Advances in Computer Vision and Pattern Recognition

Founding Editor

Sameer Singh

More information about this series at http://www.springer.com/series/4205

Nalini K. Ratha · Vishal M. Patel · Rama Chellappa
Editors

Deep Learning-Based Face Analytics

 Springer

Editors
Nalini K. Ratha
Department of Computer Science
and Engineering
University at Buffalo-SUNY
Buffalo, NY, USA

Vishal M. Patel
Department of Electrical and Computer
Engineering
Johns Hopkins University
Baltimore, MD, USA

Rama Chellappa
Departments of Electrical and Computer
Engineering (Whiting School
of Engineering) and Biomedical
Engineering (School of Medicine)
Johns Hopkins University
Baltimore, MD, USA

ISSN 2191-6586   ISSN 2191-6594   (electronic)
Advances in Computer Vision and Pattern Recognition
ISBN 978-3-030-74699-5   ISBN 978-3-030-74697-1   (eBook)
https://doi.org/10.1007/978-3-030-74697-1

This Springer imprint is published by the registered company Springer Nature Switzerland AG
The registered company address is: Gewerbestrasse 11, 6330 Cham, Switzerland

Contents

Chapter 1
Deep CNN Face Recognition: Looking at the Past and the Future

Ankan Bansal, Rajeev Ranjan, Carlos D. Castillo, and Rama Chellappa

Abstract The need for face recognition has evolved from identifying a few hundred people to identifying hundreds of thousands of people in the last decade. Most of the progress in automatic face recognition has been driven by deep networks in the past few years. In this article, we provide an overview of recent progress in this area and discuss state-of-the-art CNN-based face recognition and verification systems. We also present some open questions and discuss avenues for research in the coming years.

1.1 Synonyms

– Face verification
– Face recognition
– Face identification.

1.2 Introduction

Automatic face recognition is the problem of identifying a person from an image or a video. Due to the ubiquity of cameras and prevalence of social media networks, automatic face recognition has applications in access control, homeland security,

A. Bansal
Amazon, Pasadena, CA, USA

R. Ranjan
Amazon, Seattle, WA, USA

C. D. Castillo (✉) · R. Chellappa
Johns Hopkins University, Baltimore, MD, USA

© The Author(s), under exclusive license to Springer Nature Switzerland AG 2021
N. K. Ratha et al. (eds.), *Deep Learning-Based Face Analytics*, Advances in Computer
Vision and Pattern Recognition, https://doi.org/10.1007/978-3-030-74697-1_1

rescuing exploited children, HCI interfaces, etc. Recent years have seen significant progress in automatic face recognition technology, largely due to improvements in deep convolutional network designs and the availability of large datasets [1–4] and challenging testing standards [5–8]. In this article, we summarize recent works in automatic face recognition, focusing on methods using deep convolutional neural networks (CNNs).

The problem of face recognition can be divided into face identification and face verification. The standard approach for training a CNN for solving these problems includes four steps: face detection, alignment, representation, and classification (Fig. 1.1). Identification is the problem of assigning an identity to an image from a list of identities. From another perspective, this can be considered as trying to retrieve the best matching face from a gallery for a given probe image. On the other hand, face verification involves verifying whether two face images are of the same person. This is usually performed by computing the similarity between feature representations of the two faces. Both identification and verification have benefited immensely from developments in deep learning algorithms and more advanced CNN architectures.

In addition to improved architectures, face recognition has seen significant progress in the design of effective loss functions for training CNNs. Both face identification and verification aim to learn representations which have low intra-class variations and high inter-class variations. Several loss functions have been proposed over the past few years which encourage representations with these properties. Most of these [9–11] modify the common softmax loss using additional constraints on the features which lead to compact and discriminative representations of faces and thus, performance improvements in face recognition.

However, the first few steps of any face recognition pipeline [1, 12–14] are face detection [12, 15, 16] and fiducial landmark/key-point localization. Effective methods for face detection and key-point localization have been shown to lead to improved recognition performance [12, 17]. Face detection and landmark localization have also seen performance gains due to the availability of large datasets [18–21] and CNNs. We briefly discuss some of the methods in the upcoming sections in this article.

This article is organized as follows. First, we describe some publicly available datasets used for face recognition. Then, we discuss some recent face detection methods and also cover common loss functions used for training face recognition pipelines. Next, we describe and compare some recently proposed pipelines for face verification and identification. Finally, we discuss some open questions in face recognition.

1.3 Datasets

Deep CNNs require vast amounts of training data. Large corporations have access to hundreds of millions of proprietary images and use them to train large face recognition networks. Taigman et al. used 500 million images of 10 million subjects for training a CNN for face identification. Similarly, Schroff et al. used over 200 million images

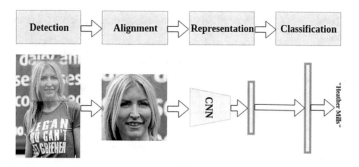

Fig. 1.1 Standard approach for training a CNN for face verification and identification

from 8 million people to train a face recognition network with 140 million parameters. However, these datasets are not publicly available. It has been shown recently [22] that pre-training models on extremely large datasets lead to better performance on other datasets. (Ironically, even the dataset used in [22] is not publicly available.) Public face recognition datasets are not close to this scale. But every year, there are larger and cleaner datasets are becoming available.

A face recognition system starts with detecting faces, then localizes landmarks which are used to align the faces to canonical views, and then classifies the detected faces. All three parts of the system require different levels of information and data types. In this section, we explore some recently released public datasets targeted for face recognition, face detection, and key-point localization. We start with a discussion on datasets for face recognition. We then give a brief overview of face detection datasets and then discuss some datasets available for training key-point detection algorithms.

Table 1.1 lists some public datasets targeted for face recognition and their properties. In-the-wild face recognition at a large scale essentially started with the release of the Labeled Faces in-the-Wild (LFW) dataset in 2008. LFW has about 13,000 face images belonging to about 5,700 subjects. It became the standard evaluation benchmark for several years [23] until the performance of methods began to saturate and there was a need for more challenging benchmarks. Also, the prevalence of deep CNNs necessitated the introduction of larger datasets for training. Recent years have seen several large datasets being released to help the training of deep networks and to provide stronger benchmarks. Some examples of such include CelebA [24], CASIA-WebFace [4], UMDFaces [2], MS-Celeb-1M [3], VGGFace [1], etc. However, these are still constrained because they only contain still images of mostly celebrities. Such photos are typically frontal and taken under bright lighting conditions. However, evaluation datasets like IJB-A [6], IJB-B [7], IJB-C [8], IJB-S [25], and Megaface [5] contain videos and images in low light conditions and profile faces. There is a clear domain shift between the aforementioned training datasets and these evaluation sets. To fill this gap, several video datasets have been proposed over the years. Among these, YoutTube Faces (YTF) [26] and UMDFaces Video [17]

Table 1.1 Recent datasets for face recognition in approximate order of number of faces available.

Face recognition

Name	#faces	#subjects
CFP [29]	7,000	500
DFW [30]	11,157 images	1,000
LFW [31]	13,233	5749
CelebA [24]	202,599	10,177
UMDFaces [2]	367,888	8,277
CASIA-WebFace [4]	494,414	10,575
PaSC [32]	2,802 videos	293
IJB-A [6]	5,712 images, 2,085 videos	500
YTF [26]	3,425 videos	1,595
IJB-B [7]	11,754 images, 7,011 videos	1,845
IJB-C [8]	31,334 images, 11,779 videos	3,531
VGGFace [1]	2.6M	2,622
VGGFaces2 [33]	3.31M	9,131
Megaface [5, 34]	4.7M	672K
UMDFaces Video [17]	22,075 videos	3,107
MS-Celeb-1M [3]	10M	100K
IJB-S [25]	>10M	202

are currently the largest publicly available annotated video datasets. Since many of the evaluation protocols contain, a mixture of still images and videos, training networks with a combination of still image datasets and video datasets should lead to performance improvements [17]. The features learned using such datasets are more robust than the features learned from networks trained with only still images. Systems trained with mixed datasets have recently been shown to achieve state-of-the-art performance on several evaluation protocols [12, 27, 28].

We list some datasets for training and testing face detection models in Table 1.2. The most popular and the largest training dataset is the WIDER FACE dataset [18]. It contains annotations for about 400,000 annotated faces half of which are in the training set and the rest are in the test set. There are about 32,000 images in the datasets. These annotations cover a large range of variations in scale, illumination, orientation, etc. It is widely used as a standard for training models and as an evaluation benchmark for face detection methods. Before WIDER FACE, the standard benchmark was FDDB [19]. It contains about 5,200 face annotations. However, due to its small size and limited variability of the faces in FDDB, the performance of the best-performing methods was saturating. This necessitated the need for larger and more difficult training and test datasets, WIDER FACE being one such. Recently, the IARPA JANUS Benchmark datasets [6–8] have also been released which contain a large number of face annotations for evaluating face detection and recognition

Table 1.2 Recent datasets for face detection with approximate number of faces and images in each

Face detection		
Name	#faces	#images
FDDB [19]	5,171	2,846 images
MALF [35]	11,931	5,250
AFLW [21]	26,000	22,000
IJB-A [6]	50,000	5,712 images, 2,085 videos
IJB-B [7]	–	11,754 images, 7,011 videos
IJB-C [8]	–	31,334 images, 11,779 videos
WIDER [18]	393,703	32,203

Table 1.3 Recent datasets for keypoint localization

Fiducial keypoint localization		
Name	#faces	Properties
AFW [20]	468 (205 images)	6 landmarks
300W [37]	600 images	68 landmarks
LFPW [36]	3,000 (1,287 images)	35 landmarks
AFLW [21]	25,993 (21,997 images)	21 landmarks, gender

in completely unconstrained settings. The IJB-C [8] dataset, for example, contains about 149,000 still images and video frames. These include about 10,000 non-face images to test operationally relevant use cases. This dataset is the superset of previously released IJB-B [7], and IJB-A [6] datasets.

Finally, datasets which contain fiducial landmark annotations are summarized in Table 1.3. Due to the difficulty in labeling and verifying facial keypoints in images, there are only a few large-scale public datasets available which include such annotations. The Annotated Face in-the-Wild (AFW) [20] dataset contains about 468 faces annotated with 6 landmarks. Some of these points might be invisible due to pose and occlusion. The 300 faces-in-the-wild dataset is a similarly sized dataset. It contains 600 images annotated with 68 facial landmarks. Half of these images are acquired indoors and the other half outdoors. The Labeled Face Parts in-the-wild (LFPW) [36] dataset contains about 3,000 faces labeled with 35 landmarks each. However, the largest dataset for keypoint detection is the Annotated Facial Landmarks in-the-Wild (AFLW) [21]. It contains 26,000 annotated faces some of which can contain significant occlusions. There are 21 landmarks annotated for each face. However, about 21% of all the landmarks are invisible due to extreme poses and occlusions. Note that the UMDFaces [2] and UMDFaces Videos [17] datasets also contain annotations for facial landmarks. However, these were automatically generated using a pre-trained model. Therefore, they will contain errors and are not usually used for training deep networks for keypoint detection.

In addition to these, there are some 3D datasets [38], age datasets [39, 40], attribute datasets [24, 41], and expression datasets [24] but a discussion on them is out of the scope of this work.

1.4 Face Detection

Face detection is the process of finding a bounding box for each face in an image. This is often the first step in any face recognition or tracking system. Counting the number of people in a crowded scene [42, 43] can also benefit from robust face and head detection. Early face detection systems like the Viola–Jones detector [44] were fast but were not effective for profile faces. Large real-world datasets like [18] and deep CNN-based representations have led to significant improvements in face detection performance. These CNN-based detection methods are largely robust to pose, illumination, clutter, and scale. Most of the popular face detection methods have been adapted from general object detectors and can be classified as either proposal-based or single-stage detectors.

Proposal-based object detection methods start with a class-agnostic object proposal generator like selective search [45], edge-boxes [46], or a region-proposal network (RPN) [47]. These proposals are then classified into object classes by a CNN. Examples of such object detectors include R-CNN [48], Fast R-CNN [49], Faster R-CNN [47], and Mask R-CNN [50]. Proposal-based face detectors follow a similar approach and generate face proposals which are then classified as face vs non-face by a CNN. We briefly describe some proposal-based methods for face detection next.

All-in-One Face [51] is a multi-task learning approach for simultaneous face detection, key-point localization, pose estimation, gender recognition, age estimation, smile detection, and face recognition. It builds upon Hyperface [52] by adding more tasks in the multi-task system. The face detector in both [51, 52], can be considered a two-stage proposal-based detector. Face proposals are first generated using selective search and are then classified as face vs non-face via one of the output branches of the multi-task network. The use of multi-task learning encourages information transfer across different tasks due to parameter sharing in the early layers of the network. This leads to performance improvements for most tasks over single-task systems. Though these models can be trained end-to-end in theory, lack of availability of a single dataset containing annotations for all tasks prevents end-to-end training in practice.

Finding Tiny Faces [15] focuses on improving detection of small faces in an image. It starts off by creating a coarse image pyramid (0.5x, 1x, and 2x). These are passed through a shared CNN which predicts detection and regression template responses at different resolutions. These templates model additional context for lower resolutions to improve small face detection. The detections from different resolutions are merged into the original scale and the final detection output is obtained after applying NMS.

Fig. 1.2 Example landmark locations from UMDFaces [2]. The figure also shows bounding boxes, and estimated pose (yaw,pitch,roll)

Other proposal-based face detectors include [53] and supervised transformer network [54]. In [53], the authors propose to train a ConvNet to estimate the 3D transformation parameters which can be used to transform a pre-defined 3D mean face model and generate face proposals and localize facial landmarks.

Unlike proposal-based detectors, single-stage object detectors do not contain an explicit proposal generation step. Such detectors typically include a single pass through a CNN and processing multi-scale image pyramid or multiple layers of a CNN. Single-shot multibox detector (SSD) [55] and YOLO [56, 57] are examples of recent Single-stage object detectors. Several recent face detectors adapt these methods. One of the most recent methods is DPSSD [12] which is described next.

DPSSD [12] adapts a standard SSD network for face detection by adding upsampling layers to create an hourglass network [58]. These layers generate contextual features and help in detecting faces at multiple scales. Anchors from six different layers in the network are classified using a small classification network which outputs the probability of an anchor being a face and the bounding box regression offsets. The authors of [12] show that the model achieves near state-of-the-art performance on multiple face detection datasets.

In addition, SSH [16], CNN Cascade [59], ScaleFace [60], and S^3 FD [61] are some other single-stage face detection methods. We refer the reader to the original papers for more details.

After a face has been detected, the step in most face recognition pipelines in facial landmark detection and face alignment. Landmarks can be considered synonymous with corners in an image. They usually determine the most discriminative locations on a face. Some examples of facial landmarks are the corners of the eyes, ear lobes, tip of the nose, corners of the mouth, etc. Fig. 1.2 shows some landmark annotations for a few faces from the UMDFaces dataset [2]. A detailed discussion on various key-point detection methods is out of the scope of this paper. We refer the reader to the brief overview in [12] and a comprehensive review in [62] for a better coverage of the topic.

1.5   Loss Functions

The loss function is an important factor in determining the performance of deep networks. For face recognition, most networks are trained to perform a $C-$way classification of faces with the hope that the learned features can be used as discriminative representations. Therefore, most existing works have used the standard cross-entropy loss with softmax for training face recognition networks. However, some variants of softmax loss have been proposed. These variants aim to address some specific issues associated with softmax loss. These issues include preference for high-quality images, early saturation, lack of margin between intra- and inter-class samples, etc. Some methods instead focus on directly optimizing the features for face verification. Metric learning approaches optimize the features to reduce intra-class separation and increase inter-class separation.

We start with a description of the standard cross-entropy-based softmax loss for training a classifier. Suppose there are M training samples in a batch. Let, $\mathbf{x}_i$ be the ith face image in the batch with the label y_i and $f(\mathbf{x}_i)$ be the feature representation of the face. The feature representation is typically a deep CNN. The feature vectors are projected into logits using weights W and bias b. The softmax loss is then given by

$$\mathcal{L}_{\text{Softmax}} = -\frac{1}{M}\sum_{i=1}^{M}\log\frac{e^{W_{y_i}^T f(\mathbf{x}_i)+b_{y_i}}}{\sum_{j=1}^{C}e^{W_j^T f(\mathbf{x}_i)+b_j}} \tag{1.1}$$

where C is the total number of classes, W_j is the jth column of the weight matrix W and b_j is the corresponding bias. Note that the bias term can be absorbed into the weights by appending 1 to $f(\mathbf{x}_i)$. Now, since $\mathbf{a}^T\mathbf{b} = \|a\|\|b\|\cos(\theta)$, where θ is the angle between $\mathbf{a}$ and $\mathbf{b}$, the above equation can be re-written as

$$\mathcal{L}_{\text{Softmax}} = -\frac{1}{M}\sum_{i=1}^{M}\log\frac{e^{\|W_{y_i}\|\|f(\mathbf{x}_i)\|\cos(\theta_{y_i})}}{\sum_{j=1}^{C}e^{\|W_j\|\|f(\mathbf{x}_i)\|\cos(\theta_j)}} \tag{1.2}$$

The network is trained with this loss till convergence. At test time, a probe face $\mathbf{x}_p$ is compared to a face in the gallery, $\mathbf{x}_g$ using cosine similarity:

$$s = \frac{f(\mathbf{x}_p)^T f(\mathbf{x}_g)^T}{\|f(\mathbf{x}_p)\|_2 \|f(\mathbf{x}_g)\|_2} \tag{1.3}$$

Crystal Loss [9] adds the following constraint:

$$\|f(\mathbf{x}_i)\|_2 = \alpha, \forall i = 1, 2, ..., M \tag{1.4}$$

to the objective in (1.1). The authors argue that the features obtained from networks trained with softmax loss strongly prefer high quality/resolution images to low-quality images. This is apparent from the L_2-norm of the features. Good quality images have a higher norm than low-quality images. Such samples are easy to classify. Near the origin (low L_2-norm), the features from different classes are easily confused. To solve these problems, [9] proposes to project all features to a hypersphere of a fixed radius α. This ensures that all features have the same norm and low-resolution faces are classified better. In addition, minimizing the softmax loss on the hypersphere is equivalent to minimizing the cosine distance between similar faces and maximizing it for dissimilar faces. This is the final target of a face verification system. Therefore, training a network with Crystal loss directly benefits face verification.

A-Softmax [63] incorporates an angular margin to the softmax formulation. This is based on the idea that at test time, we usually want dissimilar features to be angularly separated (since our distance metric is cosine distance). Adding an angular margin helps in generating features which are discriminative on a hypersphere manifold. This means that the learned features are well separated. A-Softmax starts by normalizing the weight vectors $\| W_j \| = 1$, $\forall j$. Let, $\| f(\mathbf{x}_i) \| = s$, then the A-Softmax loss is give as

$$\mathcal{L}_{\text{SphereFace}} = \frac{-1}{M} \sum_{i=1}^{M} \log \frac{e^{s \cos(m\theta_{y_i,i})}}{e^{s \cos(m\theta_{y_i,i})} + \sum_{j \neq y_i} e^{s \cos(\theta_{j,i})}} \tag{1.5}$$

where m is the size of the margin and $\theta_{y_i,i}$ is in the range $\left[0, \frac{\pi}{m}\right]$. However, training a CNN under this constraint is difficult. Therefore, the authors in [63] propose to generalize $\cos(\theta_{y_i,i})$ to a monotonic angle function $\psi(\theta_{y_i,i})$ which equals $\cos(\theta_{y_i,i})$ in $\left[0, \frac{\pi}{m}\right]$. So, A-softmax can be written as

$$\mathcal{L}_{\text{SphereFace}} = \frac{-1}{M} \sum_{i=1}^{M} \log \frac{e^{s \psi(\theta_{y_i,i})}}{e^{s \psi(\theta_{y_i,i})} + \sum_{j \neq y_i} e^{s \cos(\theta_{j,i})}} \tag{1.6}$$

where $\psi(\theta)$ is a piecewise function:

$$\psi(\theta) = (-1)^k \cos(m\theta) - 2k, \ \theta \in \left[\frac{k\pi}{m}, \frac{(k+1)\pi}{m}\right] \tag{1.7}$$
$$\text{and } k \in [0, m-1]$$

Large Margin Cosine Loss [10] uses an additive margin term instead of a multiplicative margin as used above. In addition to fixing $\| W_j \| = 1$ by L_2 normalization, the authors propose to fix $\| f(\mathbf{x}_i) \| = s$. This puts the learned features on a hypersphere were they need to be separable in the angular space. Fixing the norm of the features is a commonly used technique, e.g., [9]. Adding the margin in (1.2) thus gives the formulation:

$$\mathcal{L}_{\text{CosFace}} = -\frac{1}{M} \sum_{i=1}^{M} \log \frac{e^{s(\cos(\theta_{y_i,i})-m)}}{e^{s(\cos(\theta_{y_i,i})-m)} + \sum_{j \neq y_i} e^{s\cos(\theta_{j,i})}} \qquad (1.8)$$

The margin m and the feature scale s are inter-dependent. The feature scale should be high enough to ensure that the samples are separable for the given margin. The additive margin helps in learning more discriminative features by forcing higher inter-class variance and lower intra-class variance.

Additive Angular Margin Loss [11] also starts by normalizing W_{y_i} and scaling the feature such that
$\|f(\mathbf{x}_i)\| = s$. However, instead of directly adding an additive cosine margin as in (1.8), [11] proposes to use an additive angular margin. This is again done with the aim of increasing inter-class discrepancy and intra-class compactness. The proposed loss can be written as

$$\mathcal{L}_{\text{ArcFace}} = \frac{-1}{M} \sum_{i=1}^{M} \log \frac{e^{s(\cos(\theta_{y_i,i}+m))}}{e^{s(\cos(\theta_{y_i,i}+m))} + \sum_{j \neq y_i} e^{s\cos(\theta_{j,i})}} \qquad (1.9)$$

where m is the additive angular margin. Additionally, the authors also propose a loss which combines SphereFace (1.5), CosFace (1.8), and the proposed ArcFace (1.9):

$$\mathcal{L}_{\text{Combined}} = -\frac{1}{M} \sum_{i=1}^{M} \log \left(\frac{e^{s(\cos(m_1\theta_{y_i,i}+m_2)-m_3)}}{e^{s(\cos(m_1\theta_{y_i,i}+m_2)-m_3)} + \sum_{j \neq y_i} e^{s\cos(\theta_{j,i})}} \right) \qquad (1.10)$$

where m_1, m_2, and m_3 are the corresponding margins for SphereFace [63], ArcFace [11], and CosFace [10].

Triplet Loss [14] pushes features from different identities apart and pull features from the same identity together. A training sample consists of a triplet of faces: an anchor $\mathbf{x}^a$, a positive (same identity) face $\mathbf{x}^p$, and a negative (different identity) face $\mathbf{x}^n$. The loss can be formulated as

$$\mathcal{L}_{\text{Triplet}} = \sum_{i=1}^{|T|} \left[\left\| f(\mathbf{x_i^a}) - f(\mathbf{x_i^p}) \right\|_2^2 - \left\| f(\mathbf{x_i^a}) - f(\mathbf{x_i^n}) \right\|_2^2 + \alpha \right]_+ \qquad (1.11)$$

where α is the desired margin between positive samples and negative samples, and $[z]_+ = \max(0, z)$.

Several other loss functions have been proposed for training face recognition networks. However, space limitations do not allow a more detailed exposition of those methods. We refer the reader to the original papers for Noisy Softmax [64], Center Loss [65], Center Invariant Loss [66], Range Loss [67], Centralized Coordinate Learning [68], and Ring Loss [69].

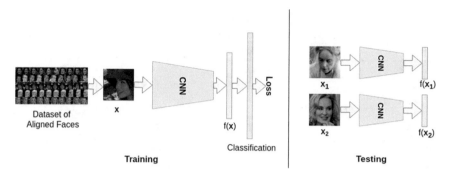

Fig. 1.3 A face verification training and testing pipeline. A dataset of aligned faces is used to train a deep CNN with a classification loss. At test time, features are extracted from two faces and their similarity is computed to determine whether the two faces are of the same person

1.6   Face Verification and Identification Using CNNs

In this section, we describe some recent face recognition pipelines which utilize some of the techniques described above. We note that both face identification and verification can be formulated as the same problem. In identification, given a probe image, the goal is to find the closest image from a gallery. This is achieved by computing the similarities between the feature representation of the probe image and feature representations of the gallery images. The image with the highest similarity with the probe images is given as output. In verification, the aim is to determine if a given pair of images belong to the same person. This is also achieved by computing the similarity between the feature representations of the two images. The basic operation in both identification and verification is to extract a feature representation and compare it with representations of the other image/images. Thus, identification and verification can be considered the same task and both follow the same pipeline: Detect, Align, Compare. We focus on methods for face verification in this section with the understanding that similar methods can be used for face identification too.

A typical face verification training and testing pipeline is shown in Fig. 1.3. A training set of aligned faces is used to train a deep network for $C-$way classification. The layer before the classification layer is used to extract a feature representation for a face at test time. Representations from two faces are compared using a similarity metric. Table 1.4 gives the performance of some recent methods for face verification on the IJB-A [6] benchmark and Table 1.5 gives performance for IJB-C [8].

DeepID [85] proposes to train a deep network on a large number of classes to obtain discriminative features which can be used for face verification. It extracts features from 60 face patches from different scales, different regions, and RGB or gray channels. Each such face patch is used to train 60 ConvNets. The final feature from each ConvNet is a 160-dimensional vector. Features for each patch and their flipped versions are extracted using these ConvNets to give a final feature representation of 19,200 dimension. The authors use Joint Bayesian [86] for face

Table 1.4 IJB-A verification performance of some recent methods

Method	True accept rate (%) @ False accept rate			
	0.0001	0.001	0.01	0.1
GOTS [6]	–	20(0.8)	41(1.4)	63(2.3)
Pose-Aware Models [70]	–	65.2(3.7)	82.6(1.8)	–
LSFS [71]	–	51.4(6)	73.3(3.4)	89.5(1.3)
Pose [72]	–	–	78.7	91.1
VGGFace [1]	–	60.4(6)	80.5(3)	93.7(1)
$DCNN_{manual}$ + metric [73]	–	–	78.7(4.3)	94.7(1.1)
$DCNN_{tpe}$ [74]	–	81.3	90.0	96.4
Chen et al. [75]	–	–	83.8(4.2)	96.7(0.9)
3d [76]	–	72.5	88.6	–
$DCNN_{fusion}$ [77]	–	76.0	88.9	96.8
$DCNN_{all}$ [51]	–	78.7	89.3	96.8
All + TPE [51]	–	82.3	92.2	97.6
TP [78]	–	–	93.9	–
NAN [79]	–	88.1	94.1	97.8
FPN [80]	77.5	85.2	90.1	–
Crystal Model C [9]	90.7(1.8)	94.7(0.4)	96.8(0.3)	98.3(0.2)
Crystal Model B [9]	91.4(1.8)	94.9(0.5)	96.9(0.3)	98.4(0.2)
Crystal Model A [9]	91.4(1.6)	94.8(0.6)	97.1(0.4)	98.5(0.2)
$RX101_{l2+tpe}$ [81]	90.9	94.3	97.0	98.4
Fast and $Accurate_A$ [12]	91.7	95.3	96.8	98.3
Fast and $Accurate_{RG1}$ [12]	91.4	94.8	97.1	98.5
Fusion [12]	92.1	95.2	96.9	98.4
TDFF [82]	87.5(1.3)	91.9(0.6)	96.1(0.7)	98.8(0.3)
TDFF + TPE [82]	87.7(1.8)	92.1(0.5)	96.1(0.7)	98.9(0.3)
TDFF* [82]	95.9(1.4)	97.9(0.4)	99.1(0.2)	99.6(0.1)

Table 1.5 IJB-C verification performance of some recent methods (* - approximate numbers from Figure 9 in [11])

Method	True accept rate (%) @ False accept rate							
	10^{-8}	10^{-7}	10^{-6}	10^{-5}	10^{-4}	10^{-3}	10^{-2}	10^{-1}
Center loss [65]	36.0	37.6	66.1	78.1	85.3	91.2	95.3	98.2
MN-vc [83]	–	–	–	–	86.2	92.7	96.8	98.9
SENet50 +DCN [84]	–	–	–	–	88.5	94.7	98.3	99.8
Fast & Acc. A [12]	16.5	19.5	43.6	77.6	91.9	95.6	97.8	99.0
Fast & Acc. $RG1$ [12]	60.6	67.4	76.4	86.2	91.9	95.7	97.9	99.2
Fusion [12]	54.1	55.9	69.5	86.9	92.5	95.9	97.9	99.2
ArcFace VGG2, R50 [11]	–	–	69*	86*	92.1	–	–	
ArcFace MS1MV2 R1000 [11]	–	–	86*	93*	95.6	–	–	

verification using this feature. All neural networks are trained with softmax loss over a training dataset containing 10,177 identities.

DeepID2 [87] combined face identification and verification signals as supervision for obtaining better feature representations. The face identification supervision pushes features from different identities apart, while the face verification supervision pulls features from the same identity closer. The identification supervision is the standard softmax loss, while the verification supervision is through the Euclidean distance between the features. Features from the same identity should have a low Euclidean distance, while features from different identities should have a high Euclidean distance. DeepID2 uses 200 ConvNets to extract features from 400 patches. To reduce redundancy among features, the authors use a forward–backward greedy algorithm to select only a small number of features which are concatenated to obtain a final 4000-dimensional feature representation. This is further reduces using PCA and final verification is again done using Joint Bayesian model.

DeepFace [13] used explicit 3D modeling, starting from 2D keypoints, to apply a piecewise affine transformation for aligning faces. This alignment starts with a 2D alignment using six fiducial keypoints. The aligned face is further warped to the image plane of a generic 3D face shape. After the alignment, DeepFace also uses a nine-layer deep network with 120 million parameters to learn the face representation.

The authors used a dataset of four million images from over 4,000 identities to train this network. The deep network used in [13] contains locally connected layers without weight sharing instead of the standard convolutional layers. This network is trained with the standard softmax cross-entropy loss.

FaceNet [14] uses a triplet loss to directly optimize the embedding instead of using the surrogate task of C−way classification. The authors claim that this leads to greater representational efficiency and this feature embedding can improve face verification and clustering performance. Each batch contains a set of triplets where each triplet consists of an anchor, a hard positive, and a hard negative sample. These are used to obtain feature embeddings which are normalized and then triplet loss over these features is used as supervision to the CNN.

Baidu [88] follows a two-step approach for training deep networks to obtain feature embeddings. Similar to DeepID [85] and DeepID2 [87], the first step uses several CNNs to extract features from different patches on an aligned face. These features are concatenated to obtain the final feature representation. Each CNN is trained with softmax loss. The second step is metric learning. The high-dimensional feature obtained from the first step is redundant and not efficient for face verification. The authors propose to use metric learning using a triplet loss to reduce the feature to low dimensions. The triplet loss ensures that the distance between the new features is small for same identity and large for different identities. Therefore, such triplet embedding directly optimizes for verification performance.

VGGFace [1] model uses a large dataset of over 2.6 million images from about 2,600 identities to train a CNN with softmax loss. The features obtained from this network are embedded using a triplet loss similar to [88].

All-in-One Face [51] proposes a multi-task learning approach for face detection, keypoint detection, pose estimation, smile detection, gender classification, age estimation, and face recognition. The network contains several heads which are responsible for learning different functionalities. The idea is that each modality will benefit from other modalities. The separate heads are trained with the corresponding losses and gradients from all heads are accumulated to train the trunk of the network. The face recognition/feature learning branch uses a standard softmax loss.

Fast and Accurate System [12] proposes a better face detector (DPSSD) and uses an ensemble of networks to extract features. The authors argue that a good face detector leads to better face verification performance by avoiding false positives and missing faces. The detected faces are aligned using facial landmarks from All-in-One Face [51]. Each network in the ensemble is a different architecture and is trained with a combination of three large datasets: UMDFaces [2], UMDFaces Videos [17], and MS-Celeb-1M [3]. Crystal loss [9] is used to train all networks. The authors also propose to do a score-level fusion, instead of feature fusion as done by previous methods. This means that the similarity scores from all networks are computed separately and then averaged to obtain the final similarities. Such ensembles were also used for recognizing disguised faces [27] and face clustering [89].

ArcFace [11] uses the Additive Angular Margin Loss and the large-scale, and clean MS1MV2 dataset to achieve state-of-the-art performance on several face recognition and verification benchmarks. The MS1MV2 dataset is a refined version of the

MS-Celebl-1M dataset and contains about 5.8M faces for 85,000 identities. ArcFace uses the popular ResNet-100 network architecture.

Some other popular methods which we could not cover due to space limitations include Masi et al. [76], LSFS [71], PAMs [70], Multi Pose Representations [72], NAN [79], FacePoseNet [80], TDFF [82], and light-CNN [90].

1.7   Open Problems

Though immense progress has been made in face recognition in the past few years, there are still several unanswered questions. In the era of deep learning, the most important questions are related to the availability of data. With the ever-increasing size of training datasets, it is not clear if there is a saturation point, i.e., if there is a point at which additional data will not lead to performance improvements. Current efforts in face recognition are focused on collecting annotated data. If we can answer this question, we will know whether these efforts are being spent in the right direction. A related problem is the use of unlabeled data. Semi-supervised learning methods which use a little labeled data along with large amounts of unlabeled data need to be developed for face recognition. This is because detecting faces in the wild is easy, but labeling them is expensive. Such semi-supervised methods will help in exploiting this unlabeled data.

Deep networks are still mostly trained with aligned faces and need aligned faces during testing. However, this cascade process introduces an unnecessary source of error. Are there ways which obviate the need for alignments? Can collecting larger datasets help? Are there better ways to align faces which do not require an additional step? These questions are important and there are no clear answers.

Another overlooked area is the presence of biases in the datasets which leads to recognition systems being biased. Most public datasets are of celebrities. These images are taken by professional photographers with good quality cameras. Networks learn to be biased to such data. Most datasets have been collected from a pre-defined list of people. Many of these people are Caucasian men. This introduces a gender and racial bias in the networks. There have been very few major attempts to remove such biases. Merler et al. [91] recently released a diversity dataset which is an important step toward this problem. An interesting question is how do we transfer knowledge from one kind of biased data to another such that we end up with a network which performs well for both categories. Transfer learning and curriculum learning methods need to be developed for faces to enable the extension of existing methods to new data instead of starting from scratch every time.

These questions and more need to be answered for even better face recognition systems in the future and will give researchers something to work on for the next few years.

Acknowledgements This research is based upon work supported by the Office of the Director of National Intelligence (ODNI), Intelligence Advanced Research Projects Activity (IARPA), via

IARPA R&D Contract No. 2014-14071600012. The views and conclusions contained herein are those of the authors and should not be interpreted as necessarily representing the official policies or endorsements, either expressed or implied, of the ODNI, IARPA, or the U.S. Government. The U.S. Government is authorized to reproduce and distribute reprints for Governmental purposes notwithstanding any copyright annotation thereon.

References

1. Parkhi OM, Vedaldi A, Zisserman A (2015) Deep face recognition. BMVC 1:6
2. Bansal A, Nanduri A, Castillo C, Ranjan R, Chellappa R (2016) Umdfaces: an annotated face dataset for training deep networks. arXiv:1611.01484
3. Guo Y, Zhang L, Hu Y, He X, Gao J (2016) Ms-celeb-1m: a dataset and benchmark for large-scale face recognition. In: European conference on computer vision. Springer, pp 87–102
4. Yi D, Lei Z, Liao S, Li SZ (2014) Learning face representation from scratch. arXiv:1411.7923
5. Kemelmacher-Shlizerman I, Seitz SM, Miller D, Brossard E (2016) The megaface benchmark: 1 million faces for recognition at scale. In: IEEE conference on computer vision and pattern recognition (CVPR), pp 4873–4882
6. Klare, BF, Taborsky E, Blanton A, Cheney J, Allen K, Grother P, Mah A, Burge M, Jain AK (2015) Pushing the frontiers of unconstrained face detection and recognition: Iarpa janus benchmark a. In: IEEE conference on computer vision and pattern recognition (CVPR), vol 13, p 4
7. Whitelam C, Taborsky E, Blanton A, Maze B, Adams JC, Miller T, Kalka ND, Jain AK, Duncan JA, Allen K et al (2017) Iarpa janus benchmark-b face dataset. In: CVPR workshops, vol 2,, p 6
8. Maze B, Adams J, Duncan JA, Kalka N, Miller T, Otto C, Jain AK, Niggel WT, Anderson J, Cheney J et al (2018) Iarpa janus benchmark—c: face dataset and protocol. In: 11th IAPR international conference on biometrics
9. Ranjan R, Bansal A, Xu H, Sankaranarayanan S, Chen J-C, Castillo CD, Chellappa R (2018) Crystal loss and quality pooling for unconstrained face verification and recognition. arXiv:1804.01159
10. Wang H, Wang Y, Zhou Z, Ji X, Gong D, Zhou J, Li Z, Liu W (2018) Cosface: large margin cosine loss for deep face recognition. In: Proceedings of the IEEE conference on computer vision and pattern recognition, pp 5265–5274
11. Deng J, Guo J, Zafeiriou S (2018) Arcface: additive angular margin loss for deep face recognition. arXiv:1801.07698
12. Ranjan R, Bansal A, Zheng J, Xu H, Gleason J, Lu B, Nanduri A, Chen J-C, Castillo CD, Chellappa R (2018) A fast and accurate system for face detection, identification, and verification. arXiv:1809.07586
13. Taigman Y, Yang M, Ranzato M, Wolf L (2014) Deepface: closing the gap to human-level performance in face verification. In: Proceedings of the IEEE conference on computer vision and pattern recognition, pp 1701–1708
14. Schroff F, Kalenichenko D, Philbin J (2015) Facenet: a unified embedding for face recognition and clustering. In: Proceedings of the IEEE conference on computer vision and pattern recognition, pp 815–823
15. Hu P, Ramanan D (2016) Finding tiny faces. arXiv:1612.04402
16. Najibi M, Samangouei P, Chellappa R, Davis L (2017) SSH: single stage headless face detector. arXiv:1708.03979
17. Bansal A, Castillo C, Ranjan R, Chellappa R (2017) The do's and don'ts for CNN-based face verification. In: Proceedings of the IEEE international conference on computer vision, pp 2545–2554

18. Yang S, Luo P, Loy C-C, Tang X (2016) Wider face: a face detection benchmark. In: IEEE conference on computer vision and pattern recognition, pp 5525–5533
19. Jain V, Learned-Miller E (2010) FDDB: a benchmark for face detection in unconstrained settings. University of Massachusetts, Amherst, Tech. Rep. UM-CS-2010-009
20. Zhu X, Ramanan D (2012) Face detection, pose estimation, and landmark localization in the wild. In: IEEE conference on computer vision and pattern recognition, pp 2879–2886
21. Kostinger M, Wohlhart P, Roth P, Bischof H (2011) Annotated facial landmarks in the wild: a large-scale, real-world database for facial landmark localization. In: IEEE international conference on computer vision workshops, pp 2144–2151
22. Sun C, Shrivastava A, Singh S, Gupta A (2017) Revisiting unreasonable effectiveness of data in deep learning era. In: 2017 IEEE international conference on computer vision (ICCV). IEEE, pp 843–852
23. Learned-Miller E, Huang GB, RoyChowdhury A, Li H, Hua G (2016) Labeled faces in the wild: a survey. In: Advances in face detection and facial image analysis, pp 189–248
24. Liu Z, Luo P, Wang X, Tang X (2015) Deep learning face attributes in the wild. In: IEEE international conference on computer vision, pp 3730–3738
25. Kalka ND, Maze B, Duncan JA, OConnor K, Elliott S, Hebert K, Bryan J, Jain AK (2018) Ijb—s: Iarpa janus surveillance video benchmark. In: 2018 IEEE 9th international conference on biometrics theory, applications and systems (BTAS). IEEE, pp 1–9
26. Wolf L, Hassner T, Maoz I (2011) Face recognition in unconstrained videos with matched background similarity. In: 2011 IEEE conference on computer vision and pattern recognition (CVPR). IEEE, pp 529–534
27. Bansal A, Ranjan R, Castillo CD, Chellappa R (2018) Deep features for recognizing disguised faces in the wild. In: 2018 IEEE/CVF conference on computer vision and pattern recognition workshops (CVPRW). IEEE, pp 10–106
28. Ranjan R, Sankaranarayanan S, Bansal A, Bodla N, Chen J-C, Patel VM, Castillo CD, Chellappa R (2018) Deep learning for understanding faces: Machines may be just as good, or better, than humans. IEEE Signal Process Mag 35(1):66–83
29. Sengupta S, Chen J-C, Castillo C, Patel VM, Chellappa R, Jacobs DW (2016) Frontal to profile face verification in the wild. In: 2016 IEEE winter conference on applications of computer vision (WACV). IEEE, pp 1–9
30. Kushwaha V, Singh M, Singh R, Vatsa M, Ratha N, Chellappa R (2018) Disguised faces in the wild. In: IEEE conference on computer vision and pattern recognition workshops, vol 8
31. Huang GB, Mattar M, Berg T, Learned-Miller E (2008) Labeled faces in the wild: a database for studying face recognition in unconstrained environments. In: Workshop on faces in real-life images: detection, alignment, and recognition
32. Beveridge JR, Phillips PJ, Bolme DS, Draper BA, Givens GH, Lui YM, Teli MN, Zhang H, Scruggs WT, Bowyer KW et al (2013) The challenge of face recognition from digital point-and-shoot cameras. In: IEEE international conference on biometrics: theory, applications and systems (BTAS). IEEE, pp 1–8
33. Cao Q, Shen L, Xie W, Parkhi OM, Zisserman A (2018) Vggface2: a dataset for recognising faces across pose and age. In: 2018 13th IEEE international conference on automatic face & gesture recognition (FG 2018). IEEE, pp 67–74
34. Nech A, Kemelmacher-Shlizerman I (2017) Level playing field for million scale face recognition
35. Yang B, Yan J, Lei Z, Li SZ (2015) Fine-grained evaluation on face detection in the wild. In: IEEE international conference on automatic face and gesture recognition
36. Belhumeur PN, Jacobs DW, Kriegman DJ, Kumar N (2011) Localizing parts of faces using a consensus of exemplars. In: IEEE conference on computer vision and pattern recognition, pp 545–552
37. Sagonas C, Tzimiropoulos G, Zafeiriou S, Pantic M (2013) 300 faces in-the-wild challenge: the first facial landmark localization challenge. In: Proceedings of the IEEE international conference on computer vision workshops, pp 397–403

38. Faltemier TC, Bowyer KW, Flynn PJ (2007) Using a multi-instance enrollment representation to improve 3D face recognition. In: First IEEE international conference on biometrics: theory, applications, and systems, BTAS 2007. IEEE, pp 1–6
39. Levi G, Hassner T (2015) Age and gender classification using convolutional neural networks. In: Proceedings of the IEEE conference on computer vision and pattern recognition workshops, pp 34–42
40. Rothe R, Timofte R, Gool LV (2015) Dex: deep expectation of apparent age from a single image. In: IEEE international conference on computer vision workshop on chalearn looking at people, pp 10–15
41. Hand EM, Castillo C, Chellappa R (2018) Doing the best we can with what we have: multi-label balancing with selective learning for attribute prediction. In: AAAI conference on artificial intelligence. AAAI
42. Bansal A, Venkatesh K (2015) People counting in high density crowds from still images. arXiv:1507.08445
43. Sindagi VA, Patel VM (2017) CNN-based cascaded multi-task learning of high-level prior and density estimation for crowd counting. In: 2017 14th IEEE international conference on advanced video and signal based surveillance (AVSS). IEEE, pp 1–6
44. Viola P, Jones MJ (2004) Robust real-time face detection. Int J Comp Vis 57(2):137–154
45. Uijlings JR, van de Sande KE, Gevers T, Smeulders AW (2013) Selective search for object recognition. Int J Comp Vis 104(2):154–171
46. Zitnick CL, Dollár P (2014) Edge boxes: locating object proposals from edges. In: European conference on computer vision. Springer, pp 391–405
47. Ren S, He K, Girshick R, Sun J (2015) Faster R-CNN: towards real-time object detection with region proposal networks. In: Advances in neural information processing systems, pp 91–99
48. Girshick R, Donahue J, Darrell T, Malik J (2014) Rich feature hierarchies for accurate object detection and semantic segmentation. In: IEEE conference on computer vision and pattern recognition, pp 580–587
49. Ross G (2015) Fast R-CNN. In: IEEE international conference on computer vision, pp 1440–1448
50. He K, Gkioxari G, Dollár P, Girshick R (2017) Mask R-CNN. In: 2017 IEEE international conference on computer vision (ICCV). IEEE, pp 2980–2988
51. Ranjan R, Sankaranarayanan S, Castillo CD, Chellappa R (2017) An all-in-one convolutional neural network for face analysis
52. Ranjan R, Patel V, Chellappa R (2016) Hyperface: a deep multi-task learning framework for face detection, landmark localization, pose estimation, and gender recognition. arXiv:1603.01249
53. Li Y, Sun B, Wu T, Wang Y (2016) Face detection with end-to-end integration of a convnet and a 3D model. In: European conference on computer vision (ECCV)
54. Chen D, Hua G, Wen F, Sun J (2016) Supervised transformer network for efficient face detection. In: European conference on computer vision. Springer, pp 122–138
55. Liu W, Anguelov D, Erhan D, Szegedy C, Reed S, Fu C-Y, Berg AC (2016) SSD: single shot multibox detector. In: European conference on computer vision (ECCV), pp 21–37
56. Redmon J, Divvala S, Girshick R, Farhadi A (2015) You only look once: unified, real-time object detection. arXiv:1506.02640
57. Redmon J, Farhadi A (2017) Yolo9000: better, faster, stronger
58. Newell A, Yang K, Deng J (2016) Stacked hourglass networks for human pose estimation. In: European conference on computer vision. Springer, pp 483–499
59. Li H, Lin Z, Shen X, Brandt J, Hua G (2015) A convolutional neural network cascade for face detection. In: IEEE conference on computer vision and pattern recognition (CVPR), pp 5325–5334
60. Yang S, Xiong Y, Loy CC, Tang X (2017) Face detection through scale-friendly deep convolutional networks. arXiv:1706.02863
61. Zhang S, Zhu X, Lei Z, Shi H, Wang X, Li SZ (2017) S^3fd: single shot scale-invariant face detector. arXiv:1708.05237

62. Wang N, Gao X, Tao D, Yang H, Li X (2017) Facial feature point detection: a comprehensive survey. Neurocomputing
63. Liu W, Wen Y, Yu Z, Li M, Raj B, Song L (2017) Sphereface: deep hypersphere embedding for face recognition
64. Chen B, Deng W, Du J (2017) Noisy softmax: improving the generalization ability of DCNN via postponing the early softmax saturation. In: The IEEE conference on computer vision and pattern recognition (CVPR)
65. Wen Y, Zhang K, Li Z, Qiao Y (2016) A discriminative feature learning approach for deep face recognition. In: European conference on computer vision (ECCV), pp 499–515
66. Wu Y, Liu H, Li J, Fu Y (2017) Deep face recognition with center invariant loss. In: Proceedings of the on thematic workshops of ACM multimedia 2017. ACM, pp 408–414
67. Zhang X, Fang Z, Wen Y, Li Z, Qiao Y (2017) Range loss for deep face recognition with long-tailed training data. In: 2017 IEEE international conference on computer vision (ICCV). IEEE, pp 5419–5428
68. X. Qi and L. Zhang, "Face recognition via centralized coordinate learning," *arXiv preprint arXiv:* 1801.05678, 2018
69. Zheng Y, Pal DK, Savvides M (2018) Ring loss: convex feature normalization for face recognition. In: Proceedings of the IEEE conference on computer vision and pattern recognition, pp 5089–5097
70. Masi I, Rawls S, Medioni G, Natarajan P (2016) Pose-aware face recognition in the wild. In: Proceedings of the IEEE conference on computer vision and pattern recognition, pp 4838–4846
71. Wang D, Otto C, Jain AK (2015) Face search at scale: 80 million gallery. arXiv:1507.07242
72. AbdAlmageed W, Wu Y, Rawls S, Harel S, Hassne T, Masi I, Choi J, Lekust J, Kim J, Natarajana P, Nevatia R, Medioni G (2016) Face recognition using deep multi-pose representations. In: IEEE winter conference on applications of computer vision (WACV)
73. Chen J-C, Ranjan R, Kumar A, Chen C-H, Patel VM, Chellappa R (2015) An end-to-end system for unconstrained face verification with deep convolutional neural networks. In: IEEE international conference on computer vision workshop on ChaLearn looking at people, pp 118–126
74. Sankaranarayanan S, Alavi A, Castillo CD, Chellappa R (2016) Triplet probabilistic embedding for face verification and clustering. In: 2016 IEEE 8th international conference on biometrics theory, applications and systems (BTAS)
75. Chen J-C, Patel VM, Chellappa R (2016) Unconstrained face verification using deep CNN features. In: 2016 IEEE winter conference on applications of computer vision (WACV). IEEE, pp 1–9
76. Masi I, Tran AT, Leksut JT, Hassner T, Medioni G (2016) Do we really need to collect millions of faces for effective face recognition? arXiv:1603.07057
77. Chen J, Ranjan R, Sankaranarayanan S, Kumar A, Chen C, Patel VM, Castillo CD, Chellappa R (2017) Unconstrained still/video-based face verification with deep convolutional neural networks. Int J Comput Vis 1–20
78. Crosswhite N, Byrne J, Parkhi OM, Stauffer C, Cao Q, Zisserman A (2017) Template adaptation for face verification and identification. In: IEEE international conference on automatic face and gesture recognition
79. Yang J, Ren P, Chen D, Wen F, Li H, Hua G (2016) Neural aggregation network for video face recognition. arXiv:1603.05474
80. Chang F-J, Tran AT, Hassner T, Masi I, Nevatia R, Medioni G (2017) Faceposenet: making a case for landmark-free face alignment. In: 2017 IEEE international conference on computer vision workshop (ICCVW). IEEE, pp 1599–1608
81. Ranjan R, Castillo CD, Chellappa R (2017) L2-constrained softmax loss for discriminative face verification. arXiv:1703.09507
82. Xiong L, Jayashree K, Zhao J, Feng J, Pranata S, Shen S (2017) A good practice towards top performance of face recognition: transferred deep feature fusion. arXiv:1704.00438
83. Xie W, Zisserman A (2018) Multicolumn networks for face recognition. arXiv:1807.09192

84. Xie W, Shen L, Zisserman A (2018) Comparator networks. In: European conference on com-
    puter vision. Springer, pp 811–826
85. Sun Y, Wang X, Tang X (2014) Deep learning face representation from predicting 10000 classes.
    In: IEEE conference on computer vision and pattern recognition (CVPR), pp 1891–1898
86. Chen D, Cao XD, Wang LW, Wen F, Sun J (2012) Bayesian face revisited: a joint formulation.
    In: European conference on computer vision, pp 566–579
87. Sun Y, Chen Y, Wang X, Tang X (2014) Deep learning face representation by joint identification-
    verification. In: Advances in neural information processing systems, pp 1988–1996
88. Liu J, Deng Y, Bai T, Wei Z, Huang C (2015) Targeting ultimate accuracy: face recognition
    via deep embedding. arXiv:1506.07310
89. Lin W-A, Chen J-C, Ranjan R, Bansal A, Sankaranarayanan S, Castillo CD, Chellappa R (2018)
    Proximity-aware hierarchical clustering of unconstrained faces. Image Vis Comput 77:33–44
90. Wu X, He R, Sun Z, Tan T (2018) A light cnn for deep face representation with noisy labels.
    IEEE Trans Inf Forensics Secur 13(11):2884–2896
91. Merler M, Ratha N, Feris RS, Smith JR (2019) Diversity in faces. arXiv:1901.10436

Chapter 2
Face Segmentation, Face Swapping, and How They Impact Face Recognition

Y. Nirkin, I. Masi, Anh Tuan Tran, T. Hassner, and G. Medioni

Abstract Face swapping refers to the task of changing the appearance of a face appearing in an image by replacing it with the appearance of a face taken from another image, in an effort to produce an authentic-looking result. We describe a method for face swapping that does not require training on faces being swapped and can be easily applied even when face images are unconstrained and arbitrarily paired. Our method offers the following contributions: (a) Instead of tailoring systems for face segmentation, as others previously proposed, we show that a standard fully convolutional network (FCN) can achieve remarkably fast and accurate segmentation, provided that it is trained on a rich enough example set. For this purpose, we describe novel data collection and generation routines which provide challenging segmented face examples. (b) We use our segmentations for robust face swapping under unprecedented conditions, without requiring subject-specific data or training. (c) Unlike previous work, our swapping is robust enough to allow for extensive quantitative tests. To this end, we use the Labeled Faces in the Wild (LFW) benchmark and measure how intra- and inter-subject face swapping affect face recognition. We show that intra-subject swapped faces remain as recognizable as their sources, testifying to the effectiveness of our method. In line with established perceptual studies, we

Y. Nirkin
Bar-Ilan University, Ramat Gan, Israel

I. Masi
Sapienza, University of Rome, Rome, Italy

A. Tuan Tran
Institute for Robotics and Intelligent Systems, USC, CA, USA

T. Hassner (✉)
The Open University of Israel, Ra'anana, Israel
e-mail: hassner@openu.ac.il

G. Medioni
University of Southern California, USC, CA, USA

© The Author(s), under exclusive license to Springer Nature Switzerland AG 2021
N. K. Ratha et al. (eds.), *Deep Learning-Based Face Analytics*, Advances in Computer Vision and Pattern Recognition, https://doi.org/10.1007/978-3-030-74697-1_2

Fig. 2.1 *Inter-subject swapping.* (Top) Bruce Willis and (Bottom) G. W. Bush's photos swapped using our method onto very different subjects and images. Unlike previous work [6, 27], we do not select convenient targets for swapping or require subject-specific training and data. Are Willis and Bush hard to recognize? We offer quantitative evidence supporting Sinha and Poggio [53] showing that faces and context are both crucial for recognition

show that better face swapping produces less recognizable inter-subject results (see, e.g., Fig. 2.1). This is the first time this effect was quantitatively demonstrated by a machine vision method. Some of the material in this chapter previously appeared in [47].

2.1   Introduction

Swapping faces means transferring facial appearance from a *source* photo onto a face appearing in a *target* photo, replacing the target face while attempting to generate realistic, authentic looking results. Although face swapping today is often associated with viral Internet memes [5, 16, 49], it is actually far more important than such practices may suggest: Face swapping can also be used for preserving privacy [6, 45], digital forensics [49] and as a potential face-specific data augmentation method [42–44] especially in applications where training data is scarce (e.g., facial emotion recognition [35]).

Going beyond particular applications, face swapping is also an excellent opportunity to develop and test essential face processing capabilities: When faces are swapped between arbitrarily selected, unconstrained images, there is no guarantee on the similarity of viewpoints [22], expressions [11], 3D face shapes [20, 56], genders [15, 34], or any other attribute that makes swapping easy [27]. In such cases, swapping requires robust and effective methods for face alignment, segmentation, 3D shape estimation (though we will later challenge this assertion), expression estimation, and more.

We describe a face swapping method and test it in settings where no control is assumed over the images *or their pairings*. We evaluate our method using extensive quantitative tests at a scale never before attempted by other face swapping methods. These tests allow us to measure the effect face swapping has on machine face recognition, providing insights from the perspectives of both security applications and face perception.

Technically, we focus on face segmentation and the design of a face swapping pipeline. Our contributions include

- *Semi-supervised labeling of face segmentation.* We provide a novel means of generating a rich image set with face segmentation labels, by using motion cues and 3D data augmentation. The data we collect is used to train an FCN to segment faces faster and more accurately than existing methods.
- *Face swapping pipeline.* We describe an image-based face swapping pipeline and show it to work well on images and image pairs of unprecedented difficulty.
- *Quantitative tests.* Despite over a decade of work and contrary to other face processing tasks (e.g., recognition), face swapping methods were never quantitatively tested. We offer the first quantitative evaluation protocols for intra- and inter-subject face swapping systems.

Our qualitative results show that our swapped faces are as compelling as those produced by others, if not more. Our quantitative tests further demonstrate that our intra-subject face swapping has little effect on face verification accuracy: our swapping does not change these images in ways which affect subject identities.

We further report inter-subject results on randomly selected pairs. These tests require facial appearance to change, sometimes substantially, in order to naturally blend source faces into their new surroundings. We show that images changed in this manner are less recognizable. Though this perceptual phenomenon was described over two decades ago by Sinha and Poggio [53] in their well-known Clinton–Gore illusion, we are unaware of previous quantitative reports on how this applies to machine face recognition.

For code and deep models, please see our project page.[1]

2.2 Related Work

2.2.1 Face Segmentation

To swap only faces, without their surrounding context or occlusions, we require per-pixel segmentation labels. Some previous methods segment individual facial regions (e.g., eyes, mouth) but not the entire face [40]. Others take example-based approaches to face segmentation [54]. More recently, faces were segmented by alter-

---

[1] https://talhassner.github.io/home/publication/2018_FG_1.

nating between segmentation and landmark localization using deformable part models [17]. This approach reported state-at-the-art performance on the Caltech Occluded Faces in the Wild (COFW) dataset [8].

Two recent methods proposed to segment faces using deep neural networks. The first trained a network to simultaneously segment multiple facial regions, including the entire face [38]. This method was used by others for face swapping [27], but this method can be computationally expensive. More recently [52], results were reported which outperformed the state-at-the-art on COFW [17] as well as demonstrating real-time processing speeds by using a deconvolutional neural network.

2.2.2   Face Swapping

Methods for swapping faces were proposed as far back as 2004 [7] with fully automatic techniques described nearly a decade ago [6]. These methods were originally offered in response to privacy preservation concerns: Face swapping can be used to obfuscate identities of subjects appearing in publicly available photos, as a substitute to face pixelation or blurring [6, 7, 45]. Since then, however, many of their applications seem to come from recreation [27] or entertainment (e.g., Alexander et al. [2, 14, 37, 57]).

Regardless of the application, previous face swapping systems often share several key aspects. First, some methods restrict the target photos used for transfer. Given an input source face, they search through large face albums to choose ones that are easy targets for face swapping [6, 13, 27]. Such targets are those which share similar appearance properties with the source, including facial tone, pose, expression, and more. Though our method can be applied in similar settings, our tests focus on more extreme conditions, where the source and target images are arbitrarily selected and can be (often are) substantially different.

Second, most previous methods estimate the structure of the face. Some estimate 3D facial shapes [2, 6, 36], by fitting 3D Morphable Face Models (3DMM) [41, 50]. Others instead estimate dense 2D active appearance models [61]. This is presumably done in order to correctly map textures across different individual facial shapes.

Finally, deep learning was recently used to transfer faces [29, 46], as if they were styles transferred between images. This method, however, requires a deep network to be trained for each source image or subject and thus can be impractical in many applications.

2.3   Swapping Faces in Unconstrained Images

Figure 2.2 summarizes our face image swapping method. When swapping a face from a source image, $\mathbf{I}_S$, to a target image, $\mathbf{I}_T$, we treat both images the same, apart from the final stage (Fig. 2.2d). Our method first localizes 2D facial landmarks in each

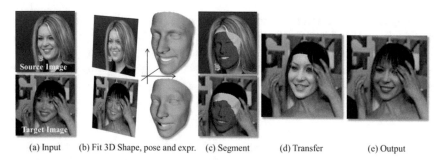

<table>
<tr><td>(a) Input</td><td>(b) Fit 3D Shape, pose and expr.</td><td>(c) Segment</td><td>(d) Transfer</td><td>(e) Output</td></tr>
</table>

Fig. 2.2 *Method overview.* **a** Source (top) and target (bottom) input images. **b** Detected facial landmarks used to establish 3D pose and facial expression for a 3D face shape (Sect. 2.3.1). **c** Our segmentation of Sect. 2.3.2 (red) overlaid on the projected 3D face (gray). **d** Source transfered onto target without blending, and the final results, **e** after blending (Sect. 2.3.3)

image (Fig. 2.2b). We use an off-the-shelf detector for this purpose [26]. Using these landmarks, we compute 3D pose (viewpoint) and modify the 3D shape to account for expression. These steps are discussed in Sect. 2.3.1.

We next segment faces from backgrounds and occlusions (Fig. 2.2c) using an FCN trained to predict per-pixel face visibility (Sect. 2.3.2). We describe how we generate rich labeled data to train our FCN. Finally, the source is efficiently warped onto the target using the two aligned 3D face shapes as proxies and blended onto the target (Sect. 2.3.3).

2.3.1 Fitting 3D Face Shapes

To enrich our set of examples for training the segmentation network (Sect. 2.3.2) we explicitly model 3D face shapes. These 3D shapes are also used as proxies to transfer textures from one face onto another, when swapping faces (Sect. 2.3.3). We experimented with two alternative methods of obtaining these shapes.

The first, inspired by [21] uses a generic 3D face, making no attempt to fit the 3D shape to the face in the image aside from performing pose (viewpoint) alignment. We, however, also estimate facial expressions and modify the 3D face accordingly.

A second approach uses the recent state-at-the-art, deep method for single image 3D face reconstruction [12, 55, 56]. This method was shown to work well on unconstrained photos such as those considered here. To our knowledge, this is the only method quantitatively shown to produce invariant, discriminative, and accurate 3D shape estimations. We have released code for that method, which regresses 3D Morphable face Models (3DMM) in neutral pose and expression. We extend it by aligning 3D shapes with input photos and modifying the 3D faces to account for facial expressions by using facial landmarks.

3D shape representation and estimation. Whether generic or regressed, we use the popular Basel Face Model (BFM) [50] to represent faces and the 3DDFA Morphable Model [62] for expressions. These are both publicly available 3DMM representations. More specifically, a 3D face shape $\mathbf{V} \subset \mathbb{R}^3$ is modeled by combining the following independent generative models:

$$\mathbf{V} = \widehat{\mathbf{v}} + \mathbf{W}_S\,\boldsymbol{\alpha} + \mathbf{W}_E\,\boldsymbol{\gamma}. \tag{2.1}$$

Here, vector $\widehat{\mathbf{v}}$ is the mean face shape, computed over aligned facial 3D scans in the Basel Faces collection and represented by the concatenated 3D coordinates of their 3D points. When using a generic face shape, we use this average face. Matrices $\mathbf{W}_S$ (shape) and $\mathbf{W}_E$ (expression) are principle components obtained from the 3D face scans. Finally, $\boldsymbol{\alpha}$ is a subject-specific 99D parameter vector estimated separately for each image and $\boldsymbol{\gamma}$ is a 29D parameter vector for expressions. To fit 3D shapes and expressions to an input image, we estimate these parameters along with camera matrices.

To estimate per-subject 3D face shapes, we regress $\boldsymbol{\alpha}$ using the deep network of [55]. They jointly estimate 198D parameters for face shape and texture. Dropping the texture components, we obtain $\boldsymbol{\alpha}$ and back-project the regressed face by $\widehat{\mathbf{v}} + \mathbf{W}_S\,\boldsymbol{\alpha}$, to get the estimated shape in 3D space.

Pose and expression fitting. Given a 3D face shape (generic or regressed) we recover its pose and adjust its expression to match the face in the input image. We use the detected facial landmarks, $\mathbf{p} = \{\mathbf{p}_i\} \subset \mathbb{R}^2$, for both purposes. To improve robustness, future implementations are planned to regress viewpoint [1, 10, 12] and expressions [11] directly using deep networks, similarly to face shape.

Specifically, we begin by solving for the pose, ignoring expression. We approximate the positions in 3D of the detected 2D facial landmarks $\widetilde{\mathbf{V}} = \{\widetilde{\mathbf{V}}_i\}$ by

$$\widetilde{\mathbf{V}} \approx f(\widehat{\mathbf{v}}) + f(\mathbf{W}_S)\,\boldsymbol{\alpha}, \tag{2.2}$$

where $f(\cdot)$ is a function selecting the landmark vertices on the 3D model. The vertices of all BFM faces are registered so that the same vertex index corresponds to the same facial feature in all faces. Hence, f need only be manually specified once, at preprocessing. From f, we get 2D-3D correspondences, $\mathbf{p}_i \leftrightarrow \widetilde{\mathbf{V}}_i$, between detected facial features and their corresponding points on the 3D shape. Similar to previous work [19], we use these correspondences to estimate 3D pose, computing 3D face rotation, $\mathbf{R} \in \mathbb{R}^3$, and translation $\mathbf{t} \in \mathbb{R}^3$ using the EPnP solver [33].

Following pose estimation, we regress expression parameters in vector $\boldsymbol{\gamma}$ by formulating expression estimation as a bounded linear problem:

$$\delta_R\Big(P(\mathbf{R},\mathbf{t})\big(f(\widehat{\mathbf{v}}) + f(\mathbf{W}_S)\,\boldsymbol{\alpha} + f(\mathbf{W}_E)\,\boldsymbol{\gamma}\big)\Big) = \delta_R(\mathbf{p}),$$
$$\text{with } |\boldsymbol{\gamma}_j| \le 3\,\sigma_j \quad \forall\, j = \{1 \ldots 29\} \tag{2.3}$$

where $\delta_R(\cdot)$ is a visibility check that removes occluded points given the head rotation $\mathbf{R}$; $P(\mathbf{R}, \mathbf{t})$ is the projection matrix, given the extrinsic parameters $(\mathbf{R}, \mathbf{t})$; and σ_j is the standard deviation of the j-th expression component in $\boldsymbol{\gamma}$. This problem can be solved using any constrained linear least-squares solver.

2.3.2 Deep Face Segmentation

Our method uses an FCN to segment the visible parts of faces from their context and occlusions. Other methods previously tailored novel network architectures for this task (e.g., Saito et al. [52]). We show that excellent segmentation results can be obtained with a standard FCN, provided that it is trained on plenty of rich and varied examples.

Obtaining enough diverse images with ground truth segmentation labels can be hard: Saito et al. [52], for example, used manually segmented LFW faces and a semi-automatic segmentation method [9] for this purpose. These labels were costly to produce and limited in their variability and number. We, instead, propose a novel means of generating numerous training examples with little manual effort and show that a *standard FCN* trained on these examples outperforms state-at-the-art face segmentation results.

Semi-supervised training data collection. We produce large quantities of segmentation labeled face images by using *motion cues* in unconstrained face videos. To this end, we process videos from the recent IARPA Janus CS2 dataset [28]. These videos include faces of different poses, ethnicities, and ages, viewed under widely varying conditions. Our training used the 1,275 videos of subjects not included in LFW, of the 2,042 CS2 videos (309 subjects out of 500).

Given a video, we produce a rough, initial segmentation using a method based on previous work [18]. Specifically, we keep a hierarchy of regions with stable region boundaries computed with dense optical flow. Though these regions may be over- or under-segmented, they are computed with temporal coherence and so these segments are consistent across frames.

We use an existing face detector [26] to detect faces and facial landmarks in each of the frames. Facial landmarks were then used to extract the face contour and extend it to include the forehead. We chose this method over simpler landmark detectors [58] because we need contour landmarks and more than five landmark positions are estimated.

All the segmented regions generated above, that did not overlap with a face contour are then discarded. All intersecting segmented regions are further processed using a simple interface which allows browsing the entire video, selecting partial segments [18], and adding or removing them from the face segmentation using simple mouse clicks. Figure 2.3a shows the interface used in the semi-supervised labeling. A selected frame is typically processed in about 5 seconds. In total, we used this method

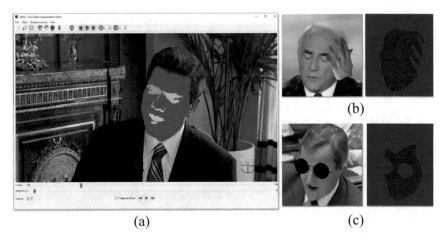

(a)   (c)

Fig. 2.3 **a** Interface used for semi-supervised labeling. **b**–**c** Augmented examples and segmentation labels for occlusions due to **b** hands and **c** synthetic sunglasses

to produce 9,818 segmented face images (frames), choosing anywhere between one to five frames per video in a little over a day of work.

Occlusion augmentation. This labeled collection is further enriched by adding synthetic occlusions. To this end, we explicitly use 3D information estimated for our example faces. Specifically, we estimate 3D face shape for our segmented faces, using the method described in Sect. 2.3.1. We then use computer graphic (CG) 3D models of various objects (e.g., sunglasses) to modify the faces. We project these CG models onto the image and record their image locations as synthetic occlusions. Each CG object added 9,500 face examples. The detector used in our system [26] failed to accurately localize facial features on the remaining 318 faces, and so this augmentation was not applied to them.

Finally, an additional source of synthetic occlusions follows previous work by overlaying hand images at various positions on our example images [52]. Hand images were taken from the EgoHands dataset [4]. Figure 2.3b shows a synthetic hand augmentation and Fig. 2.3c a sunglasses augmentation, along with their resulting segmentation labels.

2.3.3 Face Swapping and Blending

Face swapping from a source $\mathbf{I}_S$ to target $\mathbf{I}_T$ proceeds as follows. First, to avoid distortions, the source and target face horizontal rotation angles are compared, if they have a different sign and are more than $10°$ apart then the source image and its corresponding 3D shape and segmentation are flipped horizontally. The 3D shape associated with the source, $\mathbf{V}_S$, is projected down onto $\mathbf{I}_S$ using its estimated pose,

$P(\mathbf{R}_S, \mathbf{t}_S)$ (Sect. 2.3.1). We then sample the source image using bilinear interpolation, to assign 3D vertices projected onto the segmented face (Sect. 2.3.2) with intensities sampled from the image at their projected coordinates.

The shapes for both source and target, $\mathbf{V}_S$ and $\mathbf{V}_T$ correspond in the indices of their vertices. We can therefore directly transfer these sampled intensities from all vertices $\mathbf{v}_i \in \mathbf{V}_S$ to $\mathbf{v}_i \in \mathbf{V}_T$. This provides texture for the vertices corresponding to visible regions in $\mathbf{I}_S$ on the target 3D shape. We now render $\mathbf{V}_T$ onto $\mathbf{I}_T$, using the estimated target pose $(\mathbf{R}_T, \mathbf{t}_T)$, masking the rendered intensities using the target face segmentation (see Fig. 2.2d). Finally, the rendered, source face is blended in with the target context using an off the shelf method [51].

2.4   Experiments

We performed comprehensive experiments in order to test our method, both qualitatively and quantitatively. Runtimes were all measured on an Intel Core i7 4820K computer with 32GB DDR4 RAM and an NVIDIA GeForce Titan X. Our original implementation swapped faces at 1.3 fps using a GPU and taking slightly more time, 0.8 fps on the CPU [47]. A more recent, optimized implementation now runs at 25 fps on two CPUs. A similar speedup is expected by applying the same optimizations to the GPU version and is left as future work.

2.4.1   Face Segmentation Evaluations

Qualitative results. Qualitative face segmentation results are provided in Figs. 2.2, 2.4, and 2.5. Results are visualized following others [52] to show segmented regions (red) overlaying the aligned 3D face shapes, projected onto the faces (gray).

Quantitative results. We also performed quantitative tests, comparing the accuracy of our segmentation to existing methods. We follow the evaluation procedure described by previous work [17], testing the 507 face photos in the COFW dataset [8]. Previous methods include the regional predictive power (RPP) estimation [59], Structured Forest [25], segmentation-aware part model (SAPM) [17], recent deep methods [38, 52]. We provide results also for our method, trained without out occlusion augmentation (Sect. 2.3.2).

Note that Structured Forest [25] and one of the deep methods [52] used respectively 300 and 437 images for testing, without reporting which images were used. Also note that results for one of the baselines [38] were computed by us, using code released by its authors, out of the box, but optimizing for the segmentation threshold which provided the best accuracy.

Accuracy is measured using the standard *intersection over union* (IOU) metric, comparing predicted segmentations with manually annotated ground truth

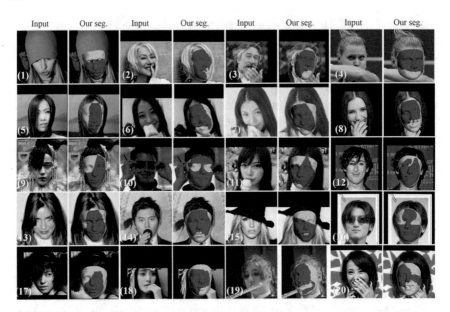

Fig. 2.4 Qualitative segmentation results from the COFW data set

Fig. 2.5 Qualitative segmentation results from the LFW data set

Table 2.1 Segmentation results on the COFW dataset [8]

Method	Mean IOU	Global	Ave(face)	FPS
Struct. Forest [25]*	–	83.9	88.6	–
RPP [59]	72.4	–	–	0.03
SAPM [17]	83.5	88.6	87.1	–
Liu et al. [38]	72.9	79.8	89.9	0.29
Saito et al. [52]* +*GraphCut*	**83.9**	88.7	92.7	43.2
Us (no occlusion augmentation at training)	81.6	87.4	93.3	**48.6**
Us	83.7	**88.8**	**94.1**	**48.6**

*These results were reported on unspecified subsets of the test set

masks [25], as well as two standard metrics [25]: *global*—overall percent of correctly labeled pixels—and *ave(face)*, the average face pixel recall. Table 2.1 reports these results along with run times. Our method is the fastest yet achieves comparable result with the state-at-the-art. Note that we use the same GPU model as the ones used by the most recent baseline [52] and report runtimes we measured ourselves for other baselines [38]. All other run times were reported in previous work.

2.4.2 Qualitative Face Swapping Results

We provide face swapping examples produced on unconstrained LFW images [24] using randomly selected targets in Figs. 2.1, 2.2, 2.6, 2.8, and 2.9. We chose these examples to demonstrate a variety of challenging settings. In particular, these results used source and target faces of widely different poses, occlusions, and facial expressions. To our knowledge, previous work never showed results for such challenging settings.

Figure 2.8 shows other qualitative results with multiple sources swapped onto multiple targets using images collected from the Web.

Our method is capable of producing high-resolution results of over 512×512 pixels as demonstrated in Fig. 2.9.

In addition, Fig. 2.7 shows a qualitative comparison with a recent baseline method [27] using the same source-target pairs. We note that this previous method used an existing segmentation approach [38] which we show in Sect. 2.4.1 to perform worse than our own. This is qualitatively evident in Fig. 2.7 by the facial hairlines. Fig. 2.7 also provides results from the publicly available code of Kowalski [30] and Hrastnik [23]. In both cases, the absence of a segmentation is clearly evident.

Fig. 2.6 *Qualitative LFW inter-subject face swapping results.* Examples were selected to represent extremely different *poses* (4, 7), *genders* (1, 2, 7, 8), *expressions* (1), *ethnicities* (1, 3, 6), *ages* (3, 4, 5, 7, 8) and *occlusions* (1, 4, 5)

2.4.3 Qualitative Ablation Study

We further performed qualitative ablation studies showing how the components employed in our face swapping affects the final result. Figures 2.10 and 2.11 provide such qualitative results for inter-subject and intra-subject face swapping: The figures show the source face transferred onto the target face using (1) a generic 3D model (2) the estimated 3D shape (3) a generic 3D model and face segmentation (4) the estimated 3D shape and face segmentation.

2.4.4 Limitations of Our System

Figure 2.12 describes a few typical failure cases and their causes. As we show earlier (Fig. 2.10), surprisingly, the differences in 3D shapes and even expressions have little effect on the quality of the generated images. Instead, swapping errors—cases where the output shows noticeable artifacts or else fails to look authentic—are typically caused by a failure of one of the preprocessing components of the system, particularly the landmark detection used for viewpoint estimation Fig. 2.12(1). We hope to solve

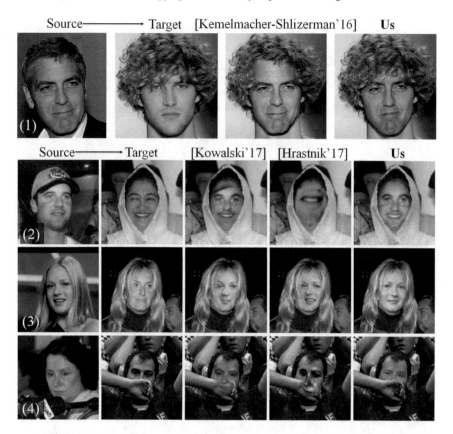

Fig. 2.7 *Comparison with previous face swap methods* (1) Result published by Kemelmacher–Shlizerman [27]. (2–4) Results obtained by the public implementations of Kowalski [30] and Hrastnik [23]

these in the future by using more robust viewpoint estimation techniques, such as direct face alignment, recently proposed [10].

Other frequent failure cases are due to severe differences in image resolutions between the source and target views, Fig. 2.12(2). Finally, the simple blending method we use can also fail to correctly merge the source face into its new target context, Fig. 2.12(3). The last two failure reasons could potentially be mitigated by smarter merging methods, possibly using deep learning as a realism filter, as some have recently proposed for face alignment [60].

Fig. 2.8 Qualitative inter-subject face swapping results

2.5 The effects of Swapping on Recognition

Similar to previous work, we offer qualitative results demonstrating our swapped faces (Sect. 2.4.2). Unlike others, however, we also offer extensive quantitative tests designed to measure the effect of swapping on the perceived identity of swapped faces. To this end, we propose two test protocols, motivated by the following assumptions.

Fig. 2.9 High-resolution intra-subject face swapping results

Assumption 1 Swapping faces between images of different subjects (i.e., *inter-subject swapping*) changes facial *context* (e.g., hair, skin tone, head shape). Effective swapping must therefore modify source faces, sometimes substantially, to blend them naturally into their new contexts, thereby producing faces that look less like the source subjects. Examples of inter-subject face swaps are provided throughout this paper, but see Fig. 2.10 and for some LFW examples.

Assumption 2 If a face is swapped between two photos of the same person (*intra-subject swapping*), the output of an effective swapping method should easily be recognizable as the two photos share the same context. See examples in Fig. 2.11.

The first assumption is based on a well-known trait of human visual perception: Face recognition requires both internal and external cues (faces and their context) to recognize faces. This idea was claimed by a seminal work [53] and extensively studied in the context of biological visual systems [3]. To our knowledge, it was never explored for machine recognition systems and never quantitatively. The robustness of our method allows us to do just that.

The second assumption is intended to verify that when the context remains the same (the same subject) swapping does not change facial appearances in a way which makes faces less recognizable. This ensures that the swapping method does not introduce artifacts or changes facial appearances.

To test these assumptions, we produce modified (face swapped) versions of the LFW benchmark [24]. We estimate how recognizable faces appear after swapping by using a publicly available, state-of-the-art face recognition system, in lieu of a large-scale human study. Though the recognition abilities of humans and machines may be different, modern systems already claim human or even super-human accuracy [39]. We therefore see the use of a state-at-the-art machine system as an adequate surrogate to human studies which often involve problems of their own [32].

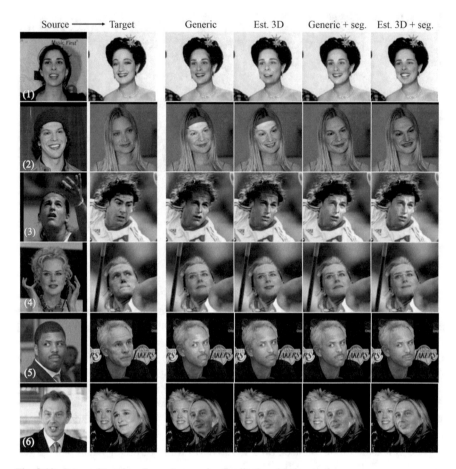

Fig. 2.10 *Inter-subject face Swapping results*. Qualitative ablation study

2.5.1 Face Verification System

We use the ResFace101 [43] face recognition system to test if faces remain recognizable after swapping. ResFace101 obtained near-perfect verification results on LFW, yet it was not optimized for that benchmark and tested also on IJB-A [28]. Moreover, it was trained on synthetic face views, not unlike the ones produced by face swapping. For these reasons, we expect ResFace101 to be well suited for our purposes. Recognition is measured by 100%-EER (Equal Error Rate), accuracy (Acc.), and normalized Area Under the Curve (nAUC). Finally, we provide ROC curves for all our tests.

Source ──────→ Target   Generic   Est. 3D   Generic + seg.   Est. 3D + seg.

Fig. 2.11 *Intra-subject face swapping results*. Qualitative ablation study

2.5.2   Inter-Subject Swapping Verification Protocols

We begin by measuring the effect of inter-subject face swapping on face verification accuracy. To this end, we process all faces in the LFW benchmark, swapping them onto photos of *other, randomly selected subjects*. We make no effort to verify the quality of the swapped results and if swapping failed, we treat the result as any other image.

We use the original LFW test protocol with its same/not-same subject pairs. Our images, however, present the original faces with possibly very different contexts. Specifically, let $(\mathbf{I}_i^1, \mathbf{I}_i^2)$ be the i-th LFW test image pair. We produce $\widehat{\mathbf{I}}_i^1$, the swapped version of $\mathbf{I}_i^1$, by randomly picking another LFW subject and image from that subject as a target, taking $\mathbf{I}_i^1$ as the source. We then do the same for $\mathbf{I}_i^2$ to obtain $\widehat{\mathbf{I}}_i^2$.

Fig. 2.12 *Face swapping failures.* (1) The most common reason for failure is facial landmark localization errors, leading to misaligned shapes or poor expression estimation. Other, less frequent reasons include (2) substantially different image resolutions (3) failures in blending very different facial hues

Matching pairs of swapped images, however, can obscure changes to both images which make the source faces equally unrecognizable: Such tests only reflect the similarity of swapped images to each other, not to their sources. We therefore test verification on benchmark pairs comparing original versus swapped images. This is done twice, once on pairs $(\widehat{\mathbf{I}}_i^1, \mathbf{I}_i^2)$, the other on pairs $(\mathbf{I}_i^1, \widehat{\mathbf{I}}_i^2)$. We then report the average results for both trials. We refer to these tests as *face preserving* tests.

We also performed *context preserving* tests: These use benchmark image pairs as *targets* rather than sources. Thus, they preserve the context of the original LFW images, not the faces. By doing so, we can measure the effect of context on recognition. This test setup is reminiscent of the *inverse mask* tests performed by others in the past [31]. Their tests were designed to measure how well humans recognize LFW faces if the face was cropped out without being replaced, and showed that doing so led to a drop in recognition. Unlike them, our images contain faces of other subjects swapped in place of the original faces, and so are more realistic.

2.5.3 Inter-Subject Swapping Results

We provide verification results for both face preserving and context preserving inter-subject face swapping in Table 2.2 and ROC curves for the various tests in Fig. 2.13. Our results include ablation studies, showing accuracy with a generic face and no segmentation (*Generic*), with an estimated 3D face shape (Sect. 2.3.1) and no segmentation (*Est. 3D*), with a generic face and segmentation (*Seg.*) and with an estimated 3D shape and face segmentation (*Est. 3D+Seg.*).

The face preserving results in Table 2.2 (bottom) are consistent with our *Assumption 1*: The more the source face is modified, by estimating 3D shape and better segmenting the face, the less it is recognizable as the original subject and the lower the verification results. Using a simple generic shape and no segmentation provides ∼8% *better* accuracy than using our the entire pipeline. Importantly, just by estimating 3D face shapes, accuracy drops by ∼3.5% compared to using a simple generic face shape.

Unsurprisingly, the context preserving results in Table 2.2 (top) are substantially lower than the face preserving tests. Unlike the face preserving tests, however, the harder we work to blend the randomly selected source faces into their contexts, the

Table 2.2 *Inter-subject face swapping.* Ablation study

Method	100%-EER	Acc.	nAUC
Baseline (ResFace101)	98.10±0.90	98.12±0.80	99.71±0.24
Context preserving (face swapped out)			
Generic	64.58±2.10	64.56±2.22	69.94±2.24
Est. 3D	69.00±1.43	68.93±1.19	75.58±2.20
Seg.	68.93±1.98	69.00±1.93	76.06± 2.15
Est. 3D+Seg.	73.17±1.59	72.94±1.39	80.77±2.22
Face preserving (face swapped in)			
Generic	92.28±1.37	92.25±1.45	97.55±0.71
Est. 3D	88.77±1.50	88.53±1.25	95.53±0.99
Seg.	89.92±1.48	89.98±1.36	96.17±0.93
Est. 3D+Seg.	86.48±1.74	86.38±1.50	93.71±1.42

Table 2.3 *Intra-subject face swapping.* Ablation study

Method	100%-EER	Acc.	nAUC
Baseline (VGGFace)	97.23±0.88	97.35±0.77	99.54±0.30
Baseline (ResFace101)	98.10±0.90	98.12±0.80	99.71±0.24
Generic	97.02±0.98	97.02±0.97	99.53±0.31
Est. 3D	97.05±0.98	97.03±1.01	99.52±0.32
Seg.	97.12±1.09	97.08±1.07	99.53±0.31
Est. 3D+Seg.	97.12±1.09	97.12±0.99	99.52±0.31
Est. 3D+Seg.	96.65±0.85	96.63±0.92	99.45±0.29

better recognition becomes. By better blending the sources into the context, more of the context is retained and the easier it is to verify the two images based on their context without the face itself misleading the match.

2.5.4 Intra-Subject Swapping Verification Protocols and Results

To test our second assumption, we again process the LFW benchmark, this time swapping faces between different images of the *same subjects* (*intra-subject* face swapping). Of course, all *same* labeled test pairs, by definition, belong to subjects that have at least two images, and so this did not affect these pairs. *Not-same* pairs, however, sometimes include images from subjects which have only a single image. To address this, we replaced them with others for which more than one photo exists.

Fig. 2.13 *Inter-subject swapping ROC curves.* Ablation study for the two experiments. Baseline shown in red

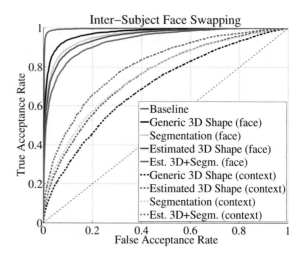

We again run our entire evaluation twice: once, swapping the first image in each test pairs keeping the second unchanged, and vice versa. Our results average these two trials. Results obtained using different components of our system are provided in Table 2.3 and Fig. 2.13. These show that even under extremely different viewing conditions, perceived subject identity remains unchanged, supporting our *Assumption 2*.

In general, accuracy drops by ∼1%, with a similar nAUC compared to the use of original LFW images. This slight drop suggests that our swapping between different images of the same subject does not alter apparent facial identities.

2.6 Conclusions

We describe a method for swapping faces between images which is robust enough to allow for large-scale, quantitative tests. From these tests, several key observations emerge. **(1)** State-of-the-art face segmentation can be obtained with a standard segmentation network, provided that the network is trained on rich and diverse examples. **(2)** Collecting such examples is easy using motion cues in video and CG augmentation techniques.

As for the effects of swapping on face recognition, **(3)** Both face and context play important roles in recognition. We offer quantitative support for the two decades old claim of Sinha and Poggio [53]. **(4)** Better swapping leads to more facial changes and a drop in recognition. Finally, **(5)**, 3D face shape estimation better blends the two faces together and so produces less recognizable faces. As these methods mature, an important goal of future research is to provide fake detection techniques. Although there is progress on that front, one challenging frontier remains *generalized* fake detection: Detecting fakes produced by methods that are unknown at the time the detector is being trained [48].

Acknowledgements This research is based upon work supported in part by the Office of the Director of National Intelligence (ODNI), Intelligence Advanced Research Projects Activity (IARPA), via IARPA 2014-14071600011. The views and conclusions contained herein are those of the authors and should not be interpreted as necessarily representing the official policies or endorsements, either expressed or implied, of ODNI, IARPA, or the U.S. Government. The U.S. Government is authorized to reproduce and distribute reprints for Governmental purpose notwithstanding any copyright annotation thereon. TH was also partly funded by the Israeli Ministry of Science, Technology and Space.

References

1. Albiero V, Chen X, Yin X, Pang G, Hassner T (2020) Img2pose: face alignment and detection via 6DoF, face pose estimation. arXiv:2012.07791
2. Alexander O, Rogers M, Lambeth W, Chiang M, Debevec P (2009) Creating a photoreal digital actor: the digital emily project. In: Conference for visual media production. IEEE, pp 176–187
3. Axelrod V, Yovel G (2010) External facial features modify the representation of internal facial features in the fusiform face area. Neuroimage 52(2):720–725
4. Bambach S, Lee S, Crandall DJ, Yu C (2015) Lending a hand: detecting hands and recognizing activities in complex egocentric interactions. In: Proceedings of the IEEE international conference on computer vision
5. Banks A (2018) What are deepfakes & why the future of porn is terrifying. https://www.highsnobiety.com/p/what-are-deepfakes-ai-porn
6. Bitouk D, Kumar N, Dhillon S, Belhumeur P, Nayar SK (2008) Face swapping: automatically replacing faces in photographs. ACM Trans Gr 27(3):39
7. Blanz V, Scherbaum K, Vetter T, Seidel H-P (2004) Exchanging faces in images. In: Computer graphics forum, vol 23. Wiley Online Library, pp 669–676
8. Burgos-Artizzu XP, Perona P, Dollár P (2013) Robust face landmark estimation under occlusion. In: Proceedings of the international conference on computer vision. IEEE, pp 1513–1520
9. Cao C, Weng Y, Zhou S, Tong Y, Zhou K (2014) Facewarehouse: a 3D facial expression database for visual computing. Trans Vis Comput Gr 20(3):413–425
10. Chang F-J, Tran AT, Hassner T, Masi I, Nevatia R, Medioni G (2017) Faceposenet: making a case for landmark-free face alignment. In: Proceedings of the international conference on computer vision workshops. IEEE, pp 1599–1608
11. Chang F-J, Tran AT, Hassner T, Masi I, Nevatia R, Medioni G (2018) Expnet: landmark-free, deep, 3D facial expressions. In: Automatic face and gesture recognition. IEEE, pp 122–129
12. Chang F-J, Tran AT, Hassner T, Masi I, Nevatia R, Medioni G (2019) Deep, landmark-free FAME: face alignment, modeling, and expression estimation. Int. J. Comput. Vision 127(6–7):930–956
13. Chen T, Tan P, Ma L-Q, Cheng M-M, Shamir A, Hu S-M (2013) Poseshop: human image database construction and personalized content synthesis. Trans Vis Comput Gr 19(5):824–837
14. De La Hunty M, Asthana A, Goecke R (2010) Linear facial expression transfer with active appearance models. In: International conference on pattern recognition. IEEE, pp 3789–3792
15. Eidinger E, Enbar R, Hassner T (2014) Age and gender estimation of unfiltered faces. Trans Inform Forensics Secur 9(12)
16. Floridi L (2018) Artificial intelligence, deepfakes and a future of ectypes. Philos Technol 31(3):317–321

17. Ghiasi G, Fowlkes CC, Irvine C (2015) Using segmentation to predict the absence of occluded parts. In: Proceedings of the British machine vision conference, pp 22–1
18. Grundmann M, Kwatra V, Han M, Essa I (2010) Efficient hierarchical graph-based video segmentation. In: Proceedings of the conference on computer vision pattern recognition. http://cpl.cc.gatech.edu/projects/videosegmentation/
19. Hassner T (2013) Viewing real-world faces in 3D. In: Proceedings of the international conference on computer vision. IEEE, pp 3607–3614. www.openu.ac.il/home/hassner/projects/poses
20. Hassner T, Basri R (2013) Single view depth estimation from examples. arXiv:1304.3915
21. Hassner T, Harel S, Paz E, Enbar R (2015) Effective face frontalization in unconstrained images. In: Proceedings of the conference on computer vision pattern recognition
22. Hassner T, Masi I, Kim J, Choi J, Harel S, Natarajan P, Medioni G (2016) Pooling faces: template based face recognition with pooled face images. In: Proceedings of the conference on computer vision pattern recognition workshops
23. Hrastnik M (2017) Faceswap code. https://github.com/hrastnik/FaceSwap
24. Huang GB, Ramesh M, Berg T, Learned-Miller E (2007) Labeled faces in the wild: a database for studying face recognition in unconstrained environments. Technical Report 07–49. UMass, Amherst
25. Jia X, Yang H, Chan K, Patras I (2014) Structured semi-supervised forest for facial landmarks localization with face mask reasoning. In: Proceedings of the British machine vision conference
26. Kazemi V, Sullivan J (2014) One millisecond face alignment with an ensemble of regression trees. In: Proceedings of the conference on computer vision pattern recognition. IEEE
27. Kemelmacher-Shlizerman I (2016) Transfiguring portraits. ACM Trans Gr 35(4):94
28. Klare BF, Klein B, Taborsky E, Blanton A, Cheney J, Allen K, Grother P, Mah A, Burge M, Jain AK (2015) Pushing the frontiers of unconstrained face detection and recognition: IARPA Janus Benchmark-A. In: Proceedings of the conference on computer vision pattern recognition
29. Korshunova I, Shi W, Dambre J, Theis L (2016) Fast face-swap using convolutional neural networks. arXiv:1611.09577
30. Kowalski M (2017) Faceswap code. https://github.com/MarekKowalski/FaceSwap
31. Kumar N, Berg AC, Belhumeur PN, Nayar SK (2009) Attribute and simile classifiers for face verification. In: 2009 IEEE 12th international conference on computer vision. IEEE, pp 365–372
32. Learned-Miller E, Huang GB, RoyChowdhury A, Li H, Hua G (2016) Labeled faces in the wild: a survey. In: Advances in face detection and facial image analysis. Springer, pp 189–248
33. Lepetit V, Moreno-Noguer F, Fua P (2009) Epnp: an accurate on solution to the PNP problem. Int J Comput Vis 81(2):155
34. Levi G, Hassner T (2015a) Age and gender classification using convolutional neural networks. In: Proceedings of the conference on computer vision pattern recognition workshops. http://www.openu.ac.il/home/hassner/projects/cnn_agegender
35. Levi G, Hassner T (2015b), Emotion recognition in the wild via convolutional neural networks and mapped binary patterns. In: International ACM, conference on multimodal interaction
36. Lin Y, Lin Q, Tang F, Wang S (2012) Face replacement with large-pose differences. In: ACM multimedia conference
37. Lin Y, Wang S, Lin Q, Tang F (2012) Face swapping under large pose variations: a 3D model based approach. In: International conference on multimedia and Expo. IEEE, pp 333–338
38. Liu S, Yang J, Huang C, Yang M-H (2015) Multi-objective convolutional learning for face labeling. In: Proceedings of the conference on computer vision pattern recognition, pp 3451–3459
39. Lu C, Tang X (2014) Surpassing human-level face verification performance on LFW with gaussianface. arXiv:1404.3840
40. Luo P, Wang X, Tang X (2012) Hierarchical face parsing via deep learning. In: Proceedings of the conference on computer vision pattern recognition. IEEE, pp 2480–2487
41. Masi I, Ferrari C, Del Bimbo A, Medioni G (2014) Pose independent face recognition by localizing local binary patterns via deformation components. In: International conference on pattern recognition

42. Masi I, Hassner T, Tran AT, Medioni G (2017) Rapid synthesis of massive face sets for improved face recognition. International conference on automatic face and gesture recognition. IEEE, pp 604–611

43. Masi I, Tran A, Hassner T, Leksut JT, Medioni G (2016) Do we really need to collect millions of faces for effective face recognition?. In: European conference on computer vision. www.openu.ac.il/home/hassner/projects/augmented_faces

44. Masi I, Tran AT, Hassner T, Sahin G, Medioni G (2019) Face-specific data augmentation for unconstrained face recognition. Int J Comput Vis 127(6–7):642–667

45. Mosaddegh S, Simon L, Jurie F (2014) Photorealistic face de-identification by aggregating donors face components. In: Asian conference on computer vision, pp 159–174

46. Nirkin Y, Keller Y, Hassner T (2019) Fsgan: subject agnostic face swapping and reenactment. In: Proceedings of the IEEE/CVF international conference on computer vision, pp 7184–7193

47. Nirkin Y, Masi I, Tuan AT, Hassner T, Medioni G (2018) On face segmentation, face swapping, and face perception. In: Automatic face and gesture recognition. IEEE, pp98–105

48. Nirkin Y, Wolf L, Keller Y, Hassner T (2020) Deepfake detection based on the discrepancy between the face and its context. arXiv:2008.12262

49. Oikawa MA, Dias Z, de Rezende Rocha A, Goldenstein S (2016) Manifold learning and spectral clustering for image phylogeny forests. Trans Inform Forensics Secur 11(1):5–18

50. Paysan P, Knothe R, Amberg B, Romhani S, Vetter T (2009) A 3D face model for pose and illumination invariant face recognition. In: International conference on advanced video and signal based surveillance

51. Prez P, Gangnet M, Blake A (2003) Poisson image editing. In: Proceedings of the ACM SIGGRAPH conference on computer graphics

52. Saito S, Li T, Li H (2016) Real-time facial segmentation and performance capture from RGB input. In: European conference on computer vision, pp 244–261

53. Sinha P, Poggio T (1996) I think I know that face. Nature 384(404)

54. Smith BM, Zhang L, Brandt J, Lin Z, Yang J (2013) Exemplar-based face parsing. In: Proceedings of the conference on computer vision pattern recognition, pp 3484–3491

55. Tran A, Hassner T, Masi I, Medioni G (2017) Regressing robust and discriminative 3D morphable models with a very deep neural network. In: Proceedings of the conference on computer vision pattern recognition, p 5, 6

56. Tran AT, Hassner T, Masi I, Paz E, Nirkin Y, Medioni G (2018) Extreme 3D face reconstruction: seeing through occlusions. In: Proceedings of the conference on computer vision pattern recognition

57. Wolf L, Freund Z, Avidan S (2010) An eye for an eye: a single camera gaze-replacement method. In: Proceedings of the conference on computer vision pattern recognition. IEEE

58. Wu Y, Hassner T, Kim K, Medioni G, Natarajan P (2017) Facial landmark detection with tweaked convolutional neural networks. Trans Pattern Anal Mach Intell

59. Yang H, He X, Jia X, Patras I (2015) Robust face alignment under occlusion via regional predictive power estimation. Trans Image Process 24(8):2393–2403

60. Zhao J, Xiong L, Jayashree PK, Li J, Zhao F, Wang Z, Pranata PS, Shen PS, Yan S, Feng J (2017) Dual-agent GANs for photorealistic and identity preserving profile face synthesis. In: Neural information processing systems, pp 66–76

61. Zhu J, Van Gool L, Hoi SC (2009) Unsupervised face alignment by robust nonrigid mapping. In: Proceedings of the international conference on computer vision. IEEE, pp 1265–1272

62. Zhu X, Lei Z, Liu X, Shi H, Li SZ (2016) Face alignment across large poses: a 3D solutio. In: Proceedings of the conference on computer vision pattern recognition

Chapter 3
Disentangled Representation Learning and Its Application to Face Analytics

Dimitris N. Metaxas, Long Zhao, and Xi Peng

3.1 Introduction

The goal of every contemporary recognition approach is to learn robust and unambiguous object representations in feature space. These learned powerful disentangled representations make it possible to build effective classifiers and are an active research topic in many fields such as face analytics.

In this chapter, we present the state-of-the-art approaches for disentangled representation learning and their application to face analytics. Most representation learning problems use an encoder/decoder approach to achieve powerful feature embeddings in representation space. Figure 3.1 illustrates this idea. The first mapping $f(\cdot)$, the encoder, projects pixels in the image space into a representation space to achieve low-dimensional feature embeddings, which could be 1D vectors, 2D maps, or multidimensional manifolds. Then, the second mapping $g(\cdot)$, the decoder, remaps the embedded feature representation into a target space to accomplish tasks such as classification labels, detection locations, and segmentation boundaries.

In face recognition, the encoder extracts image features which are then projected into a one-hot vector to represent the facial class label. In face detection, the encoder learns spatial-dependent feature maps which are followed by the decoder to generate a region of interests for face localization. In face generation, an encoder is applied to the image to achieve low-dimensional embeddings, which are then fed into a decoder to recover an image that presents pixel-wise facial information.

The representation space is usually designed to be low-dimensional with restricted variations The embeddings in this space are preferred to be informative, concise, and interpretable to final targets. To address this problem, the encoder is used to project the image space, which is high-dimensional and filled up with variations, to the representation space, which is relatively low-dimensional and disentangled

D. N. Metaxas (✉) · L. Zhao · X. Peng
Camden, USA
e-mail: dnm@cs.rutgers.edu

© The Author(s), under exclusive license to Springer Nature Switzerland AG 2021   45
N. K. Ratha et al. (eds.), *Deep Learning-Based Face Analytics*, Advances in Computer
Vision and Pattern Recognition, https://doi.org/10.1007/978-3-030-74697-1_3

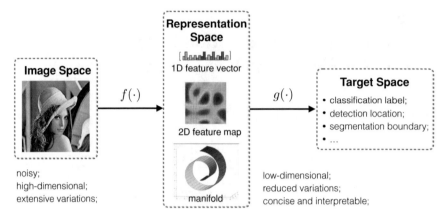

Fig. 3.1 Illustration of the encoder/decoder approach of many recognition systems. $f(\cdot)$ encodes information from the image space to achieve abstract feature embeddings in the representation space; while $g(\cdot)$ aims to decode the learned feature embeddings into the target space for the desired tasks. The representation space could consist of 1D vectors, 2D maps, or multidimensional manifolds

Table 3.1 Configurations of training datasets used to train three recently proposed face recognition networks. A large amount of labeled subjects and images are used, which is not only very expensive but also time-consuming

Method	DeepFace	FaceNet	VggFace
Training datasets	SFC	WebFace	VggFace
# of images	4.4M	200M	2.6M
# of subjects	4K	8M	2.6K

for interpretation. Learning robust and interpretable representations is a crucial and fundamental goal in computer vision to address challenging recognition applications.

However, learning robust representations is very expensive and hard given the need for annotations on large amounts of carefully selected data. Table 3.1 lists training datasets used in three recently proposed face recognition systems: DeepFace [54] developed by Google, FaceNet [51] developed by Facebook, and VggFace [42] developed by the VGG group. We can see that 2.6 and 4.4 million labeled images are used to train VggFace and DeepFace, respectively. An astonishing large dataset of 200 million labeled images of 8 million different subjects is used to train FaceNet. Undoubtedly, annotating such a large amount of images is extremely expensive, time-consuming, and tedious, whether it is done manually or semi-automatically.

The annotation cost is just one aspect of the difficulty in learning robust representations. Table 3.2 shows that even after feeding the VGGFace network with 2.6 million labeled images, the performance is still poor. For instance, the verification accuracy drops drastically to as low as 34.2% on profile faces (yaw $<75°$), although it achieves a high accuracy of 97.2% on frontal faces (yaw $<15°$). Therefore, just increasing the number of annotated data does not solve the problem.

Table 3.2 Rank-1 face recognition accuracy with respect to different head pose on MultiPIE dataset. The recognition accuracy drops significantly when the head pose changes from frontal to profile

Head pose	15° (%)	30° (%)	45° (%)	60° (%)	75° (%)	90° (%)
VggFace	97.2	96.1	92.6	84.7	62.8	34.2

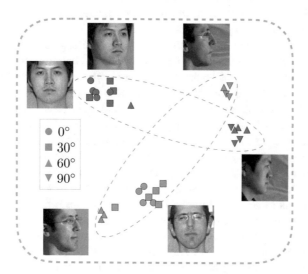

Fig. 3.2 Illustration of the feature entanglement. Two subjects (in different colors) from MultiPIE dataset are mapped into the learned representation space of VGGFace. Images in a similar head pose are embedded closer to each other even they belong to different subjects. In other words, generic data-driven features for face recognition might confound images of the same identity with others in large pose conditions

It has been demonstrated in many works that the aforementioned challenges are mainly caused by the difficulty of learning robust representations [29, 35, 48]. If we simply categorize features in the representation space as target-related and target-unrelated, we discover that in many applications, the latter may dominate the former due to feature entanglement in the representation space. Figure 3.2 illustrates the dilemma in representation learning. Two different faces (in different colors) from MultiPIE dataset [17] are mapped into the learned representation space of VGGFace [42]. We can see that generic data-driven features for face recognition might confuse images of the same faces with different faces with large pose variations. In the representation space, images of the same person may be far away from each other due to large variations in head poses, while images of different people may be close to each other when the head poses are similar. Therefore, in representation space, target-unrelated factors such as head pose may dominate target-related factors such as person identity.

A favorable representation learned from the image space should be compact, interpretable, and unambiguously disentangled for the final target. However, the image space is usually high-dimensional and noisy. Target-unrelated features may confuse target-related ones due to the nature of massive entanglement in the representation space. To achieve robust representations, the encoder, which maps the image space to the representation space, should be capable to factorize and further decouple the latent features to split target-related and target-unrelated representations. Therefore, learning a robust mapping from the image space to the representation space becomes a fundamental goal of many computer vision methods.

Learning disentangled representations is a fundamental goal in many facial analytics methods. For example, in the multi-view face recognition problem [75], we can explicitly split the latent features into *facial appearance features* (who is this person) and *viewpoint features* (which is the angle the facial image is taken). The disentangled representations can then be used to recognize the person's face consistently, regardless of viewpoint changes. In the face generation task [32], multiple latent features can be implicitly learned and decoupled from each other, such as wearing eyeglasses, having mustache, black hair. Then these decoupled features can be reorganized together to guide the generation of a novel face image that is not present in the training dataset. Overall, the essential idea in learning disentangled representations is to set apart target-related and target-unrelated features. In other words, we intend to factorize the latent representation space features into variant and invariant categories given a task, while ignoring unrelated features.

In this chapter, we show how to develop and apply disentangled representation learning to solve two classic but important face analytics problems: face landmark tracking and facial attribute inference and generation. Face landmark tracking plays a fundamental role in many computer vision tasks, such as face recognition and verification, expression analysis, person identification, and 3D face modeling. It is also the basic technology component for a wide range of applications like video surveillance, emotion recognition, augmented reality on faces, etc. In the past few years, many methods have been proposed to address this problem, with significant progress being made toward systems that work in real-world conditions ("in the wild"). Learning facial representations with deep neural networks is another essential yet challenging problem. It has broad applications in vision, graphics, and robotics. Generative adversarial networks (GANs) are widely employed to address this problem. And how to employ GANs to learn disentangled representations for faces to improve the performance of facial attribute inference and generation is still one of the hottest topics in this field.

The rest of the chapter is organized as follows. In Sect. 3.2, we split the identity embedding, which is invariant to facial landmark locations as the same subject are tracked, from the pose and expression embedding, which determines the displacements of facial landmark locations frame to frame, in the bottleneck of a recurrent encoder–decoder network [43]. The additional identity supervision significantly improves the tracking accuracy and robustness especially in large pose and partial occlusions. In Sect. 3.3, we show how disentangled representation learning is employed for facial attribute generation tasks with the help of Generative Adversarial Networks (GANs) [16]. Especially, we introduce a variant of GANs which aims to learn disentangled representations in a "complete" way and thus significantly improves the generation quality.

3.2 Application 1: Facial Landmark Tracking

In this section, we first give a brief literature review of face alignment and then introduce RED-Net proposed by Peng et al. [43], which is one of the state-of-the-art methods in this field. At last, we show the superior performance of RED-Net comparing with other methods on public benchmarks.

Face alignment has a long history of research in computer vision. Here we briefly introduce the background of face landmark detection, face landmark tracking, and recurrent encoder–decoder neural networks.

3.2.1 Related Works

3.2.1.1 Face Landmark Detection

Recently, regression-based face landmark detection methods [2, 3, 8, 25, 53, 59, 62, 63, 68, 73, 73] have achieved significant boost in the generalization performance of face landmark detection, compared to algorithms based on statistical models such as active shape models [11, 39] and active appearance models [15]. Regression-based approaches directly regress landmark locations based on features extracted from face images. Landmark models for different points are learned either in an independent manner or in a joint fashion [8]. When all the landmark locations are learned jointly, implicit shape constraints are imposed because they share the same or partially the same regressors. To compare, RED-Net performs landmark detection via both a classification model and a regression model. It deals with face alignment in a video by jointly optimizing the detection outputs of the same person in the video.

Additional accuracy improvement in face landmark detection performance can be obtained by learning cascaded regression models. Regression models from earlier cascade stages learn coarse detectors, while later cascade stages refine the result based on early predictions. Cascaded regression helps to gradually reduce the prediction variance, thus making the learning task easier for later stage detectors. Many methods have effectively applied cascade-like regression models for the face alignment task [53, 63, 68]. The supervised descent method [63] learns cascades of regression models based on SIFT features. Sun et al. [53] proposed to use three levels of neural networks to predict landmark locations. Zhang et al. [68] studied the problem via cascades of stacked auto-encoders which gradually refine the landmark position with higher resolution inputs. Compared to these efforts which explicitly define cascade structures, RED-Net learns a spatial recurrent model which implicitly incorporates the cascade structure with shared parameters. It is also more "end-to-end" compared to previous works that divide the learning process into multiple stages.

3.2.1.2 Face Landmark Tracking

Most face alignment algorithms utilize temporal information by initializing the location of landmarks with detection results from the previous frame, performing alignment in a tracking-by-detection fashion [61]. Asthana et al. [3] and Peng et al. [46] proposed to learn a person-specific model using incremental learning. However, incremental learning (or online learning) is a challenging problem [55], as the incremental scheme has to be carefully designed to prevent model drifting [56]. In RED-Net, the recurrent model is online trained to capture landmark motion correlations.

3.2.1.3 Recurrent Encoder–Decoder

Recurrent neural networks (RNNs) are widely employed in the literature of speech recognition [38] and natural language processing [37]. They have also been recently used in computer vision. For instance, in the tasks of image captioning [26] and video captioning [66], RNNs are usually employed for text generation. RNNs are also popular as a tool for action classification. As an example, Veeriah et al. [60] use RNNs to learn complex time-series representations via high-order derivatives of states for action recognition. Benefiting from the deep architecture, RNNs are naturally good alternatives to Conditional Random Fields (CRFs) [71] which are popular in image segmentation.

Encoder and decoder networks are well studied in machine translation [9] where the encoder learns the intermediate representation and the decoder generates the translation from the representation. It is also investigated in speech recognition [36] and computer vision [4, 22]. Yang et al. [65] proposed to decouple identity units and pose units in the bottleneck of the network for 3D view synthesis. However, how to fully utilize the decoupled units for correspondence regularization [33] is still unexplored. To address this issue, RED-Net takes the advantage of the encoder to learn a joint representation for identity, pose, expression as well as landmarks. The decoder translates the representation to heatmaps in order to locate the face landmarks.

3.2.2 Our Approach: RED-Net

The task is to locate facial landmarks in sequential images using an end-to-end deep neural network. Figure 3.3 shows an overview of the approach. The network consists of a series of nonlinear and multi-layered mappings, which can be functionally categorized as four modules. We now describe the details of each learning module.

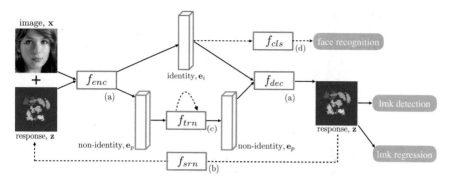

Fig. 3.3 Overview of RED-Net: **a** encoder-decoder (Sect. 3.2.2.1); **b** spatial recurrent learning (Sect. 3.2.2.2); **c** temporal recurrent learning (Sect. 3.2.2.3); and **d** supervised identity disentangling (Sect. 3.2.2.4). f_{enc}, f_{dec}, f_{srn}, f_{trn}, f_{cls} are potentially nonlinear and multi-layered mappings

3.2.2.1 Encoder–Decoder

The input of the encoder-decoder is a single video frame $\mathbf{x} \in \mathbb{R}^{W \times H \times 3}$ and the output is a response map $\mathbf{z} \in \mathbb{R}^{W \times H \times C_z}$ which indicates landmark locations. $C_z = 7$ or 68 depending on the number of landmarks to be predicted.

The *encoder* performs a sequence of convolution, pooling, and batch normalization [24] to extract a low-dimensional representation $\mathbf{e}$ from both $\mathbf{x}$ and $\mathbf{z}$:

$$\mathbf{e} = f_{enc}\big(\mathbf{x}, \mathbf{z}; \theta_{enc}\big), \quad f_{enc} : \mathbb{R}^{W \times H \times C} \rightarrow \mathbb{R}^{W_e \times H_e \times C_e}, \tag{3.1}$$

where $f_{enc}\big(\cdot; \theta_{enc}\big)$ denotes the encoder mapping with parameters θ_{enc}. We concatenate $\mathbf{x}$ and $\mathbf{z}$ along the channel dimension thus $C = 3 + C_z$. The concatenation is fed into the encoder as an updated input.

Symmetrically, the *decoder* performs a sequence of unpooling, convolution and batch normalization to upsample the representation code to the response map:

$$\mathbf{z} = f_{dec}(\mathbf{e}; \theta_{dec}), \quad f_{dec} : \mathbb{R}^{W_e \times H_e \times C_e} \rightarrow \mathbb{R}^{W \times H \times C_z}, \tag{3.2}$$

where $f_{dec}\big(\cdot; \theta_{dec}\big)$ denotes the decoder mapping with parameters θ_{dec}. $\mathbf{z}$ has the same $W \times H$ dimension as $\mathbf{x}$ but C_z channels for C_z landmarks. Each channel presents pixel-wise confidences of the corresponding landmark.

3.2.2.2 Spatial Recurrent Learning

The purpose of spatial recurrent learning is to pinpoint landmark locations in a coarse-to-fine manner. Unlike existing approaches [53, 68] that employ multiple networks in cascade, RED-Net accomplishes the coarse-to-fine search in a single network in which the parameters are jointly learned in successive recurrent steps.

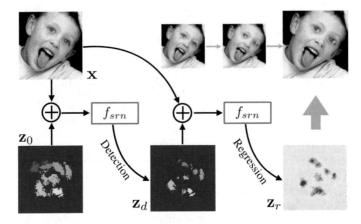

Fig. 3.4 An unrolled illustration of *spatial recurrent learning*. The response map is pretty coarse when the initial guess is far away from the ground truth if large pose and expression exist. It eventually gets refined in the successive recurrent steps

The spatial recurrent learning is performed by feeding back the previous prediction, stacked with the image as shown in Fig. 3.4, to eventually push the shape prediction from an initial guess to the ground truth:

$$\mathbf{z}_k = f_{\text{srn}}(\mathbf{x}, \mathbf{z}_{k-1}; \theta_{\text{srn}}), \quad k = 1, \ldots, K \tag{3.3}$$

where $f_{\text{srn}}(\cdot; \theta_{\text{srn}})$ denotes the spatial recurrent mapping with parameters θ_{srn}. $\mathbf{z}_0$ is the initial response map, which could be a response map generated by the mean shape or the output of the previous frame.

Specially, RED-Net carries out a two-step recurrent learning by setting $K = 2$. The first step performs *landmark detection* that aims to locate 7 major facial components (i.e., $C = 7$ in Eq. (3.2)). The second step performs *landmark regression* that refines all 68 landmarks positions (i.e. $C = 68$). This detection-followed-by-regression design [7] can largely improve the precision of landmark localization.

The landmark detection step guarantees fitting robustness especially in large pose and partial occlusions. The encoder–decoder aims to output a binary map of C_d channels, one for each major facial component. The detection step outputs:

$$\mathbf{z}_d = f_{\text{dec}}(f_{\text{enc}}(\mathbf{x}, \mathbf{z}_0; \theta_{\text{enc}}); \theta_{\text{dec}}), \quad \mathbf{z}_d \in \mathbb{R}^{W \times H \times C_d}, \tag{3.4}$$

where the detection task can be trained using pixel-wise sigmoid cross-entropy loss function:

$$\ell_d = \frac{1}{M_d} \sum_{c=1}^{C_d} \sum_{i=1}^{W} \sum_{j=1}^{H} z_{ij}^c \log y_{ij}^c + (1 - z_{ij}^c) \log(1 - y_{ij}^c) \tag{3.5}$$

where $M_d = C_d \times W \times H$. Here z_{ij}^c denotes the sigmoid output at pixel location (i, j) in $\mathbf{z}_d$ for the c-th landmark. y_{ij}^c is the ground-truth label at the same location, which is set to 1 to mark the presence of the corresponding landmark and 0 for the remaining background.

Note that this loss function is different from the N-way cross-entropy loss used in our previous conference paper [43]. It allows multiple class labels for a single pixel, which helps to tackle the landmark overlaps.

The landmark regression step improves the fitting accuracy from the outputs of the previous detection step. The encoder–decoder aims to output a heatmap of C_r channels, one for each landmark. The regression step outputs:

$$\mathbf{z}_r = f_{\text{dec}}\big(f_{\text{enc}}(\mathbf{x}, \mathbf{z}_{\text{det}}; \theta_{\text{enc}}); \theta_{\text{dec}}\big), \quad \mathbf{z}_r \in \mathbb{R}^{W \times H \times C_r}, \tag{3.6}$$

where the regression task can be trained using pixel-wise L_2 loss function:

$$\ell_r = \frac{1}{M_r} \sum_{c=1}^{C_r} \sum_{i=1}^{W} \sum_{j=1}^{H} \|z_{ij}^c - y_{ij}^c\|_2^2, \tag{3.7}$$

where $M_r = C_d \times W \times H$. Here, z_{ij}^c denotes the heatmap value of the c-th landmark at pixel location (i, j) in $\mathbf{z}_r$ for the c-th landmark. y_{ij}^c is the ground-truth value at the same location, which obeys a Gaussian distribution centered at the landmark with a pre-defined standard deviation.

Now the spatial recurrent learning (Eq. (3.3)) can be achieved by minimizing the detection loss (Eq. (3.5)) and the regression loss (Eq. (3.7)), simultaneously:

$$\text{argmin}_{\theta_{\text{enc}}, \theta_{\text{dec}}} \ell_d + \lambda \ell_r, \tag{3.8}$$

where λ balances the loss between the two tasks. Note that the spatial recurrent learning do not introduce new parameters but sharing the same parameters of the encoder-decoder network, i.e. $\theta_{\text{srn}} = \{\theta_{\text{enc}}, \theta_{\text{dec}}\}$.

3.2.2.3 Temporal Recurrent Learning

In addition to the spatial recurrent learning, RED-Net also performs temporal recurrent learning to model factors, e.g., head pose, expression, and illumination, that may change over time and affect the landmark locations significantly.

As mentioned in Sect. 3.2.2.1, the bottleneck embedding $\mathbf{e}$ can be decoupled into two parts: the identity code $\mathbf{e}_i$ and the non-identity code $\mathbf{e}_p$:

$$\mathbf{e}_i \in \mathbb{R}^{W_e \times H_e \times C_i}, \mathbf{e}_p \in \mathbb{R}^{W_e \times H_e \times C_p}, C_e = C_i + C_p, \tag{3.9}$$

where $\mathbf{e}_i$ and $\mathbf{e}_p$ model the temporal-invariant and -variant factors, respectively. We leave $\mathbf{e}_i$ to Sect. 3.2.2.4 for additional identity supervision, and exploit variations of

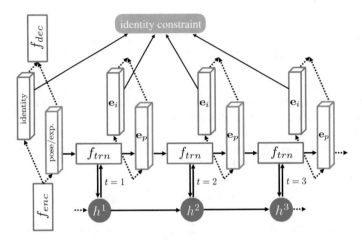

Fig. 3.5 An unrolled illustration of *temporal recurrent learning*. $\mathcal{C}_i$ encodes temporal-invariant factor which subjects to the same identity constraint. $\mathcal{C}_p$ encodes temporal-variant factors which is further modeled in f_{trn}

$\mathbf{e}_p$ via the recurrent model. Please refer to Fig. 3.5 for an unrolled illustration of the proposed temporal recurrent learning.

Mathematically, given T successive video frames $\{\mathbf{x}^t; t = 1, \ldots, T\}$, the encoder extracts a sequence of embeddings $\{\mathbf{e}_i^t, \mathbf{e}_p^t; t = 1, \ldots, T\}$. The goal is to achieve a nonlinear mapping f_{trn}, which simultaneously tracks a latent state h^t and updates $\mathbf{e}_p^t$ at time t:

$$h^t = p(\mathbf{e}_p^t, h^{t-1}; \theta_{\text{trn}}), \quad t = 1, \ldots, T$$
$$\mathbf{e}_p^{t*} = q(h^t; \theta_{\text{trn}}), \tag{3.10}$$

where $p(\cdot)$ and $q(\cdot)$ are functions of $f_{\text{trn}}(\cdot; \theta_{\text{trn}})$ with parameters θ_{trn}. $\mathbf{e}_p^{t*}$ is the update of $\mathbf{e}_p^t$.

The temporal recurrent learning is trained using T successive frames. At each frame, the detection and regression tasks are performed for the spatial recurrent learning. The recurrent learning is performed by minimizing Eq. (3.8) at every time step t:

$$\text{argmin}_{\theta_{\text{enc}}, \theta_{\text{dec}}, \theta_{\text{trn}}} \sum_{t=1}^{T} \ell_d^t + \lambda \ell_r^t, \tag{3.11}$$

where θ_{trn} denotes network parameters of the temporal recurrent learning, e.g., parameters of LSTM units.

3.2.2.4 Supervised Identity Disentangling

There is no guarantee that temporal-invariant and -variant factors can be completely decoupled in the bottleneck by simply splitting the bottleneck representation $\mathbf{e}$ into two parts. More supervised information is required to achieve the disentangling. To address this issue, RED-Net employs a side task of face recognition using identity code $\mathbf{e}_i$, in addition to the temporal recurrent learning applied on non-identity code $\mathbf{e}_p$.

The supervised identity disentangling is formulated as an N-way classification problem. N is the number of unique individuals present in the training sequences. In general, the identity code $\mathbf{e}_i$ is associated with a one-hot encoding $\mathbf{z}_i$ to indicate the score of each identity:

$$\mathbf{z}_i = f_{\text{cls}}(\mathbf{e}_i; \theta_{\text{cls}}), \ f_{\text{cls}} : \mathbb{R}^{W_e \times H_e \times C_i} \to \mathbb{R}^N, \tag{3.12}$$

where $f_{\text{cls}}(\cdot; \theta_{\text{cls}})$ is the identity classification mapping with parameters θ_{cls}. The identity task is trained using N-way cross-entropy loss:

$$\ell_{\text{cls}} = \frac{1}{N} \sum_{n=1}^{N} z^n \log y^n + (1 - z^n)\log(1 - y^n), \tag{3.13}$$

where z^n denotes the softmax activation of the n-th element in $\mathbf{z}_i$. y^n is the n-th element of the identity annotation $\mathbf{y}_i$, which is a one-hot vector with a 1 for the correct identity and all 0s for others.

All the three tasks, i.e., f_{srn}, f_{trn}, and f_{cls}, are jointly learned in an end-to-end training framework by optimizing parameters $\{\theta_{\text{enc}}, \theta_{\text{dec}}, \theta_{\text{trn}}, \theta_{\text{cls}}\}$ simultaneously. Based on Eqs. (3.11) and (3.13), RED-Net simultaneously minimize the detection and regression loss together with the identity loss at every time step t:

$$\operatorname*{argmin}_{\theta_{\text{enc}}, \theta_{\text{dec}}, \theta_{\text{trn}}, \theta_{\text{cls}}} \sum_{t=1}^{T} \ell_{\text{det}}^t + \lambda \ell_{\text{reg}}^t + \gamma \ell_{\text{cls}}^t, \tag{3.14}$$

where γ weights the identity constraint.

3.2.3 Experiments

In this section, we show the superior performance of RED-Net by validating its major components and comparing with state-of-the-art methods on widely used benchmarks. Figure 3.6 shows the fitting results of RED-Net when setting the number of tracked landmarks to 7 and 68, respectively.

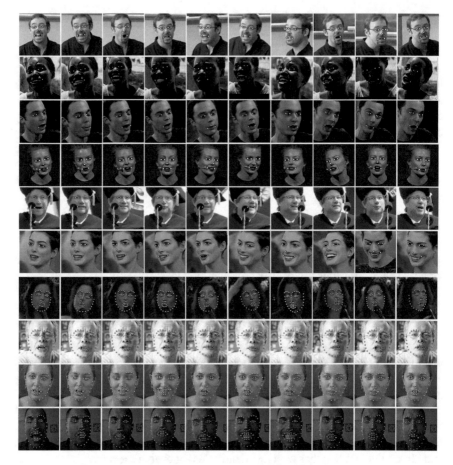

Fig. 3.6 Examples of 7-landmark (**Rows 1–6**) and 68-landmark (**Rows 7–10**) fitting results on FM and 300-VW. RED-Net achieves robust and accurate fittings when the tracked subjects suffer from large pose/expression changes (**Rows 1, 3, 4, 6, 10**), illumination variations (**Rows 2, 8**) and partial occlusions (**Rows 5, 7**)

3.2.3.1 Validation of Encoder–Decoder Network

The performance of two encoder–decoder variants on AFLW [28] and 300-VW [52] are compared: (1) VGGNet-based design with symmetrical encoder and decoder, which has been mainly investigated in our former conference paper [43]; and (2) ResNet-based design with asymmetrical encoder, i.e., the encoder is much deeper than the decoder. The results are reported in Table 3.3. The results show that the ResNet-based design outperforms the VGGNet-based variant with a substantial margin in terms of fitting accuracy (mean error) and robustness (standard deviation). Much deeper layers, as well as the proposed skipping shortcuts, contribute a lot to the improvement. In addition, the ResNet-based encoder-decoder has a very close

Table 3.3 Performance comparison of VGGNet-based and ResNet-based encoder–decoder Variants. Network configurations are described in Sect. ??. Rows 1–2: image-based results on AFLW; Rows 3–4: video-based results on 300-VW

	Mean (%)	Std (%)	Time (ms)	Memory (Mb)
VGGNet-based	6.85	4.52	43.6	184
ResNet-based	6.33	3.61	54.9	257
VGGNet-based	5.16	2.57	42.5	184
ResNet-based	4.75	2.10	56.2	257

Table 3.4 Comparison of single-step detection or regression with the proposed recurrent detection-followed-by-regression on AFLW. The proposed method (Last Row) has the best performance especially in challenging settings

	Common (%)		Challenging (%)	
	Error	Failure	Error	Failure
Single-step detection	6.05	4.62	8.14	12.4
Single-step regression	5.92	4.75	7.87	14.5
Recurrent Det.+Det.	5.86	3.44	7.33	8.20
Recurrent Det.+Reg.	5.71	3.30	6.97	8.75

computational cost to the VGGNet-based variant, e.g., the average fitting time per image/frame and the memory usage of a trained model, which should be attributed to the custom residual module design and the proposed asymmetrical encoder–decoder network.

3.2.3.2 Validation of Spatial Recurrent Learning

The spatial recurrent learning is validated on the validation set of AFLW [28]. To better investigate the benefits of spatial recurrent learning, the validation set is partitioned into two image groups according to the absolute value of the yaw angle: (1) Common settings where yaw $\in [0°–30°)$; and (2) Challenging settings where yaw $\in (30°, 90°]$. The training sets are ensembles of AFLW [28], Helen [30], and LFPW [5].

The mean fitting errors and failure rates are reported in Table 3.4. First, the results show that the two-step recurrent learning can instantly decrease the fitting error and failure rate, compared with either the single-step detection or regression. The improvement is more significant in challenging settings with large pose variations. Second, though landmark detection is more robust in challenging settings (low failure rate), it lacks the ability to predict precise locations (small fitting error) compared

Table 3.5 Comparison of cascade and recurrent learning in the challenging settings of AFLW. The latter improves accuracy with a half memory usage of the former

	Mean (%)	Std (%)	Memory (Mb)
Cascade Det. & Reg.	6.81	4.53	468
Recurrent Det. & Reg.	6.33	3.61	257

Table 3.6 Validation of temporal recurrent learning on 300-VW. f_{trn} helps to improve the tracking robustness (smaller std and lower failure rate), as well as the tracking accuracy (smaller mean error). The improvement is more significant in challenging settings of large pose and partial occlusion as demonstrated in Fig. 3.7

	Common			Challenging		
	Mean (%)	Std (%)	Fail (%)	Mean (%)	Std (%)	Failure (%)
w/o f_{trn}	4.52	2.24	3.48	6.27	5.33	13.3
f_{trn}	4.21	1.85	1.71	5.64	3.28	5.40

to landmark regression. This fact proves the effectiveness of the proposed recurrent detection-followed-by-regression.

The spatial recurrent learning and the cascade models that are widely used in other methods are compared in Table 3.5. Unsurprisingly, spatial recurrent learning can improve the fitting accuracy. The underlying reason is that the recurrent network learns the step-by-step fitting strategy jointly, while the cascade networks learn each step independently. It can better handle the challenging settings where the initial guess is usually far away from the ground truth. Moreover, the recurrent network with shared weights can instantly reduce the memory usage to one-half of the cascaded model.

3.2.3.3   Validation of Temporal Recurrent Learning

The temporal recurrent learning is validated on the validation set of 300-VW [52]. To better study the performance under different settings, the validation set is split into two groups: (1) 9 videos in common settings that roughly match "Scenario 1"; and (2) 15 videos in challenging settings that roughly match "Scenario 2" and "Scenario 3". The common, challenging, and full sets were used for evaluation. The results with and without temporal recurrent learning are compared in Table 3.6.

3.2.3.4   Benefits of Supervised Identity Disentangling

The supervised identity disentangling is proposed to better decouple the temporal-variant and -invariant factors in the bottleneck of the encoder–decoder. This facilitates the temporal recurrent training, yielding better generalization and more accurate fittings at test time (Fig. 3.8).

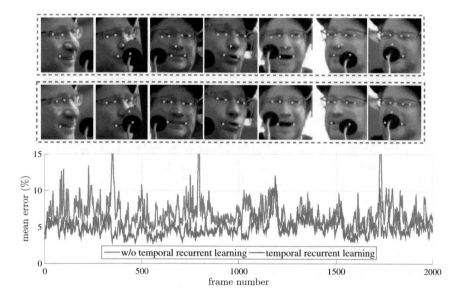

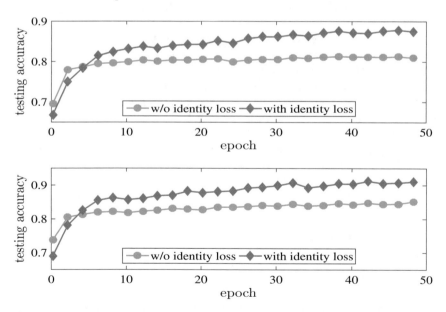

Fig. 3.7 Examples of temporal recurrent learning on 300-VW. The tracked subject undergoes intensive pose and expression variations as well as severe partial occlusions. f_{trn} substantially improves the tracking robustness (less variance) and fitting accuracy (low error), especially for landmarks on the nose tip and mouth corners

Fig. 3.8 Fitting accuracy of different facial components with respect to the number of training epochs on 300-VW. The proposed supervised identity disentangling helps to achieve a more complete factor decoupling in the bottleneck of the encoder–decoder, which yields better generalization capability and more accurate fitting results

The validation results of different facial components show similar trends: (1) The network demonstrates better generalization capability by using additional identity cues, which results in a more efficient training. For instance, after only ten training epochs, the validation accuracy for landmarks located at the left eye reaches 0.84 with identity loss compared to 0.8 without identity loss. (2) The supervised identity information can substantially boost the testing accuracy. There is an approximately 9% improvement by using the additional identity loss. It worth mentioning that, at the very beginning of the training (<5 epochs), the network has inferior testing accuracy with supervised identity disentangling. It is because the suddenly added identity loss perturbs the backpropagation process. However, the testing accuracy with identity loss increases rapidly and outperforms the one without identity loss after only a few more training epochs.

3.3 Application 2: Learning Facial Representations for Inference and Generation

In this section, we first give a brief introduction to the problem of facial representation learning and GANs, and then introduce CR-GAN proposed by Tian et al. [57], which is one of the state-of-the-art GANs in this field. At last, we show two applications of facial representation learning: multi-view facial image generation and facial attribute manipulation.

Learning a disentangled facial representation from a single image for further generations is an interesting problem with broad applications in vision, graphics, and robotics. One of its application is to generate multi-view images from a single-view input. Yet, it is a challenging problem since (1) computers need to "imagine" what a given object would look like after a 3D rotation is applied; and (2) the multi-view generations should preserve the same "identity."

Generally speaking, previous solutions to this problem include model-driven synthesis [6], data-driven generation [64, 75], and a combination of the both [49, 70, 73]. Recently, generative adversarial networks (GANs) [16] have shown impressive results in multi-view generation [58, 69].

3.3.1 Related Works

3.3.1.1 Pose-Invariant Representation Learning

Hinton et al. [21] introduced transforming auto-encoder to generate images with view variance. Yan et al. [64] proposed Perspective Transformer Nets to find the projection transformation. Zhou et al. [72] propose to synthesize views by appearance flow. Very recently, GAN-based methods usually follow a single-pathway design: an encoder–decoder network [43] followed by a discriminator network. For example, to normalize the viewpoint, e.g., face frontalization, they either combine encoder–decoder with 3DMM [6] parameters [67], or use duplicates to predict global and local details [23]. DR-GAN [58] follows the single-pathway framework to learn identity features that are invariant to viewpoints. However, it may learn "incomplete" representations due to the single-pathway framework. In contrast, CR-GAN can learn complete representations using a two-pathway network, which guarantees high-quality generations even for "unseen" inputs.

For representation learning [14, 31], early works may use *Canonical Correlation Analysis* to analyze the commonality among different pose subspaces [20, 40, 44]. Recently, deep learning-based methods use synthesized images to disentangle pose and identity factors by cross-reconstruction [45, 75], or transfer information from pose variant inputs to a frontalized appearance [74]. However, they usually use only labeled data, leading to a limited performance. We proposed a two-pathway network to leverage both labeled and unlabeled data for self-supervised learning, which can generate realistic images in challenging conditions.

3.3.1.2 Generative Adversarial Networks

Goodfellow et al. [16] introduced GANs to estimate target distribution via an adversarial process. Gulrajani et al. [19] presented a more stable approach to enforce *Lipschitz Constraint* on Wasserstein GAN [1]. AC-GAN [41] extended the discriminator by containing an auxiliary decoder network to estimate class labels for the training data. BiGANs [12, 13] try to learn an inverse mapping to project data back into the latent space. CR-GAN [57] can also find an inverse mapping, make a balanced minimax game when training data is limited.

These GAN-based methods usually have a single-pathway design: an encoder–decoder network is followed by a discriminator network. The encoder (E) maps input images into a latent space (Z), where the embeddings are first manipulated and then fed into the decoder (G) to generate novel views.

However, experiments conducted by CR-GAN indicate that this single-pathway design may have a severe issue: they can only learn "incomplete" representations, yielding limited generalization ability on "unseen" or unconstrained data. Take Fig. 3.9 as an example. During the training, the outputs of E constitute only a subspace of Z since we usually have a limited number of training samples. This would make G only "see" part of Z. During the testing, it is highly possible that E would map an "unseen" input outside the subspace. As a result, G may produce poor results due to the unexpected embedding.

To address this issue, CR-GAN is proposed to learn *Complete Representations* for multi-view generation. The main idea is, in addition to the reconstruction path, we introduce another generation path to create view-specific images from embeddings that are randomly sampled from Z. Please refer to Fig. 3.10 for an illustration. The two paths share the same G. In other words, G learned in the generation path will guide the learning of both E and D in the reconstruction path, and vice versa. E is

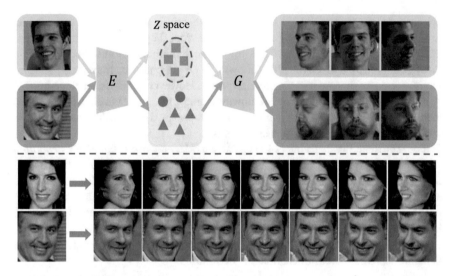

Fig. 3.9 Top: the limitation of existing GAN-based methods. They can generate good results if the input is mapped into the learned subspace (Row 1). However, "unseen" data may be mapped out of the subspace, leading to poor results (Row 2). Bottom: results of CR-GAN. By learning complete representations, CR-GAN can generate realistic, identity-preserved images from a single-view input

forced to be an inverse of G, yielding complete representations that would span the entire Z space. More importantly, the two-pathway learning can easily utilize both labeled and unlabeled data for self-supervised learning, which can largely enrich the Z space for natural generations.

3.3.2   Our Approach: CR-GAN

A single-pathway network, i.e., an encoder–decoder network followed by a discriminator network, may have the issue of learning "incomplete" representations. As illustrated in Fig. 3.10 left, the encoder E and decoder G can "touch" only a subspace of Z since we usually have a limited number of training data. This would lead to a severe issue in testing when using "unseen" data as the input. It is highly possible that E may map the novel input out of the subspace, which inevitably leads to poor generations since G has never "seen" the embedding.

A toy example is used to explain this point. We use Multi-PIE [18] to train a single-pathway network. As shown at the top of Fig. 3.9, the network can generate realistic results on Multi-PIE (the first row), as long as the input image is mapped into the learned subspace. However, when testing "unseen" images from IJB-A [27], the network may produce unsatisfactory results (the second row). In this case, the new image is mapped out of the learned subspace.

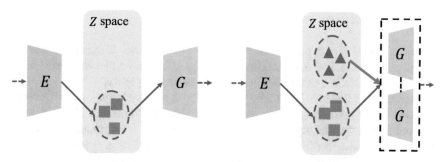

Fig. 3.10 Left: Previous methods use a single path to learn the latent representation, but it is incomplete in the whole space. Right: CR-GAN employs a two-pathway network combined with self-supervised learning, which can learn complete representations

This fact motivates CR-GAN to train E and G that can "cover" the whole Z space, so we can learn complete representations. CR-GAN achieves this goal by introducing a separate generation path, where the generator focuses on mapping the entire Z space to high-quality images. Fig. 3.10 illustrates the comparison between the single-pathway and two-pathway networks.

3.3.2.1 Generation Path

The generation path trains generator G and discriminator D. Here, the encoder E is not involved since G tries to generate from random noise. Given a view label v and random noise $\mathbf{z}$, G aims to produce a realistic image $G(v, \mathbf{z})$ under view v. D is trying to distinguish real data from G's output, which minimizes:

$$
\mathop{\mathbb{E}}_{\mathbf{z} \sim \mathbb{P}_\mathbf{z}} [D_s(G(v, \mathbf{z}))] - \mathop{\mathbb{E}}_{\mathbf{x} \sim \mathbb{P}_\mathbf{x}} [D_s(\mathbf{x})] +
$$
$$
\lambda_1 \mathop{\mathbb{E}}_{\hat{\mathbf{x}} \sim \mathbb{P}_{\hat{\mathbf{x}}}} [(\|\nabla_{\hat{\mathbf{x}}} D(\hat{\mathbf{x}})\|_2 - 1)^2] - \lambda_2 \mathop{\mathbb{E}}_{\mathbf{x} \sim \mathbb{P}_\mathbf{x}} [P(D_v(\mathbf{x}) = v)],
\tag{3.15}
$$

where $\mathbb{P}_\mathbf{x}$ is the data distribution and $\mathbb{P}_\mathbf{z}$ is the noise uniform distribution, $\mathbb{P}_{\hat{\mathbf{x}}}$ is an interpolation between pairs of points sampled from data distribution and the generator distribution [19]. G tries to fool D, it maximizes:

$$
\mathop{\mathbb{E}}_{\mathbf{z} \sim \mathbb{P}_\mathbf{z}} [D_s(G(v, \mathbf{z}))] + \lambda_3 \mathop{\mathbb{E}}_{\mathbf{z} \sim \mathbb{P}_\mathbf{z}} [P(D_v(G(v, \mathbf{z})) = v)],
\tag{3.16}
$$

where $(D_v(\cdot), D_s(\cdot)) = D(\cdot)$ denotes pairwise outputs of the discriminator. $D_v(\cdot)$ estimates the probability of being a specific view, $D_s(\cdot)$ describes the image quality, *i.e.*, how real the image is. Note that in Eq. 3.15, D learns how to estimate the correct view of a real image [41], while G tries to produce an image with that view in order to get a high score from D in Eq. 3.16.

3.3.2.2   Reconstruction Path

The reconstruction path trains E and D but keeping G fixed. E tries to reconstruct training samples, this would guarantee that E will be learned as an inverse of G, yielding complete representations in the latent embedding space.

The output of E should be identity-preserved so the multi-view images will present the same identity. We propose a cross-reconstruction task to make E disentangle the viewpoint from the identity. More specifically, we sample a real image pair $(\mathbf{x}_i, \mathbf{x}_j)$ that share the same identity but different views v_i and v_j. The goal is to reconstruct x_j from x_i. To achieve this, E takes $\mathbf{x}_i$ as input and outputs an identity-preserved representation $\bar{\mathbf{z}}$ together with the view estimation $\bar{\mathbf{v}}$: $(\bar{\mathbf{v}}, \bar{\mathbf{z}}) = (E_v(\mathbf{x}_i), E_z(\mathbf{x}_i)) = E(\mathbf{x}_i)$.

G takes $\bar{\mathbf{z}}$ and view v_j as the input. As $\bar{\mathbf{z}}$ is expected to carry the identity information of this person, with view v_j's help, G should produce $\tilde{\mathbf{x}}_j$, the reconstruction of $\mathbf{x}_j$. D is trained to distinguish the fake image $\tilde{\mathbf{x}}_j$ from the real one $\mathbf{x}_i$. Thus, D minimizes:

$$\mathop{\mathbb{E}}_{\mathbf{x}_i,\mathbf{x}_j \sim \mathbb{P}_\mathbf{x}} [D_s(\tilde{\mathbf{x}}_j) - D_s(\mathbf{x}_i)]+$$
$$\lambda_1 \mathop{\mathbb{E}}_{\hat{\mathbf{x}} \sim \mathbb{P}_{\hat{\mathbf{x}}}} [(\|\nabla_{\hat{\mathbf{x}}} D(\hat{\mathbf{x}})\|_2 - 1)^2] - \lambda_2 \mathop{\mathbb{E}}_{\mathbf{x}_i \sim \mathbb{P}_x} [P(D_v(\mathbf{x}_i) = v_i)], \tag{3.17}$$

where $\tilde{\mathbf{x}}_j = G(v_j, E_z(\mathbf{x}_i))$. E helps G to generate high-quality image with view v_j, so E maximizes:

$$\mathop{\mathbb{E}}_{\mathbf{x}_i,\mathbf{x}_j \sim \mathbb{P}_\mathbf{x}} [D_s(\tilde{\mathbf{x}}_j) + \lambda_3 P(D_v(\tilde{\mathbf{x}}_j) = v_j)-$$
$$\lambda_4 L_1(\tilde{\mathbf{x}}_j, \mathbf{x}_j) - \lambda_5 L_v(E_v(\mathbf{x}_i), v_i)], \tag{3.18}$$

where L_1 loss is utilized to enforce that $\tilde{x}_j$ is the reconstruction of x_j. L_v is the cross-entropy loss of estimated and ground-truth views, to let E be a good view estimator.

The two-pathway network learns complete representations: First, in the generation path, G learns how to produce real images from *any* inputs in the latent space. Then, in the reconstruction path, G retains the generative ability since it keeps unchanged. The alternative training details of the two pathways are summarized in Algorithm 1.

3.3.3   Results: Multi-view Facial Image Generation

CR-GAN aims to learn complete representations in the embedding space, which achieves this goal by combining the two-pathway architecture with self-supervised learning. In this section, we show results of CR-GAN when applied to multi-view facial image generation. We also show results which compare CR-GAN with DR-GAN [58], and both the visual results and t-SNE visualization in the embedding space are shown.

Algorithm 1: Supervised training algorithm of CR-GAN

Input: Sets of view labeled images X, max number of steps T, and batch size m.
Output: Trained network E, G and D.
for $t = 1$ **to** T **do**

    for $i = 1$ **to** m **do**

        1. Sample $\mathbf{z} \sim P_\mathbf{z}$ and $\mathbf{x}_i \sim P_\mathbf{x}$ with v_i;
        2. $\bar{\mathbf{x}} \leftarrow G(v_i, \mathbf{z})$;
        3. Update D by Eq. 3.15, and G by Eq. 3.16;
        4. Sample $\mathbf{x}_j \sim P_\mathbf{x}$ with v_j (where $\mathbf{x}_j$ and $\mathbf{x}_i$ share the same identity);
        5. $(\bar{\mathbf{v}}, \bar{\mathbf{z}}) \leftarrow E(\mathbf{x}_i)$;
        6. $\tilde{\mathbf{x}}_j \leftarrow G(v_j, \bar{\mathbf{z}})$;
        7. Update D by Eq. 3.17, and E by Eq. 3.18;

    end

end

3.3.3.1 Experimental Settings

We evaluate CR-GAN on datasets with and without view labels. Multi-PIE [18] is a labeled dataset collected under constrained environment. We use 250 subjects from the first session with 9 poses within $\pm 60°$, 20 illuminations, and 2 expressions. The first 200 subjects are for training and the rest 50 for testing. 300wLP [73] is augmented from 300W [50] by the face profiling approach [73], which contains view labels as well. We employ images with yaw angles ranging from $-60°$ to $+60°$, and discretize them into nine intervals.

For evaluation on unlabeled datasets, we use CelebA [32] and IJB-A [27]. CelebA contains a large amount of celebrity images with unbalanced viewpoint distributions. Thus, we collect a subset of 72,000 images from it, which uniformly ranging from $-60°$ to $+60°$. Notice that the view labels of the images in CelebA are only utilized to collect the subset, while no view or identity labels are employed in the training process. We also use IJB-A which contains 5,396 images for evaluation. This dataset is challenging since there are extensive identity and pose variations.

3.3.3.2 Multi-view Facial Image Generation

CR-GAN is employed in the following results. We use Multi-PIE and 300wLP in supervised learning. For self-supervised learning, in addition to the above datasets, CelebA is employed as well. Note that we don't use view or identity labels in CelebA during training.

Figure 3.11 shows the results generated by CR-GAN from Multi-PIE, IJB-A, and CelebA. CR-GAN produces desirable results. In each view, facial attributes are kept well, and CR-GAN consistently produces natural images with more details and fewer artifacts. These results prove that CR-GAN handles "unseen" data well by learning a complete representation in the embedding space.

| Input | -60° | -45° | -30° | -15° | 0° | 15° | 30° | 45° | 60° |

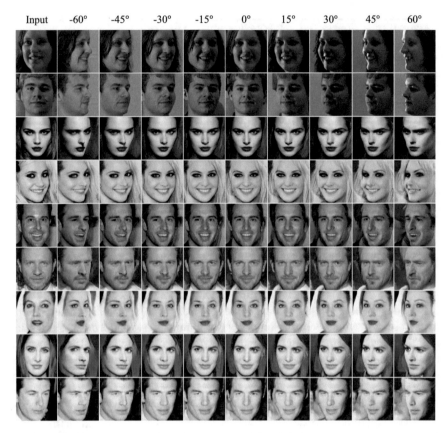

Fig. 3.11 Results generated by CR-GAN from Multi-PIE, IJB-A, and CelebA

3.3.3.3 Comparison with DR-GAN

Furthermore, we compare CR-GAN with DR-GAN [58]. We replace DC-GAN [47] network architecture used in DR-GAN with WGAN-GP for a fair comparison.

We generate nine views for each image in IJB-A both using DR-GAN and CR-GAN. Then we obtain a 128-dim feature for each view by FaceNet [51]. We evaluate the identity similarities between the real and generated images by feeding them to FaceNet. The squared L2 distances of the features directly corresponding to the face similarity: faces of the same subjects have small distances, while faces of different subjects have large distances. Table 3.7 shows the results of the average L2 distance of CR-GAN and DR-GAN in different datasets. Our method outperforms DR-GAN on all datasets, especially on IJB-A which contains unseen data. Figure 3.12 shows the t-SNE visualization in the embedding space of DR-GAN and CR-GAN respectively. For clarity, we only visualize ten randomly selected subjects along with nine generated views of each. Compared with DR-GAN, CR-GAN produces tighter clus-

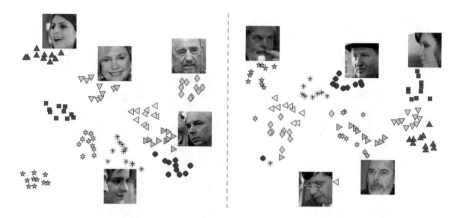

Fig. 3.12 t-SNE visualization for the embedding space of CR-GAN (**left**) and DR-GAN (**right**), with 10 subjects from IJB-A [27]. The same marker shape (color) indicates the same subject. For CR-GAN, multi-view images of the same subject are embedded close to each other, which means the identities are better preserved

Table 3.7 Identity similarities between real and generated images

	Multi-PIE	CelebA	IJB-A
DR-GAN	1.073 ± 0.013	1.281 ± 0.007	1.295 ± 0.008
CR-GAN	**1.018 ± 0.019**	**1.214 ± 0.009**	**1.217 ± 0.010**

terings: multi-view images of the same subject are embedded close to each other. It means the identities are better preserved.

We utilize DR-GAN and CR-GAN to generate images from random noises. In Fig. 3.13, CR-GAN can produce images with different styles, while DR-GAN leads to blurry results. This is because the single-pathway generator of DR-GAN learns incomplete representations in the embedding space, which fails to handle random inputs. Instead, CR-GAN produces favorable results with complete embeddings.

3.3.4 Results: Conditional Facial Attribute Manipulation

In this section, we show results of CR-GAN on CelebA [32] for facial attribute manipulation. The network implementation is modified from the residual networks in WGAN-GP [19], where E shares a similar network structure with D. 11D of our 128D embedding vector is for binary attributes while the remaining are for disentangled information.

Fig. 3.13 Generating multi-view images from the random noise. **a** DR-GAN generates blurry results and many artifacts. **b** CR-GAN generates realistic images of different styles

3.3.4.1 Single-Attribute Manipulation

For single-attribute manipulation, CR-GAN first gets the embedding through the encoder, then generates samples by feeding G with the embedding and attribute labels. One attribute is reversed each time. For example, if the input is male, the female image will be produced. The generated samples with 128×128 resolution are shown in Fig. 3.14. We can see that identities are disentangled and kept well in the results.

Fig. 3.14 Single-attribute manipulation results (128×128). Each column shows the results by reversing one attribute. Note that the identities are well preserved in each row

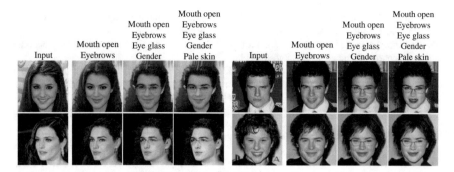

Fig. 3.15 Multiple-attribute manipulation results (128 × 128). Each column reverses multiple attributes

3.3.4.2 Multiple-attribute Manipulation

With the same settings, we further change multiple attributes at a time. The results are shown in Fig. 3.15. We can find that CR-GAN can generate high-quality images even when five attributes are changed at the same time. Besides the high visual quality, male outputs of the left female images presented in column 3 demonstrate how CR-GAN works: the model not only changes the gender but also tries to transform long hair into the background. The transformation indicates that D has learned that male-style faces should always come along with short hair. This learning ability distinguishes CR-GAN from image editing works like StarGAN [10].

3.4 Conclusion

In this chapter, we have presented the state of the art in disentanglement learning with application to face analytics. This is a fundamental and active research area due to the complex nature of visual data which results in entangled features in representation space. The data complexities are even higher in dynamic scenes and we are currently developing new feature disentanglement methods to address them.

Acknowledgements This work was funded partly by ARO-MURI-68985NSMUR and NSF 1763523, 1747778, 1733843, 1703883 to Dimitris N. Metaxas.

References

1. Arjovsky M, Chintala S, Bottou L (2017) Wasserstein generative adversarial networks. In: ICML
2. Asthana A, Zafeiriou S, Cheng S, Pantic M (2013) Robust discriminative response map fitting with constrained local models. In: The IEEE conference on computer vision and pattern recognition, pp 3444–3451

3. Asthana A, Zafeiriou S, Cheng S, Pantic M (2014) Incremental face alignment in the wild. In: The IEEE conference on computer vision and pattern recognition
4. Badrinarayanan V, Kendall A, Cipolla R (2015) Segnet: a deep convolutional encoder-decoder architecture for image segmentation. CoRR
5. Belhumeur PN, Jacobs DW, Kriegman DJ, Kumar N (2011) Localizing parts of faces using a consensus of exemplars. In: The IEEE conference on computer vision and pattern recognition
6. Blanz V, Vetter T (1999) A morphable model for the synthesis of 3D faces. In: SIGGRAPH
7. Bulat A, Tzimiropoulos G (2016) Human pose estimation via convolutional part heatmap regression. Springer International Publishing, Cham, pp 717–732
8. Cao X, Wei Y, Wen F, Sun J (2014) Face alignment by explicit shape regression. Int J Comput Vis 107(2):177–190
9. Cho K, van Merrienboer B, Bahdanau D, Bengio Y (2014) On the properties of neural machine translation: encoder-decoder approaches. CoRR. arXiv:1409.1259
10. Choi Y, Choi M, Kim M, Ha J-W, Kim S, Choo J (2018) Stargan: unified generative adversarial networks for multi-domain image-to-image translation. In: IEEE conference on computer vision and pattern recognition
11. Cootes TF, Taylor CJ (1992) Active shape models—smart snakes. In: British machine vision conference
12. Donahue J, Krähenbühl P, Darrell T (2017) Adversarial feature learning. In: ICLR
13. Dumoulin V, Belghazi I, Poole B, Lamb A, Arjovsky M, Mastropietro O, Courville A (2017) Adversarially learned inference. In: ICLR
14. Fan M, Zhou Q, Zheng TF (2016) Learning embedding representations for knowledge inference on imperfect and incomplete repositories. In: Web intelligence (WI)
15. Gao X, Su Y, Li X, Tao D (2010) A review of active appearance models. IEEE Trans Syst Man Cybern 40(2):145–158
16. Goodfellow I, Pouget-Abadie J, Mirza M, Xu B, Warde-Farley D, Ozair S, Courville A, Bengio Y (2014) Generative adversarial nets. In: NIPS
17. Gross R, Matthew I, Cohn J, Kanade T, Baker S (2009) Multiple. Image Vis Comput
18. Gross R, Matthews I, Cohn J, Kanade T, Baker S (2010) Multiple. Image Vis Comput 28(5):807–813
19. Gulrajani I, Ahmed F, Arjovsky M, Dumoulin V, Courville A (2017) Improved training of wasserstein GANs. In: NIPS
20. Hardoon DR, Szedmak S, Shawe-Taylor J (2004) Canonical correlation analysis: an overview with application to learning methods. Neural Comput 16(12):2639–2664
21. Hinton GE, Krizhevsky A, Wang SD (2011) Transforming auto-encoders. In: ICANN
22. Hong S, Noh H, Han B (2015) Decoupled deep neural network for semi-supervised semantic segmentation. CoRR. arXiv:1506.04924
23. Huang R, Zhang S, Li T, He R (2017) Beyond face rotation: global and local perception GAN for photorealistic and identity preserving frontal view synthesis. In: ICCV
24. Ioffe S, Szegedy C (2015) Batch normalization: accelerating deep network training by reducing internal covariate shift. CoRR. arXiv:1502.03167
25. Jourabloo A, Liu X (2016) Large-pose face alignment via CNN-based dense 3D model fitting. In: The IEEE conference on computer vision and pattern recognition
26. Karpathy A, Fei-Fei L (2015) Deep visual-semantic alignments for generating image descriptions. In: The IEEE conference on computer vision and pattern recognition
27. Klare BF, Klein B, Taborsky E, Blanton A, Cheney J, Allen K, Grother P, Mah A, Jain AK (2015) Pushing the frontiers of unconstrained face detection and recognition: Iarpa janus benchmark-a. In: CVPR
28. Koestinger M, Wohlhart P, Roth PM, Bischof H (2011) Annotated facial landmarks in the wild: a large-scale, real-world database for facial landmark localization. In: Workshop on benchmarking facial image analysis technologies
29. Krizhevsky A, Sutskever I, Hinton GE (2012) Imagenet classification with deep convolutional neural networks. In: Proceedings of the international conference on neural information processing systems

30. Le V, Brandt J, Lin Z, Bourdev L, Huang TS (2012) Interactive facial feature localization. In: European conference on computer vision, pp 679–692
31. Li Y, Yang M, Zhang Z (2016) Multi-view representation learning: a survey from shallow methods to deep methods. arXiv:1610.01206
32. Liu Z, Luo P, Wang X, Tang X (2015) Deep learning face attributes in the wild. In: Proceedings of the IEEE international conference on computer vision, pp 3730–3738
33. Long JL, Zhang N, Darrell T (2014a) Do convnets learn correspondence?. In: Proceedings of the international conference on neural information processing systems, pp 1601–1609
34. Long J, Shelhamer E, Darrell T (2014b) Fully convolutional networks for semantic segmentation. CoRR. arXiv:1411.4038
35. Lu L, Zhang X, Cho K, Renals S (2015) A study of the recurrent neural network encoder-decoder for lar GE vocabulary speech recognition. In: INTERSPEECH
36. Mikolov T, Joulin A, Chopra S, Mathieu M, Ranzato M (2014) Learning longer memory in recurrent neural networks. CoRR. arXiv:1412.7753
37. Mikolov T, Karafiát M, Burget L, Černocký J, Khudanpur S (2010) Recurrent neural network based language model. In: INTERSPEECH
38. Milborrow S, Nicolls F (2008) Locating facial features with an extended active shape model. In: European conference on computer vision, pp 504–513
39. Nielsen AA (2002) Multiset canonical correlations analysis and multispectral, truly multitemporal remote sensing data. IEEE Trans Image Process 11(3):293–305
40. Odena A, Olah C, Shlens J (2017) Conditional image synthesis with auxiliary classifier GANS. In: ICML
41. Parkhi OM, Vedaldi A, Zisserman A (2015) Deep face recognition. In: British machine vision conference
42. Peng X, Feris RS, Wang X, Metaxas DN (2016) A recurrent encoder-decoder network for sequential face alignment. In: European conference on computer vision. Springer International Publishing, pp 38–56
43. Peng X, Huang J, Hu Q, Zhang S, Elgammal A, Metaxas D (2015) From circle to 3-sphere: head pose estimation by instance parameterization. Comput Vis Image Underst 136:92–102
44. Peng X, Yu X, Sohn K, Metaxas DN, Chandraker M (2017) Reconstruction-based disentanglement for pose-invariant face recognition. In: Proceedings of the IEEE international conference on computer vision
45. Peng X, Zhang S, Yang Y, Metaxas DN (2015) Piefa: personalized incremental and ensemble face alignment. In: Proceedings of the IEEE international conference on computer vision workshops, pp 3880–3888
46. Radford A, Metz L, Chintala S (2016) Unsupervised representation learning with deep convolutional generative adversarial networks. In: ICLR
47. Ren S, He K, Girshick R, Sun J (2015) Faster R-CNN: Towards real-time object detection with region proposal networks. In: Cortes C, Lawrence ND, Lee DD, Sugiyama M, Garnett R (eds) Advances in neural information processing systems 28. Curran Associates, Inc., pp 91–99
48. Rezende DJ, Eslami SA, Mohamed S, Battaglia P, Jaderberg M, Heess N (2016) Unsupervised learning of 3D structure from images. In: NIPS
49. Sagonas C, Tzimiropoulos G, Zafeiriou S, Pantic M (2013) 300 faces in-the-wild challenge: The first facial landmark localization challenge. In: ICCVW
50. Schroff F, Kalenichenko D, Philbin J (2015) Facenet: a unified embedding for face recognition and clustering. In: The IEEE conference on computer vision and pattern recognition, pp 815–823
51. Shen J, Zafeiriou S, Chrysos G, Kossaifi J, Tzimiropoulos G, Pantic M (2015) The first facial landmark tracking in-the-wild challenge: benchmark and results. In: Proceedings of the IEEE international conference on computer vision workshops
52. Sun Y, Wang X, Tang X (2013) Deep convolutional network cascade for facial point detection. In: The IEEE conference on computer vision and pattern recognition, pp 3476–3483
53. Taigman Y, Yang M, Ranzato M, Wolf L (2014) Deepface: closing the gap to human-level performance in face verification. In: The IEEE conference on computer vision and pattern recognition

54. Tang M, Peng X (2012) Robust tracking with discriminative ranking lists. IEEE Trans Image Process 21(7):3273–3281
55. Tang M, Peng X, Chen D (2010) Robust tracking with discriminative ranking lists. In: Asian conference on computer vision. Springer, pp 283–295
56. Tian Y, Peng X, Zhao L, Zhang S, Metaxas DN (2018) CR-GAN: learning complete representations for multi-view generation. In: Proceedings of the international joint conference on artificial intelligence (IJCAI), pp 942–948
57. Tran L, Yin X, Liu X (2017) Disentangled representation learning GAN for pose-invariant face recognition. In: CVPR
58. Tzimiropoulos G (2015) Project-out cascaded regression with an application to face alignment. In: The IEEE conference on computer vision and pattern recognition, pp 3659–3667
59. Veeriah V, Zhuang N, Qi GJ (2015) Differential recurrent neural networks for action recognition. In: Proceedings of the IEEE international conference on computer vision
60. Wang X, Yang M, Zhu S, Lin Y (2015) Regionlets for generic object detection. IEEE Trans Pattern Anal Mach Intell 37(10):2071–2084
61. Wu Y, Ji Q (2016) Constrained joint cascade regression framework for simultaneous facial action unit recognition and facial landmark detection. In: The IEEE conference on computer vision and pattern recognition
62. Xuehan-Xiong & De la Torre F (2013) Supervised descent method and its application to face alignment. In: The IEEE conference on computer vision and pattern recognition
63. Yan X, Yang J, Yumer E, Guo Y, Lee H (2016) Perspective transformer nets: learning single-view 3D object reconstruction without 3D supervision. In: NIPS
64. Yang J, Reed S, Yang M-H, Lee H (2015) Weakly-supervised disentangling with recurrent transformations for 3D view synthesis. In: Proceedings of the international conference on neural information processing systems
65. Yao L, Torabi A, Cho K, Ballas N, Pal C, Larochelle H, Courville A (2015) Describing videos by exploiting temporal structure. In: Proceedings of the IEEE international conference on computer vision
66. Yin X, Yu X, Sohn K, Liu X, Chandraker M (2017) Towards large-pose face frontalization in the wild. In: ICCV
67. Zhang J, Shan S, Kan M, Chen X (2014) Coarse-to-fine auto-encoder networks (CFAN) for real-time face alignment. In: European conference on computer vision, pp 1–16
68. Zhao B, Wu X, Cheng Z-Q, Liu H, Feng J (2017) Multi-view image generation from a single-view. arXiv:1704.04886
69. Zhao L, Peng X, Tian Y, Kapadia M, Metaxas D (2018) Learning to forecast and refine residual motion for image-to-video generation. In: European conference on computer vision (ECCV), pp 387–403
70. Zheng S, Jayasumana S, Romera-Paredes B, Vineet V, Su Z, Du D, Huang C, Torr PHS (2015) Conditional random fields as recurrent neural networks. In: Proceedings of the IEEE international conference on computer vision
71. Zhou T, Tulsiani S, Sun W, Malik J, Efros AA (2016) View synthesis by appearance flow. In: ECCV
72. Zhu S, Li C, Change Loy C, Tang X (2015) Face alignment by coarse-to-fine shape searching. In: The IEEE conference on computer vision and pattern recognition
73. Zhu X, Lei Z, Liu X, Shi H, Li S (2016) Face alignment across large poses: a 3D solution. In: The IEEE conference on computer vision and pattern recognition
74. Zhu Z, Luo P, Wang X, Tang X (2013) Deep learning identity-preserving face space. In: ICCV
75. Zhu Z, Luo P, Wang X, Tang X (2014) Multi-view perceptron: a deep model for learning face identity and view representations. In: Proceedings of the international conference on neural information processing systems

Chapter 4
Learning 3D Face Morphable Model from In-the-Wild Images

Luan Tran and Xiaoming Liu

As a classic statistical model of 3D facial shape and albedo, 3D Morphable Model (3DMM) is widely used in facial analysis, e.g., model fitting, and image synthesis. Conventional 3DMM is learned from a set of 3D face scans with associated well-controlled 2D face images, and represented by two sets of PCA basis functions. Due to the type and amount of training data, as well as, the linear bases, the representation power of 3DMM can be limited. To address these problems, this chapter presents an innovative framework to learn a nonlinear 3DMM model from a large set of in-the-wild face images, without collecting 3D face scans. Specifically, given a face image as input, a network encoder estimates the projection, lighting, shape, and albedo parameters. Two decoders serve as the nonlinear 3DMM to map from the shape and albedo parameters to the 3D shape and albedo, respectively. With the projection parameter, lighting, 3D shape, and albedo, a novel analytically- differentiable rendering layer is designed to reconstruct the original input face. The entire network is end-to-end trainable with only weak supervision. We demonstrate the superior representation power of our nonlinear 3DMM over its linear counterpart, and its contribution to face alignment, 3D reconstruction, and face editing.

L. Tran · X. Liu (✉)
East Lansing, MI, USA
e-mail: liuxm@cse.msu.edu

L. Tran
e-mail: tranluan@msu.edu

© The Author(s), under exclusive license to Springer Nature Switzerland AG 2021   73
N. K. Ratha et al. (eds.), *Deep Learning-Based Face Analytics*, Advances in Computer Vision and Pattern Recognition, https://doi.org/10.1007/978-3-030-74697-1_4

4.1 Introduction

The 3D Morphable Model (3DMM) is a statistical model of 3D facial shape and texture in a space where there are explicit correspondences [4]. The morphable model framework provides two key benefits: firstly, a point-to-point correspondence between the reconstruction and all other models, enabling morphing, and secondly, modeling underlying transformations between types of faces (male to female, neutral to smile, etc.). 3DMM has been widely applied in numerous areas including, but not limited to, computer vision [4, 61, 72], computer graphics [1, 51, 57, 58], human behavioral analysis [2, 71] and craniofacial surgery [54].

Traditionally, 3DMM is learnt through *supervision* by performing dimension reduction, typically Principal Component Analysis (PCA), on a training set of co-captured 3D face scans and 2D images. To model highly variable 3D face shapes, a large amount of high-quality 3D face scans is required. However, this requirement is expensive to fulfill as acquiring face scans is very laborious, in both data capturing and post-processing stage. The first 3DMM [4] was built from scans of 200 subjects with a similar ethnicity/age group. They were also captured in well-controlled conditions, with only neutral expressions. Hence, it is fragile to large variances in the face identity. The widely used Basel Face Model (BFM) [42] is also built with only 200 subjects in neutral expressions. Lack of expression can be compensated using expression bases from FaceWarehouse [11] or BD-3FE [70], which are learned from the offsets to the neutral pose. After more than a decade, almost all existing models use no more than 300 training scans. Such small training sets are far from adequate to describe the full variability of human faces [9]. Until recently, with a significant effort as well as a novel automated and robust model construction pipeline, Booth et al. [9] build the first large-scale 3DMM from scans of ∼10,000 subjects.

Second, the texture model of 3DMM is normally built with a small number of 2D face images *co-captured* with 3D scans, under well-controlled conditions. Despite there is a considerable improvement of 3D acquisition devices in the last few years, these devices still cannot operate in arbitrary in-the-wild conditions. Therefore, all the current 3D facial datasets have been captured in the laboratory environment. Hence, such models are only learnt learned to represent the facial texture in similar, rather than in-the-wild, conditions. This substantially limits its application scenarios.

Finally, the representation power of 3DMM is limited by not only the size or type of training data but also its *formulation*. The facial variations are nonlinear in nature., Ee.g., the variations in different facial expressions or poses are nonlinear, which violates the linear assumption of PCA-based models. Thus, a PCA model is unable to interpret facial variations sufficiently well. This is especially true for facial texture. For all current 3DMM models, their low-dimensional albedo subspace faces the same problem of lacking facial hair, e.g., beards. To reduce the fitting error, it compensates unexplainable texture by alternating surface normal, or shrinking the face shape [75]. Either way, linear 3DMM-based applications often degrade their performances when handling out-of-subspace variations.

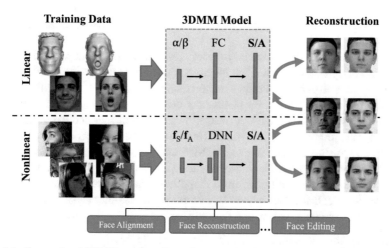

Fig. 4.1 Conventional 3DMM employs linear bases models for shape/albedo, which are trained with 3D face scans and associated controlled 2D images. We propose a nonlinear 3DMM to model shape/albedo via deep neural networks (DNNs). It can be trained from in-the-wild face images without 3D scans, and also better reconstruct the original images due to the inherent nonlinearity

Given the barrier of 3DMM in its data, supervision and linear bases, this paper aims to revolutionize the paradigm of learning 3DMM by answering a fundamental question:

Whether and how can we learn a nonlinear 3D Morphable Model of face shape and albedo from a set of in-the-wild 2D face images, without collecting 3D face scans?

If the answer were yes, this would be in sharp contrast to the conventional 3DMM approach, and remedy all aforementioned limitations. Fortunately, we have developed approaches to offer positive answers to this question. With the recent development of deep neural networks, we view that it is the right time to undertake this new paradigm of 3DMM learning. Therefore, the core of this paper is regarding how to learn this new 3DMM, what is the representation power of the model, and what is the benefit of the model to facial analysis.

We propose a novel paradigm to *learn a nonlinear 3DMM model from a large in-the-wild 2D face image collection, without acquiring 3D face scans*, by leveraging the power of deep neural networks captures variations and structures in complex face data. As shown in Fig. 4.1, starting with an observation that the linear 3DMM formulation is equivalent to a single- layer network, using a deep network architecture naturally increases the model capacity. Hence, we utilize two convolution neural network decoders, instead of two PCA spaces, as the shape and albedo model components, respectively. Each decoder will take a shape or albedo parameter as input and output the dense 3D face mesh or a face skin reflectant. These two decoders are essentially the nonlinear 3DMM.

Further, we learn the fitting algorithm to our nonlinear 3DMM, which is formulated as a CNN encoder. The encoder network takes a face image as input and generates the shape and albedo parameters, from which two decoders estimate shape and albedo.

The 3D face and albedo would *perfectly* reconstruct the input face, if the fitting algorithm and 3DMM are well ~~learnt~~learned. Therefore, we design a differentiable rendering layer to generate a reconstructed face by fusing the 3D face, albedo, lighting, and the camera projection parameters estimated by the encoder. Finally, the end-to-end learning scheme is constructed where the encoder and two decoders are learnt learned jointly to minimize the difference between the reconstructed face and the input face. Jointly learning the 3DMM and the model fitting encoder allows us to leverage the large collection of *in-the-wild* 2D images without relying on 3D scans. We show significantly improved shape and facial texture representation power over the linear 3DMM. Consequently, this also benefits other tasks such as 2D face alignment, 3D reconstruction, and face editing.

In summary, this work makes the following contributions:

- We learn a *nonlinear* 3DMM model, fully models shape, albedo, and lighting, that has greater representation power than its traditional linear counterpart.
- Both shape and albedo are represented as 2D images, which help to maintain spatial relations as well as leverage CNN power in image synthesis.
- We jointly learn the model and the model fitting algorithm via *weak supervision*, by leveraging a large collection of 2D images without 3D scans. The novel rendering layer enables the end-to-end training.
- The new 3DMM further improves performance in related tasks: face alignment, face reconstruction, and face editing.

4.2 Prior Work

4.2.1 Linear 3DMM

Blanz and Vetter [4] propose the first generic 3D face model learned from scan data. They define a linear subspace to represent shape and texture using ~~principal~~Principal component Component analysis Analysis (PCA) and show how to fit the model to data. Since this seminal work, there has been a large amount of effort on improving 3DMM modeling mechanism. In [4], the dense correspondence between facial mesh is solved with a ~~regularised~~ regularized form of optical flow. However, this technique is only effective in a constrained setting, where subjects share similar ethnicities and ages. To overcome this challenge, Patel and Smith [41] employ a Thin-Plate Splines (TPS) warp [7] to register the meshes into a common reference frame. Alternatively, Paysan et al. [42] use a Nonrigid nonrigid Iterative iterative Closest closest Point point [3] to directly align 3D scans. In a different direction, Amberg et al. [2] extended Blanz and Vetter's PCA-based model to emotive facial shapes by adopting an additional PCA modeling of the residuals from the neutral pose. This results in a single linear model of both identity and expression variation of 3D facial shape. Vlasic et al. [66] use a multilinear model to represent the combined effect of identity and expression variation on the facial shape. Later, Bolkart and Wuhrer [6] show how such a multilinear model can be estimated directly from the 3D scans using a joint optimization over the model parameters and groupwise registration of 3D scans.

4.2.2 Improving Linear 3DMM

With PCA bases, the statistical distribution underlying 3DMM is Gaussian. Koppen et al. [28] argue that single-mode Gaussian can't cannot well represent real-world distribution. They introduce the Gaussian Mixture 3DMM that models the global population as a mixture of Gaussian subpopulations, each with its own mean, but shared covariance. Booth et al. [8] aim to improve texture of 3DMM to go beyond controlled settings by learning in-the-wild feature-based texture model. On another direction, Tran et al. [60] learn to regress robust and discriminative 3DMM representation, by leveraging multiple images from the same subject. However, all works are still based on statistical PCA bases. Duong et al. [38] address the problem of linearity in face modeling by using Deep Boltzmann Machines. However, they only work with 2D face and sparse landmarks; and hence cannot handle faces with large-pose variations or occlusion well. Concurrent to our work, Tewari et al. [55] learn a (potentially non-linear) corrective model on top of a linear model. The final model is a summation of the base linear model and the learned corrective model, which contrasts to with our unified model. Furthermore, our model has thean advantage of using 2D representation of both shape and albedo, which maintains spatial relations between vertices and leverages CNN power for image synthesis. Finally, thanks for to our novel rendering layer, we are able to employ perceptual, adversarial loss to improve the reconstruction quality.

4.2.3 2D Face Alignment

2D Face Alignment [30, 68] can be cast as a regression problem where 2D landmark locations are regressed directly [14]. For large-pose or occluded faces, strong priors of 3DMM face shape have been shown to be beneficial [?]. Hence, there is increasing attention in conducting face alignment by fitting a 3D face model to a single 2D image [24, 25, 74]. Among the prior works, iterative approaches with a cascade of regressors tend to be preferred. At each cascade, there is a single [65] or even two regressors [69] used to improve its prediction. Recently, Jourabloo and Liu [25] propose a CNN architecture that enables the end-to-end training ability of their network cascade. Contrasted to aforementioned works that use a fixed 3DMM model, our model and model fitting are learned jointly. This results in a more powerful model:

a single-pass encoder, which is learned jointly with the model, achieves state-of-the-art face alignment performance on AFLW2000 [74] benchmark dataset.

4.2.4   3D Face Reconstruction

Face reconstruction creates a 3D face model from an image collection [47] or even with a single image [45, 50]. This long-standing problem draws a lot of interest because of its wide applications. 3DMM also demonstrates its strength in face reconstruction, especially in the monocular case. This problem is a highly under-constrained, as with a single image, present information about the surface is limited. Hence, 3D face reconstruction must rely on prior knowledge like 3DMM [48]. Statistical PCA linear 3DMM is the most commonly used approach. Besides 3DMM fitting methods [5, 15, 19, 29, 56, 73], recently, Richardson et al. [46] design a refinement network that adds facial details on top of the 3DMM-based geometry. However, this approach can only learn 2.5D depth map, which loses the correspondence property of 3DMM. The follow- up work by Sela et al. [50] try to overcome this weakness by learning a correspondence map. Despite having some impressive reconstruction results, both these methods are limited by training data synthesized from the linear 3DMM model. Hence, they fail to handle out-of-subspace variations, e.g., facial hair.

4.2.5   Unsupervised Learning in 3DMM

Collecting large-scale 3D scans with detailed labels for learning 3DMM is not an easy task. A few works try to use large-scale synthetic data as in [27, 45], but they don't generalize well as there still be a domain gap with real images. Tewari et al. [56] is are among the first works attempting to learn 3DMM fitting from unlabeled images. They use an unsupervised loss which compares projected textured face mesh with the original image itself. The sparse landmark alignment is also used as an auxiliary loss. Genova et al. [18] further improve this approach by comparing reconstructed images and original input using higher- level features from a pretrained face recognition network. Compared to these work, our work has a different objective of learning a *nonlinear* 3DMM.

4.3   The Proposed Nonlinear 3DMM

In this section, we start by introducing the traditional linear 3DMM and then present our novel nonlinear 3DMM model.

4.3.1 Conventional Linear 3DMM

The 3D Morphable Model (3DMM) [4] and its 2D counterpart, Activeactive Appearance appearance Modelmodel [13, 31, 32], provide parametric models for synthesizing faces, where faces are modeled using two components: shape and albedo (skin reflectant). In [4], Blanz et al. propose to describe the 3D face space with PCA:

$$\mathbf{S} = \bar{\mathbf{S}} + \mathbf{G}, \tag{4.1}$$

where $\mathbf{S} \in \mathbb{R}^{3Q}$ is a 3D face mesh with Q vertices, $\bar{\mathbf{S}} \in \mathbb{R}^{3Q}$ is the mean shape, $\alpha \in \mathbb{R}^{l_s}$ is the shape parameter corresponding to a 3D shape bases $\mathbf{G}$. The shape bases can be further split into $\mathbf{G} = [\mathbf{G}_{id}, \mathbf{G}_{exp}]$, where $\mathbf{G}_{id}$ is trained from 3D scans with neutral expression, and $\mathbf{G}_{exp}$ is from the offsets between expression and neutral scans.

The albedo of the face $\mathbf{A} \in \mathbb{R}^{3Q}$ is defined within the mean shape $\bar{\mathbf{S}}$, which describes the R, G, B colors of Q corresponding vertices. $\mathbf{A}$ is also formulated as a linear combination of basis functions:

$$\mathbf{A} = \bar{\mathbf{A}} + \mathbf{Rfi}, \tag{4.2}$$

where $\bar{\mathbf{A}}$ is the mean albedo, $\mathbf{R}$ is the albedo bases, and $\mathbf{fi} \in \mathbb{R}^{l_T}$ is the albedo parameter.

The 3DMM can be used to synthesize novel views of the face. Firstly, a 3D face is projected onto the image plane with the weak perspective projection model:

$$\mathbf{V} = \mathbf{R} * \mathbf{S}, \tag{4.3}$$

$$g(\mathbf{S}, \mathbf{m}) = \mathbf{V}^{2D} = f * \mathbf{Pr} * \mathbf{V} + \mathbf{t}_{2d} = M(\mathbf{m}) * \begin{bmatrix} \mathbf{S} \\ \mathbf{1} \end{bmatrix}, \tag{4.4}$$

where $g(\mathbf{S}, \mathbf{m})$ is the projection function leading to the 2D positions $\mathbf{V}^{2D}$ of 3D rotated vertices $\mathbf{V}$, f is the scale factor, $\mathbf{Pr} = \begin{bmatrix} 1 & 0 & 0 \\ 0 & 1 & 0 \end{bmatrix}$ is the orthographic projection matrix, $\mathbf{R}$ is the rotation matrix constructed from three rotation angles (pitch, yaw, roll), and $\mathbf{t}_{2d}$ is the translation vector. While the project matrix M is of the size of 2×4, it has six degrees of freedom, which is parameterized by a 6-dim vector $\mathbf{m}$. Then, the 2D image is rendered using texture and an illumination model such as Phong reflection model [43] or Spherical Harmonics [44].

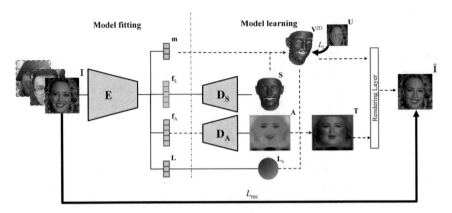

Fig. 4.2 Jointly learning a nonlinear 3DMM and its fitting algorithm from unconstrained 2D in-the-wild face image collection, in a weakly supervised fashion. L_S is a visualization of shading on a sphere with lighting parameters **L**

4.3.2 Nonlinear 3DMM

As mentioned in Sect. 4.1, the linear 3DMM has the problems such as requiring 3D face scans for supervised learning, unable to leverage massive in-the-wild face images for learning, and the limited representation power due to the linear bases. We propose to learn a nonlinear 3DMM model using only large-scale in-the-wild 2D face images.

4.3.2.1 Problem Formulation

In linear 3DMM, the factorization of each of the components (shape, albedo) can be seen as a matrix multiplication between coefficients and bases. From a neural network's perspective, this can be viewed as a shallow network with only *one fully connected layer* and no activation function. Naturally, to increase the model's representation power, the shallow network can be extended to a deep architecture. In this work, we design a novel learning scheme to joint learn a deep 3DMM model and its inference (or fitting) algorithm.

Specifically, as shown in Fig. 4.2, we use two deep networks to decode the shape, albedo parameters into the 3D facial shape and albedo, respectively. To make the framework end-to-end trainable, these parameters are estimated by an encoder network, which is essentially the fitting algorithm of our 3DMM. Three deep networks join forces for the ultimate goal of reconstructing the input face image, with the assistant of a physically- based rendering layer. Figure 4.2 visualizes the architecture of the proposed framework. Each component will be present in the following sections.

Formally, given a set of K 2D face images $\{\mathbf{I}_i\}_{i=1}^{K}$, we aim to learn an encoder $\mathcal{E}$: $\mathbf{I} \rightarrow \mathbf{m}, \mathbf{L}, \mathbf{f}_S, \mathbf{f}_A$ that estimates the projection $\mathbf{m}$, lighting parameter $\mathbf{L}$, shape param-

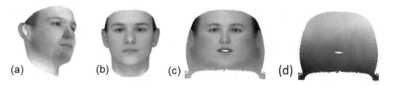

Fig. 4.3 Three albedo representations: **a** Albedo value per vertex, **b** Albedo as a frontal face, **c** UV space 2D unwarped albedo; and **d** UV space 2D unwarped shape

eters $\mathbf{f}_S \in \mathbb{R}^{l_S}$, and albedo parameter $\mathbf{f}_A \in \mathbb{R}^{l_A}$, a 3D shape decoder $\mathcal{D}_S: \mathbf{f}_S \to \mathbf{S}$ that decodes the shape parameter to a 3D shape $\mathbf{S} \in \mathbb{R}^{3Q}$, and an albedo decoder $\mathcal{D}_A: \mathbf{f}_A \to \mathbf{A}$ that decodes the albedo parameter to a realistic albedo $\mathbf{A} \in \mathbb{R}^{3Q}$, with the objective that the rendered image with $\mathbf{m}$, $\mathbf{L}$, $\mathbf{S}$, and $\mathbf{A}$ can well approximate the original image. Mathematically, the objective function is:

$$\underset{\mathcal{E}, \mathcal{D}_S, \mathcal{D}_A}{\arg\min} \sum_{i=1}^{K} \left\| \hat{\mathbf{I}}_i - \mathbf{I}_i \right\|_1 ,$$

$$\hat{\mathbf{I}} = \mathcal{R}\left(E_m(\mathbf{I}), E_L(\mathbf{I}), \mathcal{D}_S(E_S(\mathbf{I})), \mathcal{D}_A(E_A(\mathbf{I})) \right), \tag{4.5}$$

where $\mathcal{R}(\mathbf{m}, \mathbf{L}, \mathbf{S}, \mathbf{A})$ is the rendering layer (Sect. 4.3.2.3).

4.3.2.2 Albedo and Shape Representation

Figure 4.3 illustrates three possible albedo representations. In traditional 3DMM, albedo is defined per vertex (Fig. 4.3a). This representation is also adopted in recent work such as [55, 56]. There is an albedo intensity value corresponding to each vertex in the face mesh. Despite widely used, this representation has its limitations. Since 3D vertices are not defined on a 2D grid, this representation is mostly parameterized as a vector, which not only loses the spatial relation of its vertices, but also prevents it to leverage the convenience of deploying CNN on 2D albedo. In contrast, given the rapid progress in image synthesis, it is desirable to choose a 2D image, e.g., a frontal-view face image in Fig. 4.3b, as an albedo representation. However, frontal faces contain little information of two sides, which would lose many albedo information for side-view faces.

In light of these considerations, we use an unwrapped 2D texture as our texture representation (Fig. 4.3c). Specifically, each 3D vertex $\mathbf{v}$ is projected onto the UV space using cylindrical unwrap. Assuming that the face mesh has the top pointing up the y- axis, the projection of $\mathbf{v} = (x, y, z)$ onto the UV space $\mathbf{v}^{\mathrm{uv}} = (u, v)$ is computed as:

$$v \to \alpha_1 . \arctan\left(\frac{x}{z}\right) + \beta_1, \quad u \to \alpha_2 \cdot y + \beta_2, \tag{4.6}$$

where α_1, α_2, β_1, β_2 are constant scale and translation scalars to place the unwrapped face into the image boundaries. Here, per-vertex albedo $\mathbf{A} \in \mathbb{R}^{3Q}$ could be easily computed by sampling from its UV space counterpart $\mathbf{A}^{uv} \in \mathbb{R}^{U \times V}$:

$$\mathbf{A}(\mathbf{v}) = \mathbf{A}^{uv}(\mathbf{v}^{uv}). \tag{4.7}$$

Usually, it involves sub-pixel sampling via bilinear interpolation:

$$\mathbf{A}(\mathbf{v}) = \sum_{\substack{u' \in \{\lfloor u \rfloor, \lceil u \rceil\} \\ v' \in \{\lfloor v \rfloor, \lceil v \rceil\}}} \mathbf{A}^{uv}(u', v')(1 - |u - u'|)(1 - |v - v'|), \tag{4.8}$$

where $\mathbf{v}^{uv} = (u, v)$ is the UV space projection of $\mathbf{v}$ via Eq. 4.6.

Albedo information is naturally expressed in the UV space but spatial data can be embedded in the same space as well. Here, a 3D facial mesh can be represented as a 2D image with three channels, one for each spatial dimension, x, y, and z. Figure 4.3 gives an example of this UV space shape representation $\mathbf{S}^{uv} \in \mathbb{R}^{U \times V}$.

Representing 3D face shape in UV space allows us to use a CNN for shape decoder $\mathcal{D}_S$ instead of using a multi-layer perceptron (MLP) as in our preliminary version [62]. Avoiding using wide fully-connected layers allows us to use a deeper network for $\mathcal{D}_S$, potentially model more complex shape variations. This results in better fitting results as being demonstrated in our experiment (Sec. spsrefsec:ablsps1dsps2d).

Also, it is worth to note that different from our preliminary version [62] where the reference UV space, for texture, is build upon a projection of the mean shape with neutral expression; in this version, the reference shape used has the mouth open. This change helps the network to avoid learning a large gradient near the two lips' borders in the vertical direction when the mouth is open.

To regress these 2D representations of shape and albedo, we can employ CNNs as shape and albedo networks, respectively. Specifically, $\mathcal{D}_S$, $\mathcal{D}_A$ are CNN constructed by multiple fractionally- strided convolution layers. After each convolution is batch-norm and eLU activation, except the last convolution layers of encoder and decoders. The output layer has a *tanh* activation to constrain the output to be in the range of $[-1, 1]$.

4.3.2.3   In-Network Physically-Based Face Rendering

To reconstruct a face image from the albedo $\mathbf{A}$, shape $\mathbf{S}$, lighting parameter $\mathbf{L}$, and projection parameter $\mathbf{m}$, we define a rendering layer $\mathcal{R}(\mathbf{m}, \mathbf{L}, \mathbf{S}, \mathbf{A})$ to render a face image from the above parameters. This is accomplished in three steps, as shown in Fig. 4.4. Firstly, the facial texture is computed using the albedo $\mathbf{A}$ and the surface normal map of the rotated shape $N(\mathbf{V}) = N(\mathbf{m}, \mathbf{S})$. Here, following [67], we assume distant illumination and a purely *Lambertian* surface reflectance. Hence, the incoming radiance can be approximated using spherical harmonics (SH) basis functions $H_b : \mathbb{R}^3 \to \mathbb{R}$, and controlled by coefficients $\mathbf{L}$. Specifically, the texture in UV space

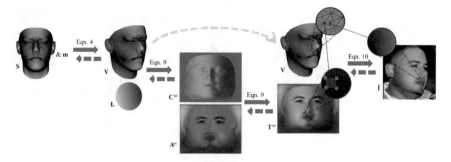

Fig. 4.4 Forward and backward passes of the physically-based rendering layer

$\mathbf{T}^{uv} \in \mathbb{R}^{U \times V}$ is composed of albedo $\mathbf{A}^{uv}$ and shading $\mathbf{C}^{uv}$:

$$\mathbf{T}^{uv} = \mathbf{A}^{uv} \odot \mathbf{C}^{uv} = \mathbf{A}^{uv} \odot \sum_{b=1}^{B^2} L_b H_b(N(\mathbf{m}, \mathbf{S}^{uv})), \qquad (4.9)$$

where B is the number of spherical harmonics bands. We use $B = 3$, which leads to $B^2 = 9$ coefficients in $\mathbf{L}$ for each of three color channels. Secondly, the 3D shape/mesh $\mathbf{S}$ is projected to the image plane via Eq. 4.4. Finally, the 3D mesh is then rendered using a Z-buffer renderer, where each pixel is associated with a single triangle of the mesh,

$$\hat{\mathbf{I}}(m, n) = \mathcal{R}(\mathbf{m}, \mathbf{L}, \mathbf{S}^{uv}, \mathbf{A}^{uv})_{m,n} = \mathbf{T}^{uv}\left(\sum_{\mathbf{v}_i \in \Phi^{uv}(g,m,n)} \lambda_i \mathbf{v}_i \right), \qquad (4.10)$$

where $\Phi(g, m, n) = \{\mathbf{v}_1, \mathbf{v}_2, \mathbf{v}_3\}$ is an operation returning three vertices of the triangle that encloses the pixel (m, n) after projection g; $\Phi^{uv}(g, m, n)$ is the same operation with resultant vertices mapped into the referenced UV space using Eq. 4.6. In order to handle occlusions, when a single- pixel resides in more than one triangle, the triangle that is closest to the image plane is selected. The final location of each pixel is determined by interpolating the location of three vertices via barycentric coordinates $\{\lambda_i\}_{i=1}^3$.

There are alternative designs to our rendering layer. If the texture representation is defined per vertex, as in Fig. 4.3a, one may warp the input image $\mathbf{I}_i$ onto the vertex space of the 3D shape $\mathbf{S}$, whose distance to the per-vertex texture representation can form a reconstruction loss. This design is adopted by the recent work of [55, 56]. In comparison, our rendered image is defined on a 2D grid, while the alternative is on top of the 3D mesh. As a result, our rendered image can enjoy the convenience of applying the perceptual loss or adversarial loss, which is shown to be critical in improving the quality of synthetic texture. Another design for rendering layer is image warping based on the spline interpolation, as in [12]. However, this warping is continuous: every pixel in the input will map to the output. Hence, this warping

Fig. 4.5 Rendering with
segmentation masks. Left to
right: segmentation results,
naive rendering,
occulusion-aware rendering

operation fails in the occluded region. As a result, Cole et al. [12] limit their scope
to only synthesizing frontal-view faces by warping from normalized faces.

4.3.2.4 Occlusion-Aware Rendering

Very often, in-the-wild faces are occluded by glasses, hair, hands, etc. Trying to
reconstruct abnormal occluded regions could make the model learning more difficult
or result in an model with external occlusion baked in. Hence, we propose to use a
segmentation mask to exclude occluded regions in the rendering pipeline:

$$\hat{\mathbf{I}} \leftarrow \hat{\mathbf{I}} \odot \mathbf{M} + \mathbf{I} \odot (1 - \mathbf{M}). \tag{4.11}$$

As a result, these occluded regions won't affect our optimization process. The
foreground mask $\mathbf{M}$ is estimated using the segmentation method given by Nirkin
et al. [39]. Examples of segmentation masks and rendering results can be found in
Fig. 4.5.

4.3.2.5 Model Learning

The entire network is end-to-end trained to reconstruct the input images, with the
loss function:

$$\mathcal{L} = \mathcal{L}_{\text{rec}}(\hat{\mathbf{I}} + \mathcal{L}_{\text{lan}} + \mathcal{L}_{\text{reg}}, \tag{4.12}$$

where the reconstruction loss L_{rec} enforces the rendered image $\hat{\mathbf{I}}$ to be similar to the
input $\mathbf{I}$, the landmark loss $Lland$ enforces geometry constraint, and the regularization
loss $\mathcal{L}_{\text{reg}}$ encourages plausible solutions.

Reconstruction Loss The main objective of the network is to reconstruct the original face via disentangle representation. Hence, we enforce the reconstructed image to be similar to the original input image:

$$\mathcal{L}^i_{\text{rec}} = \frac{1}{|\mathcal{V}|} \sum_{q \in \mathcal{V}} ||\hat{\mathbf{I}}(q) - \mathbf{I}(q)||_2 \qquad (4.13)$$

where $\mathcal{V}$ is the set of all pixels in the images covered by the estimated face mesh. There are different norms can be used to measure the closeness. To better handle outliers, we adopt the robust $l_{2,1}$, where the distance in the 3D RGB color space is based on l_2 and the summation over all pixels enforces sparsity based on l_1-norm [58, 59].

To improve from blurry reconstruction results of l_p losses, in our preliminary work [62], thanks for to our rendering layer, we employ adversarial loss to enhance the image realism. However, adversarial objectives only encourage the reconstruction to be close to the real image distribution but not necessarily the input image. Also, it's known to be not stable to optimize. Here, we propose to use a perceptual loss to enforce the closeness between images $\hat{\mathbf{I}}$ and $\mathbf{I}$, which overcomes both of adversarial loss's weaknesses. Besides encouraging the pixels of the output image $\hat{\mathbf{I}}$ to exactly match the pixels of the input $\mathbf{I}$, we encourage them to have similar feature representations as computed by the loss network φ.

$$\mathcal{L}^f_{\text{rec}} = \frac{1}{|\mathcal{C}|} \sum_{j \in \mathcal{C}} \frac{1}{W_j H_j C_j} ||\varphi_j(\hat{\mathbf{I}}) - \varphi_j(\mathbf{I})||^2_2. \qquad (4.14)$$

We choose VGG-Face [40] as our φ to leverage its face-related features and also because of simplicity. The loss is summed over $\mathcal{C}$, a subset of layers of φ. Here, $\varphi_j(\mathbf{I})$ is the activations of the j-th layer of φ when processing the image $\mathbf{I}$ with dimension $W_j \times H_j \times C_j$. This feature reconstruction loss is one of the perceptual losses widely used in different image processing tasks [23].

The final reconstruction loss is a weighted sum of two terms:

$$\mathcal{L}_{\text{rec}} = \mathcal{L}^i_{\text{rec}} + \lambda_f \mathcal{L}^f_{\text{rec}} \qquad (4.15)$$

Sparse Landmark Alignment To help achieving better model fitting, which in turn helps to improve the model learning itself, we employ the landmark alignment loss, measuring Euclidean distance between estimated and groundtruth landmarks, as an auxiliary task,

$$\mathcal{L}_{\text{lan}} = \left\| M(\mathbf{m}) * \begin{bmatrix} \mathbf{S}(:, \mathbf{d}) \\ \mathbf{1} \end{bmatrix} - \mathbf{U} \right\|^2_2, \qquad (4.16)$$

where $\mathbf{U} \in \mathbb{R}^{2 \times 68}$ is the manually labeled 2D landmark locations, $\mathbf{d}$ is a constant 68-dim vector storing the indexes of 68 3D vertices corresponding to the labeled 2D

Fig. 4.6 Effect of albedo
regularizations: albedo
symmetry ($\mathcal{L}_{\text{sym}}$) & albedo
constancy ($\mathcal{L}_{\text{con}}$)

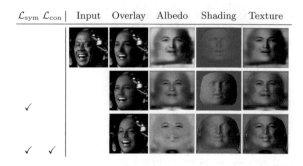

landmarks. Different from traditional face alignment work where the shape bases
are fixed, our work jointly learns the base's functions (i.e., the shape decoder $\mathcal{D}_S$)
as well. Minimizing the landmark loss while updating $\mathcal{D}_S$ only moves a tiny subsets
of vertices. If the shape $\mathbf{S}$ is represented as a vector and $\mathcal{D}_S$ is an MLP consisting of
fully connected layers, vertices are independent. Hence, L_L only adjusts 68 vertices.
In case $\mathbf{S}$ is represented in the UV space and $\mathcal{D}_S$ is a CNN, the local neighbor
region could also be modified. In both cases, updating $\mathcal{D}_S$ based on L_L only moves a
subsets of vertices, which could lead to implausible shapes. Hence, when optimizing
the landmark loss, we fix the decoder $\mathcal{D}_S$ and only update the encoder.

Also, note that different from some prior work [17], our network only requires
ground-truth landmarks during training. It is able to predict landmarks via $\mathbf{m}$ and $\mathbf{S}$
during the test time.

Regularizations To ensure plausible reconstruction, we add a few regularization
terms:

$$\mathcal{L}_{\text{reg}} = \mathcal{L}_{\text{sym}}(\mathbf{A}) + \lambda_{\text{con}}\mathcal{L}_{\text{con}}(\mathbf{A}) + \lambda_{\text{smo}}\mathcal{L}_{\text{smo}}(\mathbf{S}). \tag{4.17}$$

Albedo Symmetry: As faces are symmetry, we enforce the symmetry constraint:

$$\mathcal{L}_{\text{sym}} = \left\| \mathbf{A}^{\text{uv}} - \text{flip}(\mathbf{A}^{\text{uv}}) \right\|_1. \tag{4.18}$$

Employing on 2D albedo, this constraint can be easily implemented via a horizontal image flip operation flip().

Albedo Constancy: Using symmetry constraint can help to correct the global shading. However, symmetrical details, i.e., dimples, can still be embedded in the albedo
channel. To further remove shading from the albedo channel, following Retinex theory, which assumes albedo to be piecewise constant, we enforce sparsity in two
directions of its gradient, similar to [36, 53]:

$$\mathcal{L}_{\text{con}} = \sum_{\mathbf{v}_j^{\text{uv}} \in \mathcal{N}_i} \omega(\mathbf{v}_i^{\text{uv}}, \mathbf{v}_j^{\text{uv}}) \left\| \mathbf{A}^{\text{uv}}(\mathbf{v}_i^{\text{uv}}) - \mathbf{A}^{\text{uv}}(\mathbf{v}_j^{\text{uv}}) \right\|_2^p, \tag{4.19}$$

where $\mathcal{N}_i$ denotes a set of 4-pixel neighborhood of pixel $\mathbf{v}_i^{\text{uv}}$. With the assumption that
pixels with the same chromaticity (i.e., $\mathbf{c}(x) = \mathbf{I}(x)/|\mathbf{I}(x)|$) are more likely to have the

Fig. 4.7 Effect of shape
smoothness regularization
($\mathcal{L}_{\mathrm{smo}}$) on learned models

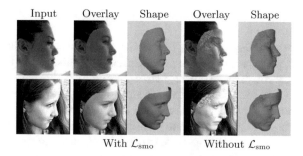

Input   Overlay   Shape   Overlay   Shape

With $\mathcal{L}_{\mathrm{smo}}$ Without $\mathcal{L}_{\mathrm{smo}}$

same albedo, we set the constant weight $\omega(\mathbf{v}_i^{\mathrm{uv}}, \mathbf{v}_j^{\mathrm{uv}}) = \exp\left(-\alpha \left\| \mathbf{c}(\mathbf{v}_i^{\mathrm{uv}}) - \mathbf{c}(\mathbf{v}_j^{\mathrm{uv}}) \right\| \right)$,
where the color is referenced from the input image using the current estimated projection. Following [36], we set $\alpha = 15$ and $p = 0.8$ in our experiment.

Effects of above albedo regularizations are demonstrated in Fig. 4.6. Learning without any constraints results in the lighting is totally explained by the albedo, meanwhile is the shading is almost constant (Fig. 4.6a). Using symmetry helps to correct the global lighting. However, symmetric geometry details are still baked into the albedo (Fig. 4.6b). Enforcing albedo constancy helps to further remove shading from it (Fig. 4.6c). Combining these two regularizations helps to learn plausible albedo and lighting, which improves the shape estimation.

Shape Smoothness: For shape component, we impose the smoothness by adding the Laplacian regularization on the vertex locations for the set of all vertices.

$$\mathcal{L}_{\mathrm{smo}} = \sum_{\mathbf{v}_i^{\mathrm{uv}} \in \mathbf{S}^{\mathrm{uv}}} \left\| \mathbf{S}^{\mathrm{uv}}(\mathbf{v}_i^{\mathrm{uv}}) - \frac{1}{|\mathcal{N}_i|} \sum_{\mathbf{v}_j^{\mathrm{uv}} \in \mathcal{N}_i} \mathbf{S}^{\mathrm{uv}}(\mathbf{v}_j^{\mathrm{uv}}) \right\|_2 . \tag{4.20}$$

Figure 4.7 shows visual comparisons between our model and its variant without the shape smoothness constraint. Without the smoothness term, the learned shape becomes noisy especially on two sides of the face. The reason is that, the hair region is not completely excluded during training because of imprecise segmentation estimation.

Intermediate Semi-Supervised Training Fully unsupervised training using only the reconstruction and adversarial loss on the rendered images could lead to a degenerate solution, since the initial estimation is far from ideal to render meaningful images. Therefore, we introduce intermediate loss functions to guide the training in the early iterations.

With the face profiling technique, Zhu et al. [74] expand the 300W dataset [49] into 122, 450 images with fitted 3DMM shapes $\widetilde{\mathbf{S}}$ and projection parameters $\widetilde{\mathbf{m}}$. Given $\widetilde{\mathbf{S}}$ and $\widetilde{\mathbf{m}}$, we create the pseudo- groundtruth texture $\widetilde{\mathbf{T}}$ by referring every pixel in the UV space back to the input image, i.e., the backward of our rendering layer. With $\widetilde{\mathbf{m}}$, $\widetilde{\mathbf{S}}$, $\widetilde{\mathbf{T}}$, we define our intermediate loss by:

$$\mathcal{L}_0 = \mathcal{L}_S + \lambda_T \mathcal{L}_T + \lambda_m \mathcal{L}_m + \mathcal{L}_{lan} + \mathcal{L}_{reg}, \qquad (4.21)$$

where: $\mathcal{L}_S = ||\mathbf{S} - \tilde{\mathbf{S}}||_2^2, \mathcal{L}_T = ||\mathbf{T} - \tilde{\mathbf{T}}||_1$, and $\mathcal{L}_m = ||\mathbf{m} - \tilde{\mathbf{m}}||_2^2$.

It's also possible to provide pseudo- groundtruth to the SH coefficients $\mathbf{L}$ and followed by albedo $\mathbf{A}$ using least square optimization with a constant albedo assumption, as has been done in [53, 67]. However, this estimation is not reliable for in-the-wild images with occlusion regions. Also empirically, with proposed regularizations, the model is able to explore plausible solutions for these components by itself. Hence, we decide to refrain from supervising $\mathbf{L}$ and $\mathbf{A}$ to simplify our pipeline.

Due to the pseudo- groundtruth, using L_0 may run into the risk that our solution learns to mimic the linear model. Thus, we switch to the loss of Eq. 4.12 after L_0 converges. Note that the estimated groundtruth of $\tilde{\mathbf{m}}, \tilde{\mathbf{S}}, \tilde{\mathbf{T}}$ and the landmarks are the only supervision used in our training, for which our learning is considered as *weakly* supervised.

4.4 Improving Model Fidelity

4.4.1 Nonlinear 3DMM with Proxy and Residual

Proxy and Residual Learning Strong regularization has been shown to be critical in ensuring the plausibility of the learned models [55]. However, the strong regularization also prevents the model from recovering high-level details in either shape or albedo. Hence, this prevents us from achieving the ultimate goal of learning a high-fidelity 3DMM model.

In this work, we propose to learn additional **proxy shape** ($\tilde{\mathbf{S}}$) and **proxy albedo** ($\tilde{\mathbf{A}}$), on which we can apply the regularization. All presented regularizations, as in Eq. 4.17, will now be moved to proxies:

$$\mathcal{L}_{reg}^* = \mathcal{L}_{sym}(\tilde{\mathbf{A}}) + \lambda_{con}\mathcal{L}_{con}(\tilde{\mathbf{A}}) + \lambda_{smo}\mathcal{L}_{smo}(\tilde{\mathbf{S}}). \qquad (4.22)$$

There will be no regularization applied directly to the actual shape $\mathbf{S}$ and albedo $\mathbf{A}$, other than a weak regularization encouraging each to be close to its proxy:

$$\mathcal{L}_{res} = \|\Delta \mathbf{S}\|_1 + \|\Delta \mathbf{A}\|_1 = \left\| \mathbf{S} - \tilde{\mathbf{S}} \right\|_1 + \left\| \mathbf{A} - \tilde{\mathbf{A}} \right\|_1. \qquad (4.23)$$

By pairing two shapes $\mathbf{S}, \tilde{\mathbf{S}}$ and two albedos $\mathbf{A}, \tilde{\mathbf{A}}$, we can render four different output images (Fig. 4.2). Any of them can be used to compare with the original input image. We rewrite our reconstruction loss as:

$$\mathcal{L}_{rec}^* = \mathcal{L}_{rec}(\hat{\mathbf{I}}(\tilde{\mathbf{S}}, \tilde{\mathbf{A}}), \mathbf{I}) + \mathcal{L}_{rec}(\hat{\mathbf{I}}(\tilde{\mathbf{S}}, \mathbf{A}), \mathbf{I}) + \mathcal{L}_{rec}(\hat{\mathbf{I}}(\mathbf{S}, \tilde{\mathbf{A}}), \mathbf{I}). \qquad (4.24)$$

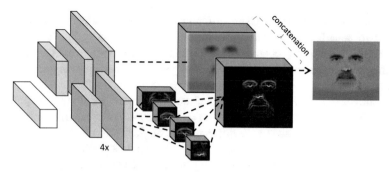

Fig. 4.8 The proposed global–local-based network architecture

Pairing strongly regularized proxies and weakly regularized components is a critical point in our approach. Using proxies allows us to learn high-fidelity shape and albedo without sacrificing the quality of either component. This pairing is inspired by the observation that Shape from Shading techniques are able to recover detailed face mesh by assuming over regularized albedo or even using the mean albedo [46]. Here, $\mathcal{L}_{rec}(\hat{\mathbf{I}}(\mathbf{S}, \tilde{\mathbf{A}}), \mathbf{I})$ loss promotes $\mathbf{S}$ to recover more details as $\tilde{\mathbf{A}}$ is constrained by piece-wise constant $\mathcal{L}_{con}(\tilde{\mathbf{A}})$ objective. Vice versa, $\mathcal{L}_{rec}(\hat{\mathbf{I}}(\tilde{\mathbf{S}}, \mathbf{A}), \mathbf{I})$ aims to learn better albedo. In order for these two losses to work as desired, proxies $\tilde{\mathbf{S}}$ and $\tilde{\mathbf{A}}$ should perform well enough to approximate the input images by themselves. Without $\mathcal{L}_{rec}(\hat{\mathbf{I}}(\tilde{\mathbf{S}}, \tilde{\mathbf{A}}), \mathbf{I})$, a valid solution that minimizes $\mathcal{L}_{rec}(\hat{\mathbf{I}}(\mathbf{S}, \tilde{\mathbf{A}}), \mathbf{I})$ is combination of a constant albedo proxy and noisy shape creating surface normal with dark shading in necessary regions, i.e., eyebrows.

Another notable design choice is that we intentionally left out the loss function on $\hat{\mathbf{I}}(\mathbf{S}, \mathbf{A})$, even though this is theoretically is the most important objective. This is to avoid the case that the shape $\mathbf{S}$ learns an in-between solution that works well with both $\tilde{\mathbf{A}}, \mathbf{A}$ and vice versa.

4.4.2 Global–Local-Based Network Architecture

While global-based models are usually robust to noise and mismatches, they are over-constrained and do not provide sufficient flexibility to represent high-frequency deformations as local-based models. To take the best of both worlds, we propose to use dual-pathway networks for our shape and albedo decoders.

Here, we transfer the success of combining local and global models in image synthesis [21, 37] to 3D face modeling. The general architecture of a decoder is shown in Fig. 4.8. From the latent vector, there is a global pathway focusing on inferring the global structure and a local pathway with four small sub-networks generating details of different facial parts, including eyes, nose, and mouth. The global pathway is built from fractional strided convolution layers with five up-sampling steps. Meanwhile, each sub-network in the local pathway has the similar architecture but shallower with only three up-sampling steps. Using different small sub-networks for each facial part offers two benefits: (i) with less up-sampling steps, the network is better able to represent high-frequency details in early layers; (ii) each sub-network can learn part-specific filters, which is more computationally efficient than applying across global face.

As shown in Fig. 4.8, to fuse two pathways' features, we firstly integrate four local pathways' outputs into one single feature tensor. Different from other works that synthesize face images with different yaw angles [26, 63, 64] with no fixed keypoints' locations, our 3DMM generates facial albedo as well as 3D shape in UV space with predefined topology. Merging these local feature tensors is efficiently done with the zero- padding operation. The max-pooling fusion strategy is also used to reduce the stitching artifacts on the overlapping areas. Then the resultant feature is simply concatenated with the global pathway's feature, which has the same spatial resolution. Successive convolution layers integrate information from both pathways and generate the final albedo/shape (or their proxies).

4.5 Experimental Results

The experiments study three aspects of the proposed nonlinear 3DMM, in terms of its expressiveness, representation power, and applications to facial analysis. Using facial mesh triangle definition by Basel Face Model (BFM) [42], we train our 3DMM using 300W-LP dataset [74], which contains 122, 450 in-the-wild face images, in a wide pose range from $-90°$ to $90°$. Images are loosely square cropped around the face and scale to 256×256. During training, images of size 224×224 are randomly cropped from these images to introduce translation variations.

The model is optimized using Adam optimizer with a learning rate of 0.001 in both training stages. We set following parameters: $Q = 53, 215, U = 192, V = 224$, $l_S = l_T = 160$. λ values are set to make losses to have similar magnitudes.

4.5.1 Ablation Study

Reconstruction Loss Functions. We study effects of different reconstruction losses on quality of the reconstructed images (Fig. 4.9). As expected, the model trained with $l_{2,1}$ loss only results in blurry reconstruction, similar to other l_p loss. To make the reconstruction more realistic, we explore other options such as gradient differ-ence [35] or perceptual loss [23]. While adding the gradient difference loss creates more details in the reconstruction, combining perceptual loss with $l_{2,1}$ gives the best results with high level of details and realism. For the rest of the paper, we will refer to the model trained using this combination.

Understanding image pairing Figure 4.10 shows fitting results of our model on a 2D face image. By using the proxy or the final components (shape or albedov), we can

Fig. 4.9 Reconstruction results with different loss functions

Input

$l_{2,1}$

$l_{2,1}+$ Grad. diff.

$l_{2,1}+$ Perceptual

Fig. 4.10 Image reconstruction using proxies and the true shape and albedo. Our shape and albedo can faithfully recover details of the face

Input $\hat{\mathbf{I}}(\tilde{\mathbf{S}}, \tilde{\mathbf{A}})$ $\hat{\mathbf{I}}(\mathbf{S}, \tilde{\mathbf{A}})$ $\hat{\mathbf{I}}(\tilde{\mathbf{S}}, \mathbf{A})$ $\hat{\mathbf{I}}(\mathbf{S}, \mathbf{A})$

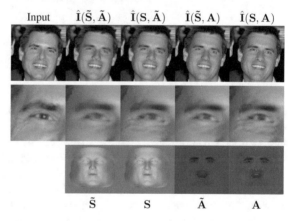

$\tilde{\mathbf{S}}$ $\mathbf{S}$ $\tilde{\mathbf{A}}$ $\mathbf{A}$

render four different reconstructed images with different quality and characteristics. The image generated by two proxies $\tilde{\mathbf{S}}$, $\tilde{\mathbf{A}}$ is quite blurry but is still be able to capture major variations in the input face. By pairing $\mathbf{S}$ and the proxy $\tilde{\mathbf{A}}$, S is enforced to capture high level of details to bring the image closer to the input. Similarly, $\mathbf{A}$ is also encouraged to capture more details by pairing with the proxy $\tilde{\mathbf{S}}$. The final image $\hat{\mathbf{I}}(\mathbf{S}, \mathbf{A})$ inherently achieves high level of details and realism even without direct optimization.

4.5.2 Expressiveness

Exploring feature space We feed the entire CelebA dataset [34] with $\sim$200k images to our network to obtain the empirical distribution of our shape and texture parameters.

Fig. 4.11 Each column shows shape changes when varying one element of $\mathbf{f}_S$, by 10 times standard deviations, in opposite directions

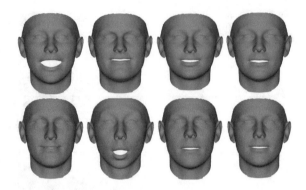

Fig. 4.12 Each column shows albedo changes when varying one element of $\mathbf{f}_A$

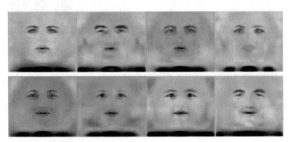

By varying the mean parameter along each dimension proportional to its standard deviation, we can get a sense of how each element contribute to the final shape and texture. We sort elements in the shape parameter $\mathbf{f}_S$ based on their differences to the mean 3D shape. Figure 4.11 shows four examples of shape changes, whose differences rank No.1, 40, 80, and 120 among 160 elements. Most of the top changes are expression- related. Similarly, in Fig. 4.12, we visualize different texture changes by adjusting only one element of $\mathbf{f}_A$ off the mean parameter $\bar{\mathbf{f}}_A$. The elements with the same 4 ranks as the shape counterpart are selected.

Attribute Embedding To better understand different shapes and albedos instances embedded in our two decoders, we dig into their attribute meaning. For a given attribute, e.g., male, we feed images with that attribute $\{\mathbf{I}_i\}_{i=1}^n$ into our encoder E to obtain two sets of parameters $\{\mathbf{f}_S^i\}_{i=1}^n$ and $\{\mathbf{f}_A^i\}_{i=1}^n$. These sets represent corresponding empirical distributions of the data in the low-dimensional spaces. Computing the mean parameters $\bar{\mathbf{f}}_S$, $\bar{\mathbf{f}}_A$ and feed into their respective decoders, also using the mean lighting parameter, we can reconstruct the mean shape and texture with that attribute. Figure 4.13 visualizes the reconstructed textured 3D mesh related to some attributes. Differences among attributes present in both shape and texture. Here, we can observe the power of our nonlinear 3DMM to model small details such as "bag under eyes,", or "rosy cheeks,", etc.

Fig. 4.13 Nonlinear 3DMM generate shape and albedo embedded with different attributes

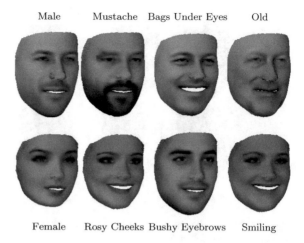

Male   Mustache   Bags Under Eyes   Old

Female   Rosy Cheeks   Bushy Eyebrows   Smiling

Fig. 4.14 Qualitative comparisons on texture representation power. Our model can better reconstruct in-the-wild facial texture

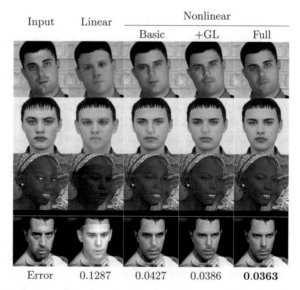

Input	Linear	Nonlinear		
		Basic	+GL	Full

| Error | 0.1287 | 0.0427 | 0.0386 | **0.0363** |

4.5.3 Representation Power

We compare the representation power of the proposed nonlinear 3DMM versus traditional linear 3DMM.

Texture We evaluate our model's power to represent in-the-wild facial texture on AFLW2000-3D dataset [74]. Given a face image, also with the groundtruth geometry and camera projection, we can jointly estimate an albedo parameter $\mathbf{f}_A$ and a lighting parameter $\mathbf{L}$ whose decoded texture can reconstruct the original image. To accomplish this, we use SGD on $\mathbf{f}_A$ and $\mathbf{L}$ with the initial parameters estimated by our encoder $\mathcal{E}$. For the linear model, Zhu et al. [74] fitting results of Basel albedo using

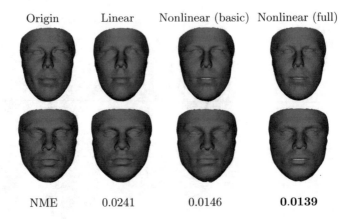

Origin   Linear   Nonlinear (basic) Nonlinear (full)

NME   0.0241   0.0146   **0.0139**

Fig. 4.15 Shape representation power comparison. Given a 3D shape, we optimize the feature $\mathbf{f}_S$ to approximate the original one

Phong illumination model [43] is used. As in Fig. 4.14, nonlinear models significantly outperforms the Basel Face model. Despite, being close to the original image, the basic nonlinear model (without proxy learning and global–local architecture) 's reconstruction results are still blurry. Using global–local-based network architecture ("+GL") with the same loss functions helps to bring the image closer to the input. However, these models are still constrained by regularizations on the albedo. By learning using proxy technique, our full model can learn more realistic albedo with more high- frequency details on the face. This conclusion is further supported with quantitative comparison in Fig. 4.14. We report the averaged $l_{2,1}$ reconstruction error over the face portion of each image. The full nonlinear model achieves the lowest averaged reconstruction error among four models, 0.0363, which is a 15% error reduction of the basic nonlinear 3DMM.

Shape Similarly, we also compare models' power to represent real-world 3D scans. Using ten 3D face meshes provided by [42], which share the same triangle topology with us, we can optimize the shape parameter to generate, through the decoder, shapes matching the groundtruth scans. The optimization objective is defined based on vertex distances (Euclidean) as well as surface normal direction (cosine distance), which empirically improves reconstructed meshes' fidelity compared to optimizing the former only. Figure 4.15 shows the visual comparisons between different reconstructed meshes from the linear 3DMM, from our basic nonlinear 3DMM and our full nonlinear 3DMM with proxy and global–local architecture. Our reconstructions closely match the face shapes details. To quantify the difference, we use NME, averaged per-vertex errors between the recovered and groundtruth shapes, normalized by inter-ocular distances. The proposed model has a significantly smaller reconstruction error than the linear model, and is also smaller than the basic nonlinear model (0.0139 vs. 0.0146, and 0.0241 [42]).

Input Overlay Albedo Shape Shading   Input Overlay Albedo Shape Shading

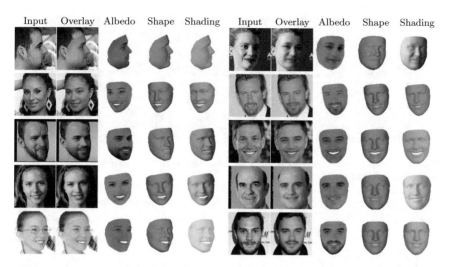

Fig. 4.16 3DMM fits to faces with diverse skin color, pose, expression, lighting, facial hair, and faithfully recovers these cues. Left half shows results from AFLW2000 dataset, and right half shows results from CelebA

4.5.4 Applications

Having shown the capability of our nonlinear 3DMM (i.e., two decoders), now we demonstrate the applications of our entire network, which has the additional encoder. Many applications of 3DMM are centered on its ability to fit to 2D face images. Our nonlinear 3DMM can be utilized for model fitting, which decomposes a 2D face into its shape, albedo and lighting. Figure 4.16 visualizes our 3DMM fitting results on AFLW2000 and CelebA dataset. Our encoder estimates the shape S, albedo A as well as lighting L and projection parameter m. We can recover personal facial characteristics in both shape and albedo. Our albedo can present facial hair, which is normally hard to be recovered by linear 3DMM.

4.5.4.1 Face Alignment

Face alignment is a critical step for many facial analysis tasks such as face recognition [63, 64]. With enhancement in the modeling, we hope to improve this task (Fig. 4.17). We compare face alignment performance with state-of-the-art methods, 3DDFA [74], DeFA [33], 3D-FAN [10] and PRN [16], on AFLW2000 dataset on both 2D and 3D settings.

The accuracy is evaluated using Normalized Mean Error (NME) as the evaluation metric with bounding box size as the normalization factor [10]. For a fair comparison with these methods in terms of computational complexity, for this comparison, we use ResNet18 [20] as our encoder. Here, 3DDFA and DeFA use the linear 3DMM model

Fig. 4.17 Our face alignment results. Invisible landmarks are marked as red. We can well handle extreme pose, lighting, and expression

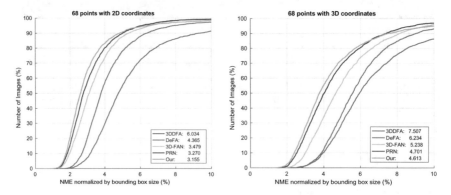

Fig. 4.18 Face alignment cumulative errors distribution curves on AFLW2000 on 2D (left) and 3D landmarks (right). NMEs are shown in legend boxes

(BFM). Even though being trained with a larger training corpus (DeFA) or having a cascade of CNNs iteratively refines the estimation (3DDFA), these methods are still significantly outperformed by our nonlinear model (Fig. 4.18). Meanwhile, 3D-FAN and PRN achieve competitive performances by by-passing the linear 3DMM model. 3D-FAN uses heat map representation. PRN uses the position map representation which shares a similar spirit to our UV representation. Not only outperforms these methods in term of regressing landmark locations (Fig. 4.18), our model also directly provides head pose information as well as the facial albedo and environment lighting condition.

4.5.4.2 3D Face Reconstruction

Using our model $\mathcal{D}_S$, $\mathcal{D}_A$, together with the model fitting CNN $\mathcal{E}$, we can decompose a 2D photograph into different components: 3D shape, albedo and lighting (Fig. 4.16). Here we compare our 3D reconstruction results with different lines of works: linear 3DMM fitting [56], nonlinear 3DMM fitting [55] and approaches beyond 3DMM [22, 50]. Comparisons are made on CelebA dataset [34].

Input   Our   Tewari17

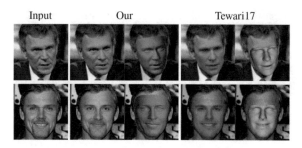

Fig. 4.19   3D reconstruction comparison to Tewari et al. [56]

Input   Our   Tewari18

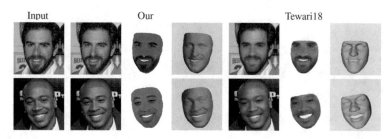

Fig. 4.20   3D reconstruction comparisons to nonlinear 3DMM approaches by Tewari et al. [55].
Our model can reconstruct face images with higher level of details

For linear 3DMM model, the representative work, MoFA [56], learns to regress
3DMM parameters in an unsupervised fashion. Even being trained on in-the-wild
images, it is still limited to the linear subspace, with limited power to recovering in-
the-wild texture. This results in the surface shrinkage when dealing with challenging
texture, i.e., facial hair as discussed in [55, 62]. Besides, even with regular skin texture
their reconstruction is still blurry and has less details compared to ours (Fig. 4.19).

The most related work to our proposed model is Tewari et al. [55], in which 3DMM
bases are embedded in neural networks. With more representation power, these models can recover details that the traditional 3DMM usually can't, i.e. make-up, facial
hair. However, the model learning process is attached with strong regularization,
which limits their ability to recover high-frequency details of the face. Our proposed model enhances the learning process in both learning objective and network
architecture to allow higher-fidelity reconstructions (Fig. 4.20).

To improve 3D reconstruction quality, many approaches also try to move beyond
the 3DMM such as Richardson et al. [46], Sela et al. [50] or Tran et al. [61]. The
current state-of-the-art 3D monocular face reconstruction method by Sela et al. [50]
using a fine detail reconstruction step to help reconstructing high- fidelity meshes.
However, their first depth map regression step is trained on synthetic data generated
by the linear 3DMM. Besides domain gap between synthetic and real, it faces a
more serious problem of lacking facial hair in the low-dimensional texture. Hence,
this network's output tends to ignore these unexplainable regions, which leads to
failure in later steps. Our network is more robust in handling these in-the-wild vari-

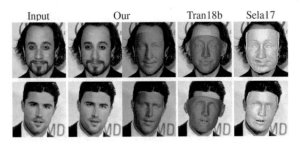

Fig. 4.21 3D reconstruction comparisons to Sela et al. [50] or Tran et al. [61], which go beyond latent representations

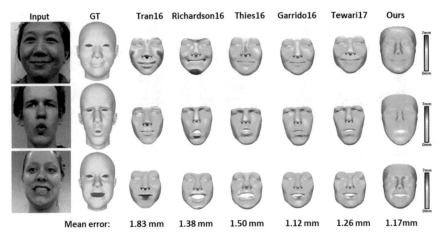

Fig. 4.22 Quantitative evaluation of 3D reconstruction. We obtain a low error that is comparable to optimization based methods

ations (Fig. 4.21). The approach of Tran et al. [61] shares a similar objective with us to be both robust and maintain high level of details in 3D reconstruction. However, they use an over-constrained foundation, which loses the personal characteristics of the each face mesh. As a result, the 3D shapes look similar across different subjects (Fig. 4.21).

Following the same setting in [56], we also quantitatively compare our method with prior works on 9 subjects of FaceWarehouse database [11]. Visual and quantitative comparisons are shown in Fig. 4.22. We achieve on-par results with Garrido et al. [17], an offline optimization method, while surpass all other regression methods [46, 56, 60].

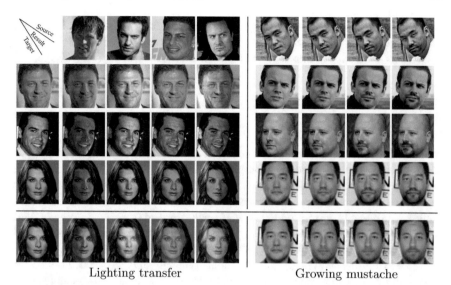

Lighting transfer   Growing mustache

Fig. 4.23 Face editing results. For lighting transfer, we transfer from of sources (first row) to target images (first column). We have similar performance compare to [52] (last row) despite being orders of magnitude faster (150 ms vs. 3 min per image). For mustache growing, the first collumn shows original images, the following collumns show edited images with increasing magnitudes. Comparing to [53] results (last row), our edited images are more realistic and identity-preserved

4.5.4.3   Face Editing

Decomposing face image into individual components gives us the ability to edit the face by manipulating any component. Here, we show two examples of face editing.

Relighting Firstly, we show an application to replacing the lighting of a target face image using lighting from a source face (Fig. 4.23). After estimating the lighting parameters $\mathbf{L}_{\text{source}}$ of the source image, we render the transfer shading using the target shape $\mathbf{S}_{\text{target}}$ and the source lighting $\mathbf{L}_{\text{source}}$. This transfered shading can be used to replace the original source shading. Also, here we use the original texture instead of the output of our decoder to maintain image details.

Attribute Manipulation Given faces fitted by 3DMM model, we can edit images by naive modifying one or more elements in the albedo or shape representation. More interestingly, we can even manipulate the semantic attribute, such as growing beard, smiling, etc. The approach is similar to learning attribute embedding in Sect. 4.5.2. Assuming, we would like to edit appearance only. For a given attribute, e.g., beard, we feed two sets of images with and without that attribute $\{\mathbf{I}_i^p\}_{i=1}^n$ and $\{\mathbf{I}_i^n\}_{i=1}^n$ into our encoder to obtain two average parameters $\mathbf{f}_A^p$ and $\mathbf{f}_A^n$. Their difference $\Delta\mathbf{f}_A = \mathbf{f}_A^p - \mathbf{f}_A^n$ is the direction to move from the distribution of negative images to positive ones. By adding $\Delta\mathbf{f}_A$ with different magnitudes, we can generate modified images with different degrees of changes. To achieve high-quality editing with identity-preserved, the final editing result is obtained by adding the residual, the different difference between the modified image and our reconstruction, to the original input image. This is a critical difference to Shu et al. [53] to improve results quality (Fig. 4.23).

4.6 Conclusions

Since its debut in 1999, 3DMM has became becamo a cornerstone of facial analysis research with applications to many problems. Despite its impact, it has drawbacks in requiring training data of 3D scans, learning from controlled 2D images, and limited representation power due to linear bases for both shape and texture. These drawbacks could be formidable when fitting 3DMM to unconstrained faces, or learning 3DMM for generic objects such as shoes. This paper demonstrates that there exists an alternative approach to 3DMM learning, where a nonlinear 3DMM can be learned from a large set of in-the-wild face images without collecting 3D face scans. Further, the model fitting algorithm can be learnt jointly with 3DMM, in an end-to-end fashion.

Our experiments cover a diverse aspects of our learnt learned model, some of which might need the subjective judgment of the readers. We hope that both the judgment and quantitative results could be viewed under the context that, unlike linear 3DMM, no genuine 3D scans are used in our learning. Finally, we believe that unsupervisedly or weak-supervisedly learning 3D models from large-scale in-the-wild 2D images is one promising research direction. This work is one step along this direction.

References

1. Aldrian O, Smith WA (2013) Inverse rendering of faces with a 3D morphable model. In: TPAMI
2. Amberg B, Knothe R, Vetter T (2008) Expression invariant 3D face recognition with a morphable model. In: FG
3. Amberg B, Romdhani S, Vetter T (2007) Optimal step nonrigid ICP algorithms for surface registration. In: CVPR
4. Blanz V, Vetter T (1999) A morphable model for the synthesis of 3D faces. In: Proceedings of the 26th annual conference on Computer graphics and interactive techniques
5. Blanz V, Vetter T (2003) Face recognition based on fitting a 3D morphable model. In: TPAMI
6. Bolkart T, Wuhrer S (2015) A groupwise multilinear correspondence optimization for 3D faces. In: ICCV
7. Bookstein FL (1989) Principal warps: thin-plate splines and the decomposition of deformations. In: TPAMI
8. Booth J, Antonakos E, Ploumpis S, Trigeorgis G, Panagakis Y, Zafeiriou S (2017) 3D face morphable models "In-the-wild". In: CVPR
9. Booth J, Roussos A, Zafeiriou S, Ponniah A, Dunaway D (2016) A 3D morphable model learnt from 10,000 faces. In: CVPR
10. Bulat A, Tzimiropoulos G (2017) How far are we from solving the 2D & 3D face alignment problem? (and a dataset of 230,000 3D facial landmarks). In: ICCV
11. Cao C, Weng Y, Zhou S, Tong Y, Zhou K (2014) Facewarehouse: a 3D facial expression database for visual computing. In: TVCG

12. Cole F, Belanger D, Krishnan D, Sarna A, Mosseri I, Freeman WT (2017) Face synthesis from facial identity features. In: CVPR
13. Cootes TF, Edwards GJ, Taylor CJ (2001) Active appearance models. In: TPAMI
14. Dollár P, Welinder P, Perona P (2010) Cascaded pose regression. In: CVPR
15. Dou P, Shah SK, Kakadiaris IA (2017) End-to-end 3D face reconstruction with deep neural networks. In: CVPR
16. Feng Y, Wu F, Shao X, Wang Y, Zhou X (2018) Joint 3D face reconstruction and dense alignment with position map regression network. In: ECCV
17. Garrido P, Zollhöfer M, Casas D, Valgaerts L, Varanasi K, Pérez P, Theobalt C (2016) Reconstruction of personalized 3D face rigs from monocular video. In: ACM TOG
18. Genova K, Cole F, Maschinot A, Sarna A, Vlasic D, Kyle W (2018) Unsupervised training for 3D morphable model regression. In: CVPR
19. Gu L, Kanade T (2008) A generative shape regularization model for robust face alignment. In: ECCV
20. He K, Zhang X, Ren S, Sun J (2016) Deep residual learning for image recognition. In: CVPR
21. Huang R, Zhang S, Li T, He R, et al (2017) Beyond face rotation: global and local perception GAN for photorealistic and identity preserving frontal view synthesis. In: ICCV
22. Jackson AS, Bulat A, Argyriou V, Tzimiropoulos G (2017) Large pose 3D face reconstruction from a single image via direct volumetric CNN regression. In: ICCV
23. Johnson J, Alahi A, Fei-Fei L (2016) Perceptual losses for real-time style transfer and super-resolution. In: ECCV. Pose-invariant 3D face alignment, ICCV
24. Jourabloo A, Liu X (2016) Large-pose face alignment via CNN-based dense 3D model fitting. In: CVPR
25. Jourabloo A, Liu X, Ye M, Ren L (2017) Pose-invariant face alignment with a single CNN. In: ICCV
26. Karras T, Aila T, Laine S, Lehtinen J (2018) Progressive growing of GANS for improved quality, stability, and variation. In: ICLR
27. Kim H, Zollhöfer M, Tewari A, Thies J, Richardt C, Theobalt C (2018) Inversefacenet: deep single-shot inverse face rendering from a single image. In: CVPR'
28. Koppen P, Feng Z-H, Kittler J, Awais M, Christmas W, Wu X-J, Yin H-F (2017) Gaussian mixture 3D morphable face model. Pattern Recogn
29. Liu F, Zhu R, Zeng D, Zhao Q, Liu X (2018) Disentangling features in 3D face shapes for joint face reconstruction and recognition. In: CVPR
30. Liu X (2009) Discriminative face alignment. In: TPAMI
31. Liu X (2010) Video-based face model fitting using adaptive active appearance model. Image Vis Comput
32. Liu X, Tu P, Wheeler F (2006) Face model fitting on low resolution images. In: BMVC
33. Liu Y, Jourabloo A, Ren W, Liu X (2017) Dense face alignment. In: ICCVW
34. Liu Z, Luo P, Wang X, Tang X (2015) Deep learning face attributes in the wild. In: ICCV
35. Mathieu M, Couprie C, LeCun Y (2015) Deep multi-scale video prediction beyond mean square error. arXiv:1511.05440
36. Meka A, Zollhöfer M, Richardt C, Theobalt C (2016) Live intrinsic video. In: ACM TOG
37. Mohammed U, Prince SJ, Kautz J (2009) Visio-lization: generating novel facial images. TOG
38. Nhan Duong C, Luu K, Gia Quach K, Bui TD (2015) Beyond principal components: deep Boltzmann machines for face modeling. In: CVPR
39. Nirkin Y, Masi I, Tran AT, Hassner T, Medioni GM (2018) On face segmentation, face swapping, and face perception. In: FG
40. Parkhi OM, Vedaldi A, Zisserman A (2015) Deep face recognition. In: BMVC
41. Patel A, Smith WA (2009) 3D morphable face models revisited. In: CVPR
42. Paysan P, Knothe R, Amberg B, Romdhani S, Vetter T (2009) A 3D face model for pose and illumination invariant face recognition. In: AVSS
43. Phong BT (1975) Illumination for computer generated pictures. Commun ACM
44. Ramamoorthi R, Hanrahan P (2001) An efficient representation for irradiance environment maps. In: Proceedings of the 28th annual conference on Computer graphics and interactive techniques

45. Richardson E, Sela M, Kimmel R (2016) 3D face reconstruction by learning from synthetic data. In: 3DV
46. Richardson E, Sela M, Or-El R, Kimmel R (2017) Learning detailed face reconstruction from a single image. In: CVPR
47. Roth J, Tong Y, Liu X (2015) Unconstrained 3D face reconstruction In; CVPR
48. Roth J, Tong Y, Liu X (2017) Adaptive 3D face reconstruction from unconstrained photo collections. In: TPAMI
49. Sagonas C, Antonakos E, Tzimiropoulos G, Zafeiriou S, Pantic M (2016) 300 faces in-the-wild challenge: database and results. Image Vis Comput
50. Sela M, Richardson E, Kimmel R (2017) Unrestricted facial geometry reconstruction using image-to-image translation. In: ICCV
51. Shi F, Wu H-T, Tong X, Chai J (2014) Automatic acquisition of high-fidelity facial performances using monocular videos. ACM TOG
52. Shu Z, Hadap S, Shechtman E, Sunkavalli K, Paris S, Samaras D (2018) Portrait lighting transfer using a mass transport approach. TOG
53. Shu Z, Yumer E, Hadap S, Sunkavalli K, Shechtman E, Samaras D (2017) Neural face editing with intrinsic image disentangling. In: CVPR
54. Staal FC, Ponniah AJ, Angullia F, Ruff C, Koudstaal MJ, Dunaway D (2015) Describing crouzon and pfeiffer syndrome based on principal component analysis. J Cranio-Maxillof Surg
55. Tewari A, Zollhöfer M, Garrido P, Bernard F, Kim H, Pérez P, Theobalt C (2018) Self-supervised multi-level face model learning for monocular reconstruction at over 250 Hz. In: CVPR
56. Tewari A, Zollhöfer M, Kim H, Garrido P, Bernard F, Pérez P, Theobalt C (2017) MoFA: model-based deep convolutional face autoencoder for unsupervised monocular reconstruction. In: ICCV
57. Thies J, Zollhöfer M, Nießner M, Valgaerts L, Stamminger M, Theobalt C (2015) Real-time expression transfer for facial reenactment. ACM Trans Graph 34(6):183:1–183:14
58. Thies J, Zollhöfer M, Stamminger M, Theobalt C, Nießner M (2016a) Face2face: real-time face capture and reenactment of RGB videos. In: CVPR
59. Thies J, Zollhöfer M, Stamminger M, Theobalt C, Nießner M (2016b) FaceVR: real-time facial reenactment and eye gaze control in virtual reality. arXiv:1610.03151
60. Tran AT, Hassner T, Masi I, Medioni G (2017) Regressing robust and discriminative 3D morphable models with a very deep neural network. In: CVPR
61. Tran AT, Hassner T, Masi I, Paz E, Nirkin Y, Medioni G (2018) Extreme 3D face reconstruction: looking past occlusions. In: CVPR
62. Tran L, Liu X (2018) Nonlinear 3D morphable model. In: CVPR
63. Tran L, Yin X, Liu X (2017) Disentangled representation learning GAN for pose-invariant face recognition. In: CVPR
64. Tran L, Yin X, Liu X (2018) Representation learning by rotating your faces. TPAMI
65. Tulyakov S, Sebe N (2015) Regressing a 3D face shape from a single image. In: ICCV
66. Vlasic D, Brand M, Pfister H, Popović J (2005) Face transfer with multilinear models. In: TOG
67. Wang Y, Zhang L, Liu Z, Hua G, Wen Z, Zhang Z, Samaras D (2009) Face relighting from a single image under arbitrary unknown lighting conditions. TPAMI
68. Wu H, Liu X, Doretto G (2008) Face alignment via boosted ranking models. In: CVPR
69. Wu Y, Ji Q (2015) Robust facial landmark detection under significant head poses and occlusion. In: ICCV
70. Yin L, Wei X, Sun Y, Wang J, Rosato MJ (2006) A 3D facial expression database for facial behavior research. In: FGR
71. Yin X, Yu X, Sohn K, Liu X, Chandraker M (2017) Towards large-pose face frontalization in the wild. In: ICCV
72. Yu R, Saito S, Li H, Ceylan D, Li H (2017) Learning dense facial correspondences in unconstrained images. In: ICCV
73. Zhang L, Samaras D (2006) Face recognition from a single training image under arbitrary unknown lighting using spherical harmonics. TPAMI

74. Zhu X, Lei Z, Liu X, Shi H, Li SZ (2016) Face alignment across large poses: a 3D solution. In: CVPR
75. Zollhöfer M, Thies J, Bradley D, Garrido P, Beeler T, Péerez P, Stamminger M, Nießner M, Theobalt C (2018) State of the art on monocular 3D face reconstruction, tracking, and applications. Eurographics

Chapter 5
Deblurring Face Images Using Deep Networks

Rajeev Yasarla, Federico Perazzi, and Vishal M. Patel

Image deblurring entails the recovery of an unknown true image from a blurry image. Image deblurring is an ill-posed problem, and therefore, it is crucial to leverage additional properties of the data to successfully recover the lost facial details in the deblurred image. Priors such as sparsity [2, 13, 16], low-rank [14], manifold [10], and patch similarity [22] have been proposed in the literature to obtain a regularized solution. In recent years, deep learning-based methods have also gained some traction [9, 11, 26, 30].

The inherent semantic structure of faces is an important information that can be exploited to improve the deblurring results. In this chapter, we provide an overview of deep CNN-based methods that make use of the facial semantic information to deblur face images.

5.1 Deep Semantic Face Deblurring

One of the first approaches that made use of facial semantic cues via deep learning for face deblurring was proposed in [18]. Their approach is based on the fact that face images are highly structured and they share several key semantic components such as mouth and eyes. As a result, the semantic information of a face can provide a strong prior for restoration. In their approach, the authors use incorporate global semantic priors as input and impose local structure losses to regularize the output within a multi-scale deep convolutional neural network (CNN). In addition, the proposed

R. Yasarla · V. M. Patel (✉)
Baltimore, USA
e-mail: vpatel36@jhu.edu

F. Perazzi
San Francisco, USA

© The Author(s), under exclusive license to Springer Nature Switzerland AG 2021
N. K. Ratha et al. (eds.), *Deep Learning-Based Face Analytics*, Advances in Computer Vision and Pattern Recognition, https://doi.org/10.1007/978-3-030-74697-1_5

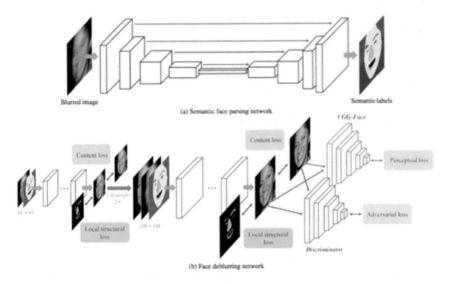

Fig. 5.1 Semantic face deblurring network proposed in [18]. The overall approch consists of two sub-networks: a semantic face parsing network and a multi-scale deblurring network

network was trained using the perceptual and adversarial losses to generate photo-realistic deblurred images. Furthermore, an incremental training strategy was also introduced to handle random blur kernels seen in practice.

Figure 5.1 gives an overview of the semantic face deblurring network proposed in [18]. The overall approach consists of two sub-networks—a semantic face parsing network and a multi-scale deblurring network. The face parsing network generates the semantic labels of the input blurry image. On the other hand, the multi-scale deblurring network uses the extracted facial semantic labels to deblur the face image. In particular, the blurred image and semantic labels are concatenated as the input to the first scale of the multi-scale deblurring network. The upsampled deblurred image from the first scale, the blurred image, and the corresponding semantic labels are then fed into the second scale of the network. Each scale of the deblurring network receives the supervision from the pixel-wise content loss and local structural losses. In addition, the the perceptual and adversarial losses are also imposed at the output of the second scale.

The following pixel-wise L1 loss was used as the content loss for the face deblurring network

$$L_c = \|G(B, P(B)) - I\|_1, \tag{5.1}$$

where P and G are the face parsing and deblurring networks. Here, B and I are blurry and ground truth clear images, respectively. To better incorporate local facial structures such as eyes, lips, and mouth, the following local structural loss is imposed

$$L_s = \sum_{i=1}^{K} \| M_k(P(B)) \odot (G(B, P(B)) - I) \|_1, \tag{5.2}$$

where M_k denotes the structural mask of the k-th component and $\odot$ denotes element-wise multiplication. In addition, the perceptual loss on the Pool2 and Pool5 layers of the pretrained VGG-Face network was used to obtained high-quality images. The perceptual loss is defined as follows:

$$L_p = \sum_{l} \| \phi_l(G(B)) - \phi_l(I) \|_2^2, \tag{5.3}$$

where ϕ_l denotes the activation at the l-th layer of the network ϕ. Finally, treating the deblurring network as a generator network, a discriminator network is also constructed to incorporate adversarial training. The adversarial training is formulated as solving the following min-max problem

$$\min_{G} \max_{D} E[\log D(I)] + E[\log(1 - D(G(B)))]. \tag{5.4}$$

In particular, when updating the generator, the following adversarial loss is used

$$L_{\text{adv}} = -\log(D(G(B))). \tag{5.5}$$

The semantic face parsing network was trained using the Helen dataset [6]. On the other hand, a set of training images were collected using the Helen dataset [6], CMU PIE dataset [19], and CelebA dataset [8] to train the deblurring network.

Various quantitative and qualitative evaluations were conducted in [18] and it was shown that the semantic face deblurring algorithm restores sharp images with more facial details and performs favorably against state-of-the-art deblurring methods in terms of restoration quality, face recognition, and execution speed. Sample results corresponding to this method are shown in Fig. 5.2. As can be seen from this figure, the use of facial semantic information clearly improves the deblurring performance.

5.2 Deblurring Via Structure Generation and Detail Enhancement

Another CNN-based approach that made use of facial semantic maps for restoring high resolution images from blurry low-resolution inputs was recently proposed in [21]. In their approach, the restored image is formulated using a base layer and a detail layer. The base layer is learned by using a CNN guided by facial components. On the other hand, the detail layer is generated by an exemplar-based texture synthesis module. First, their facial structure generation network takes the upsampled face image and its facial semantic maps as the inputs and generates base images. Then a patch-

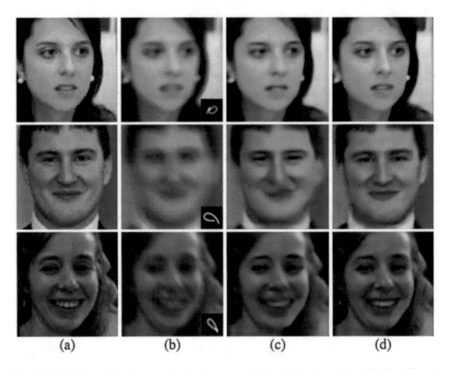

(a)   (b)   (c)   (d)

Fig. 5.2 Sample results corresponding to the semantic face deblurring network [18]. **a** Ground truth images. **b** Blurry face images. **c** Deblurring results without using the facial semantic maps. **d** Deblurring results with the use of semantic maps

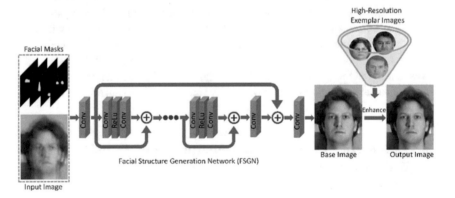

Fig. 5.3 An overview of the joint face hallucination and deblurring method proposed in [21]. Input blurry low-resolution face image along with facial masks are used in the facial structure generation network to produce a base image. Then, a detail enhancement algorithm is developed to estimate the missing details in the base image using the high resolution exemplar images

wise K-Nearest Neighbor (KNN) method is used to search between the intermediate face image and exemplar images. As a result, one can establish the correspondences on the HR training images, which ensures that the fine-grained facial structures from the high resolution exemplar images are effectively extracted. Finally, the details from these structures are transferred into the base image through an edge-aware image filtering procedure. Figure 5.3 gives an overview of their approach. Various properties of this method was analyzed and it was shown that this new approach performs favorably against state-of-the-art face hallucination and deblurring methods on the public benchmarks.

5.3 Uncertainty Guided Multi-stream Semantic Networks

Image restoration methods described above [18, 21] make use of prior information in the form of semantic labels. However, these methods do not account for the class imbalance of semantic maps corresponding to faces. Interior parts of a face like eyes, nose, and mouth are less represented as compared to face skin, hair, and background labels. Depending on the pose of the face, some of the interior parts may even disappear. Without re-weighting the importance of less represented semantic regions, the method proposed in [18] fails to reconstruct the eyes and the mouth regions as shown in Fig. 5.4. Similar observations can also be made regarding the method proposed in [21] (Fig. 5.4).

To address the imbalance of different semantic classes, a novel CNN architecture called Uncertainty guided Multi-stream Semantic Networks (UMSN) was recently proposed in [28], which learns class-specific features independently and combine them to deblur the whole face image. Class-specific features are learned by subnetworks trained to reconstruct a single semantic class. Nested residual learning paths are used to improve the propagation of semantic features. Additionally, a classbased confidence measure is proposed to train the network. The confidence measure describes how well the network is likely to deblur each semantic class. This measure is incorporated in the loss to train the network. Figure 5.4 shows sample results from the UMSN network, where one can clearly see that UMSN is able to provide better results as compared to the state-of-the-art techniques [18, 21].

5.3.1 Image Deblurring Network

A blurry face image y can be modeled as the convolution of a clean image x with a blur kernel k, as

$$y = k * x + \eta,$$

where $*$ denotes the convolution operation and η is noise. Given y, in blind deblurring, the objective is to estimate the underlying clean face image x. 11 semantic face labels

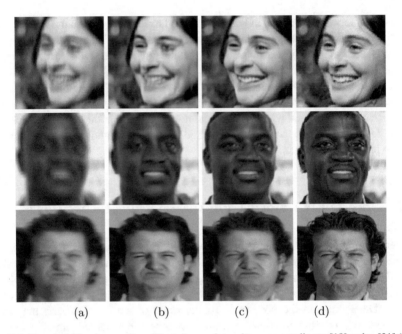

Fig. 5.4 Sample deblurring results: **a** Blurry image. **b** results corresponding to [18] and to [21] (last row); **c** Results corresponding to Uncertainty Guided Multi-Stream Semantic Network (UMSN); **d** ground truth. Our approach recovers more details and better preserves fine structures like eyes and hair

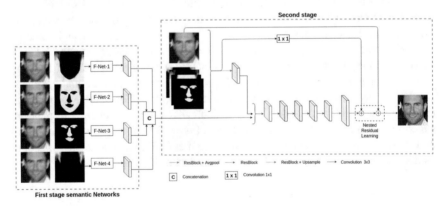

Fig. 5.5 An overview of the proposed UMSN network. First stage semantic networks consist of F-Net-i. Second stage is constructed using the base network (BN), where outputs of the F-Net-i's are concatenated with the output of the first ResBlock layer in BN

are grouped into 4 classes as follows: $m_1 = \{background\}$, $m_2 = \{face\ skin\}$, $m_3 = \{left\ eyebrow,\ right\ eyebrow,\ left\ eye,\ right\ eye,$
$nose,\ upper\ lip,\ lower\ lip,\ teeth\}$ and $m_4 = \{hair\}$. Thus, the semantic class mask of a clean image x is the union $m = m_1 \cup m_2 \cup m_3 \cup m_4$. Similarly one can define the semantic class masks of a blurry image $\hat{m}$.

Semantic class masks for blurry image, $\hat{m}$ are generated using the semantic segmentation network (SN) (Fig. 5.6), and given together with the blurry image as input to the deblurring network, UMSN. This is important in face deblurring as some parts like face, skin, and hair are easy to reconstruct, while face parts like eyes, nose, and mouth are difficult to reconstruct and require special attention while deblurring a face image. This is mainly due to the fact that parts like eyes, nose, and mouth are small in size and contain high frequency elements compared to the other components. Different from [12] that uses edge information and [18] that feed the semantic map to a single-stream deblurring network, this method addresses this problem by proposing a multi-stream semantic network, in which individual branches F-net-i learn to reconstruct different parts of the face image separately. Figure 5.5 gives an overview of the UMSN method.

As can be seen from Fig. 5.5, the UMSN network consists of two stages. The semantic class maps, $\hat{m}$, of a blurry face image are generated using the SN network. The semantic maps are used as the global masks to guide each stream of the first stage network. These semantic class maps $\hat{m}$ are further used to learn class-specific residual feature maps with nested residual learning paths (NRL). In the first stage of the network, the weights are learned to deblur the corresponding class of the face image. In the second stage of the network, the outputs from the first stage are fused to learn the residual maps that are added to the blurry image to obtain the final deblurred image. The network is trained with a confidence guided class-based loss.

5.3.2 Semantic Segmentation Network (SN)

The semantic class maps $\hat{m}_i$ of a face, are extracted using the SN network as shown in Fig. 5.6. The residual blocks (ResBlock) are used as the building module for the segmentation network. A ResBlock consist of a 1×1 convolution layer, a 3×3 convolution layer and two 3×3 convolution layers with dilation factor of 2 as shown in Fig. 5.7.

5.3.3 Base Network (BN)

The base network is constructed using a combination of UNet [15] and DenseNet [3] architectures with the ResBlock as the basic building block. To increase the receptive field size, smoothed dilation convolutions are introduced in the ResBlock as shown in Fig. 5.7. BN is a sequence of eight ResBlocks similar to the first stage semantic

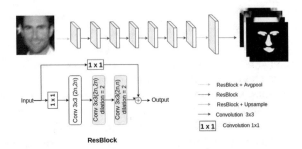

Fig. 5.6 An overview of the segmentation network. Conv $l \times l$ (p, q) contains instance normalization [23], Rectified Linear Unit (ReLU), Conv $(l \times l)$—convolutional layer with kernel of size $l \times l$, where p and q are the number of input and output channels, respectively

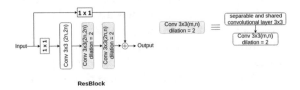

Fig. 5.7 An overview of the ResBlock. Conv $l \times l(p, q)$ contains Instance Normalization [23], ReLU—Rectified Linear Units, Conv $(l \times l)$—convolutional layer with kernel of size $l \times l$, where p and q are number of input and output channels respectively. In the right side of the figure, we show smoothed dilation convolutions introduced in ResBlock which is similar to [25]

network as shown in Fig. 5.8. Note that all convolutional layers are densely connected [3]. Residual-based learning is followed in estimating the deblurred image for the base network as shown in Fig. 5.8.

5.3.4 UMSN Network

The UMSN network is a two-stage network. The first stage network is designed to obtain deblurred outputs from the semantic class-wise blurry inputs. These outputs are further processed by the second stage network to obtain the final deblurred image. The first stage semantic network contains a sequence of five ResBlocks with residual connections, as shown in Fig. 5.8. The set of all convolution layers of the first stage network excluding the last ResBlock and Conv3 × 3 are referred to as F-Net.

The blurry image y and the semantic masks $\hat{m}_i$ are fed to F-Net-i to obtain the corresponding class-specific deblurred features which are concatenated with the output of the first layer (ResBlock-Avgpool) in Base Network(BN) for constructing the UMSN network. The Nested Residual Learning (NRL) is also used in UMSN network where class-specific residual feature maps are learned and further used in estimating the residual feature maps that are added to the blurry image for obtaining

Fig. 5.8 An overview of the first stage semantic network (F-Net)

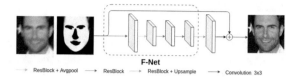

the final output. For example, one can observe a residual connection between the last layer of UMSN and class-specific feature maps obtained from Conv 1×1 using y and $\hat{m}$. Output of this residual connection is further processed as the input to the residual connection with the input blurry image, y. In this way, one can define NRL and obtain the final deblurred image. A class-based loss function is proposed to train the UMSN network.

5.3.5   Loss for UMSN

The network parameters Θ are learned by minimizing a loss $\mathcal{L}$ as follows,

$$\hat{\Theta} = \operatorname{argmin}_\Theta \mathcal{L}(f_\Theta(y, \hat{m}), x) = \operatorname{argmin}_\Theta \mathcal{L}(\hat{x}, x), \qquad (5.6)$$

where $f_\Theta(.)$ represents the UMSN network, $\hat{x}$ is the deblurred result, $\hat{m}$ is the semantic map obtained from SN. The reconstruction loss is defined as $\mathcal{L} = \|x - \hat{x}\|_1$. A face image can be expressed as the sum of masked images using the semantic maps as

$$x = \sum_{i=1}^{M} m_i \odot x,$$

where $\odot$ is the element-wise multiplication and M is the total number of semantic maps. As the masks are independent of one another, Eq. (5.6) can be re-written as

$$\hat{\Theta} = \operatorname{argmin}_\Theta \sum_{i=1}^{M} \mathcal{L}(m_i \odot \hat{x}, m_i \odot x). \qquad (5.7)$$

In other words, the loss is calculated for every class independent and summed up in order to obtain the overall loss as follows:

$$\mathcal{L}(\hat{x}, x) = \sum_{i=1}^{M} \mathcal{L}(m_i \odot \hat{x}, m_i \odot x). \qquad (5.8)$$

5.3.6   Uncertainty Guidance

A confidence measure is introduced for every class and use it to re-weight the contribution of the loss from each class to the total loss. By introducing a confidence measure and re-weighting the loss, one can benefit in two ways. If the network is giving less importance to a particular class by not learning appropriate features of it, then the Confidence Network (CN) helps UMSN to learn those class features by estimating low confidence values and higher gradients for those classes through CN network. Additionally, by re-weighting the contribution of loss from each class, it counters for the imbalances in the error estimation from different classes. The loss function can be written as

$$\mathcal{L}_c(\hat{x}, x) = \sum_{i=1}^{M} C_i \mathcal{L}(m_i \odot \hat{x}, m_i \odot x) - \lambda \log(C_i), \tag{5.9}$$

where $\log(C_i)$ acts as a regularizer that prevents the value of C_i going to zero and λ is a constant. The confidence measure C_i for each class are estimated by passing $m_i \odot \hat{x}, m_i \odot x$ as inputs to CN as shown in Fig. 5.9. C_i represents how confident UMSN is in deblurring the ith class components of the face image. Note that, $C_i (\in [0, 1])$, confidence measure is used only in the loss function while training the weights of UMSN, and it is not used (or estimated) during inference.

Inspired by the benefits of the perceptual loss in style transfer [4, 29] and image super-resolution [7], it is also used to train this network. Let $\Phi(.)$ denote the features obtained using the VGG16 model [20], then the perceptual loss is defined as follows:

$$\mathcal{L}_p = \|\Phi(\hat{x}) - \Phi(x)\|_2^2. \tag{5.10}$$

The features from layer $relu1\_2$ of a pretrained VGG-16 network [20] are used to compute the perceptual loss. The total loss used to train UMSN is as follows:

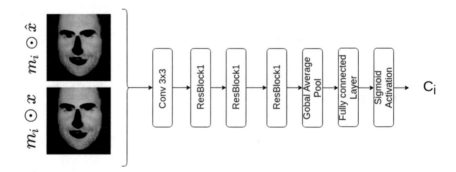

Fig. 5.9 An overview of the confidence network (CN). x is ground truth image. $\hat{x}$ is deblurred image obtained from UMSN. m semantic maps of x

$$\mathcal{L}_{\text{total}} = \mathcal{L}_c + \lambda_1 \mathcal{L}_p, \tag{5.11}$$

where λ_1 is a constant.

5.3.7 Experimental Results

The networks were evaluated using the images provided by the authors of [18], which consists of 8000 blurry images generated using the Helen dataset [6], and 8000 blurry images generated using the CelebA dataset [8]. Furthermore, the network was tested on a test dataset called PubFig, provided by the authors of [21], which contains 192 blurry images. The results were also quantitatively evaluated using the Peak-Signal-to-Noise Ratio (PSNR) and the Structural Similarity index (SSIM) [24]. Results are shown in Table 5.1. As can be seen from this table, the UMSN method is able to deblur the face images much better than the previous methods including [18, 21]. This can also be clearly seen by comparing the deblurringv results on real-world blurry images corresponding to different methods as shown in Fig. 5.10.

5.4 Conclusion

In this chapter, we provided an overview of recent face deblurring methods that make sure of semantic facial information for restoring blurred images. Various deep CNN-based methods were reviewed. The performance of different methods was also compared on synthetic as well as real-world datasets.

Table 5.1 PSNR and SSIM comparision of UMSN against state-of-the-art methods

Deblurring method	Helen		CelebA	
	PSNR	SSIM	PSNR	SSIM
Krishnan et al. [5] (CVPR'11)	19.30	0.670	18.38	0.672
Pan et al. [12] (ECCV 2014)	20.93	0.727	18.59	0.677
Shan et al. [17] (SIGGRAPH'08)	19.57	0.670	18.43	0.644
Xu et al. [27] (CVPR'13)	20.11	0.711	18.93	0.685
Cho et al. [1] (SIGGRAPH'09)	16.82	0.574	13.03	0.445
Zhong et al. [31] (CVPR'13)	16.41	0.614	17.26	0.695
Nah et al. [9] (CVPR'17)	24.12	0.823	22.43	0.832
Ziyi et al. [18] (CVPR'18) w/GAN	25.58	0.861	24.34	0.860
Ziyi et al. [18] (CVPR'18)	25.99	0.871	25.05	0.879
UMSN (ours w/$\mathcal{L}_{\text{total}}$)	26.93	0.897	25.90	0.906

(a)   (b)   (c)   (d)   (e)

Fig. 5.10 Sample results on real blurry images. **a** Blurry, **b** Xu et al. [27], **c** Zhong et al. [31], **d** Shen et al. [18], **e** UMSN [28]

References

1. Cho S, Lee S (2009) Fast motion deblurring. ACM Trans Grap (TOG) 28(5):145
2. Fergus R, Singh B, Hertzmann A, Roweis ST, Freeman WT (2006) Removing camera shake from a single photograph. ACM Trans Graph (TOG) 25(3):787–794
3. Huang G, Liu Z, Van Der Maaten L, Weinberger KQ (2017) Densely connected convolutional networks. In: Proceedings of the IEEE conference on computer vision and pattern recognition, pp 4700–4708
4. Johnson J, Alahi A, Fei-Fei L (2016) Perceptual losses for real-time style transfer and super-resolution. ECCV
5. Krishnan D, Tay T, Fergus R (2011) Blind deconvolution using a normalized sparsity measure. CVPR 2011:233–240
6. Le V, Brandt J, Lin Z, Bourdev L, Huang TS (2012) Interactive facial feature localization. In: European conference on computer vision. Springer, pp 679–692
7. Ledig C, Theis L, Huszár F, Caballero J, Cunningham A, Acosta A, Aitken A, Tejani A, Totz J, Wang Z, et al (2017) Photo-realistic single image super-resolution using a generative adversarial network, pp 4681–4690
8. Liu Z, Luo P, Wang X, Tang X (2015) Deep learning face attributes in the wild. In: Proceedings of the IEEE international conference on computer vision, pp 3730–3738
9. Nah S, Hyun Kim T, Mu Lee K (2017) Deep multi-scale convolutional neural network for dynamic scene deblurring. In: Proceedings of the IEEE conference on computer vision and pattern recognition, pp 3883–3891
10. Ni J, Turaga P, Patel VM, Chellappa R (2011) Example-driven manifold priors for image deconvolution. IEEE Trans Image Process 20(11):3086–3096
11. Nimisha TM, Kumar Singh A, Rajagopalan AN (2017) Blur-invariant deep learning for blind-deblurring. In: Proceedings of the IEEE international conference on computer vision, pp 4752–4760
12. Pan J, Hu Z, Su Z, Yang M-H (2014) Deblurring face images with exemplars, pp 47–62
13. Patel VM, Easley GR, Healy DM (2009) Shearlet-based deconvolution. IEEE Trans Image Process 18(12):2673–2685
14. Ren W, Cao X, Pan J, Guo X, Zuo W, Yang M-H (2016) Image deblurring via enhanced low-rank prior. IEEE Trans Image Process 25(7):3426–3437
15. Ronneberger O, Fischer P, Brox T (2015) U-net: convolutional networks for biomedical image segmentation. In: Medical image computing and computer-assisted intervention (MICCAI), vol 9351. LNCS, pp 234–241
16. Schuler CJ, Christopher Burger H, Harmeling S, Scholkopf B (2013) A machine learning approach for non-blind image deconvolution. In: Proceedings of the IEEE conference on computer vision and pattern recognition, pp 1067–1074
17. Shan Q, Jia J, Agarwala A (2008) High-quality motion deblurring from a single image. ACM Trans Graph (TOG) 27(3):73
18. Shen Z, Lai W-S, Xu T, Kautz J, Yang M-H (2018) Deep semantic face deblurring. In: Proceedings of the IEEE conference on computer vision and pattern recognition, pp 8260–8269
19. Sim T, Baker S, Bsat M (2002) The CMU pose, illumination, and expression (pie) database. In: Proceedings of fifth IEEE international conference on automatic face gesture recognition, pp 53–58
20. Simonyan K, Zisserman A (2014) Very deep convolutional networks for large-scale image recognition. arXiv:1409.1556
21. Song Y, Zhang J, Gong L, He S, Bao L, Pan J, Yang Q, Yang M-H (2019) Joint face hallucination and deblurring via structure generation and detail enhancement. Int J Comput Vis 127(6–7):785–800
22. Sun L, Cho S, Wang J, Hays J (2013) Edge-based blur kernel estimation using patch priors. In: IEEE international conference on computational photography (ICCP). IEEE, pp 1–8
23. Ulyanov D, Vedaldi A, Lempitsky V (2016) Instance normalization: the missing ingredient for fast stylization. arXiv:1607.08022

24. Wang Z, Bovik AC, Sheikh HR, Simoncelli EP (2004) Image quality assessment: from error visibility to structural similarity. IEEE Trans Image Process 13(4):600–612
25. Wang Z, Ji S (2018) Smoothed dilated convolutions for improved dense prediction. In: Proceedings of the 24th ACM SIGKDD international conference on knowledge discovery & data mining. ACM, pp 2486–2495
26. Xu L, Ren JS, Liu C, Jia J (2014) Deep convolutional neural network for image deconvolution. In: Advances in neural information processing systems, pp 1790–1798
27. Xu L, Zheng S, Jia J (2013) Unnatural l0 sparse representation for natural image deblurring. In: Proceedings of the IEEE conference on computer vision and pattern recognition, pp 1107–1114
28. Yasarla R, Perazzi F, Patel VM (2019) Deblurring face images using uncertainty guided multi-stream semantic networks. arXiv:1907.13106
29. Zhang H, Dana K (2017) Multi-style generative network for real-time transfer. arXiv:1703.06953
30. Zhang S, Shen X, Lin Z, Měch R, Costeira JP, Moura JM (2018) Learning to understand image blur. In: Proceedings of the IEEE conference on computer vision and pattern recognition. pp 6586–6595
31. Zhong L, Cho S, Metaxas D, Paris S, Wang J (2013) Handling noise in single image deblurring using directional filters. In: Proceedings of the IEEE conference on computer vision and pattern recognition, pp 612–619

Chapter 6
Blind Super-resolution of Faces for Surveillance

T. M. Nimisha and A. N. Rajagopalan

6.1 Introduction

Super-resolution (SR) refers to a class of techniques that derive a high resolution image from its low resolution (LR) counterpart. A vast amount of literature exists on SR spanning both multi and single image approaches. The classical approaches in SR use sub-pixel motion across multiple low resolution (LR) frames. These works [3, 11] typically assume that the blur encountered in the LR images is only due to downsampling and that the camera is static while capturing LR frames. The only motion addressed in these frameworks is the *inter*-frame motion which is used to infer the underlying high resolution (HR) image.

While multi-frame approaches supplement missing information in one frame from another, availability of multiple frames cannot always be assured. Single image SR [12, 17] is a lot more ill-posed and works by hallucinating the missing data or by exploiting patch-recurrences within an image across different scales. Of-late, many deep learning approaches have been proposed [9, 22, 24] that address the single image SR problem. However, all these methods assume that the blur encountered in the LR frame is only due to downsampling action.

Estimating an HR frame directly from a single motion blurred LR frame is highly ill-posed and is of great relevance in surveillance scenarios. Motion blur is an inevitable phenomenon that co-occurs with long exposure photography. Blur is considered as a nuisance in many image processing algorithms and inverting it is a difficult proposition. Many works exist [6, 23, 40, 49] that focus on the issue of removing motion blur due to camera shake from images. All these works aim for

T. M. Nimisha
Indian Institute of Technology Madras, Chennai, India
e-mail: nimiviswants@gmail.com

A. N. Rajagopalan (✉)
Electrical Engineering, Indian Institute of Technology Madras, Chennai, India
e-mail: raju@ee.iitm.ac.in

© The Author(s), under exclusive license to Springer Nature Switzerland AG 2021
N. K. Ratha et al. (eds.), *Deep Learning-Based Face Analytics*, Advances in Computer Vision and Pattern Recognition, https://doi.org/10.1007/978-3-030-74697-1_6

deblurring as the main goal and do not really consider resolution enhancement. SR and deblurring are well-studied problems but are treated as independent topics. Only a few works [33, 41, 47, 49] exist in the literature that addresses both SR and motion deblurring.

The challenge in arriving at an SR image escalates when the underlying LR frames have motion blur artifacts. These situations arise when the subject of interest is far away from the camera and the subject/camera is moving. In these situations, the observed images will be degraded both by motion blur and the downsampling action. Since face recognition (FR) systems are of great use nowadays and are employed as biometric in many areas, a motion distorted LR probe image that deviates significantly from that of the gallery image reduces recognition accuracy. This necessitates the need for single image blind SR. The class of algorithms that estimates an HR image from LR irrespective of artifacts due to motion blur are referred to as blind SR algorithms. It is interesting to note that motion blur occurs due to averaging of several warped instances of the clean frame during exposure. Thus, a single blurred LR frame by default aggregates information from multiple clean frames. Hence, scope exists to harness this aggregated information for deblurring as well as super-resolving.

Performing blind SR sequentially can lead to poor results. The error from the first stage (either SR or deblurring) can propagate to the second and worsen the final output. We propose here a blind SR framework that jointly deblurs and upsamples the probe images to help in achieving better recognition rates for FR systems. Prior works that have addressed the blind SR problem [26, 33, 41, 49], for instance, assumed a multi-frame approach. In contrast, ours is a single image blind SR specifically aimed at improving the accuracy of face recognition systems in surveillance applications.

In this work, we explore invariant feature learning for the purpose of single image SR from motion blurred frames. We employ a deep learning framework for achieving the task at hand. With the underlying idea that natural images follow a sparse distribution and that a shallow dictionary can capture invariance in a sparse domain, we attempt to generalize this invariance to deep non-linear networks. Our network consists of an Encoder-Decoder pair that learns the clean high resolution data domain. This is followed by a Generative Adversarial Network (GAN) that is trained to produce *blur and resolution invariant* features from LR blurred frames. The learned representations are processed by the Decoder to get the final result. We deploy this framework for face surveillance applications where the collected probe images are highly distorted.

6.2   Related Works

Deblurring and SR, though two extensively studied topics, have mostly been dealt with independently. SR frameworks assume static camera leading to LR images affected by downsampling alone. These methods neglect the effect of motion artifacts. Similarly, deblurring approaches assume the availability of high resolution frames and do not work well at a lower resolution. Hence, the performance of these methods drops considerably when the assumptions of blur/resolution do not hold. We discuss here in brief conventional and deep learning based works on deblurring and SR. These can be mainly classified as single image and multi-image approaches.

Super-resolution: Existing works in SR can be broadly classified into two categories (1) Multi-image approaches: Methods [3, 11] that utilize inter-frame sub-pixel motion in the LR frames to restore the HR image and (2) Single image based: These techniques either resort to exemplars or patch-recurrence (also termed 'image hallucination') [12, 17] or patch-based learning [48, 50] to create the HR image. *Single image SR techniques* (which is the focus of our work) employ a database of LR and HR image pairs to learn the correspondences between LR and HR image patches [48, 50]. The patch correspondences thus learned are used during testing to map an LR image to its most likely HR version. However, these techniques are known to hallucinate HR details that may not even be present in the true HR image. Based on the observation that patches in a natural image tend to recur within the same image, both at the same as well as at different scales, the works in [12, 17] sought to combine the strengths of both traditional multi-image SR as well as example-based SR. Recently, deep learning and generative networks have also made forays into computer vision and image processing, and their influence and impact are growing rapidly by the day. Single image SR with deep networks [9, 22, 24] have shown remarkable results that outperform traditional methods. Dong et al. [9] introduce a skip connection-based network that learns residual features for SR. The work in [22] uses a GAN architecture to produce photo-realistic SR outputs from a single LR frame. It is important to note that state-of-the-art SR techniques achieve remarkable results of resolution enhancement only when there is no motion blur in the LR input.

Deblurring: Many methods exist [7, 19, 49] that rely on information from multiple frames captured using video or burst mode and work by harnessing the information from these frames to solve for the underlying original (latent) image. Single image blind deblurring is considerably more challenging as the blur kernel, as well as the latent image, must be estimated from just one observation. Works in [6, 23, 40] perform an iterative approach to solve for the latent image and blur kernel. Most of these methods employ priors on the underlying clean image and motion to stabilize the optimization process. The most widely used priors are total variational regularizer [5, 35], sparsity prior on image gradients, l_1 / l_2 image regularization [21], the unnatural l_0 prior [46], and the very recent dark channel prior [32] for images. Even though such prior-based optimization schemes have shown promise, the extent to which a prior is able to perform under general conditions is questionable [21]. Some priors (such as the sparsity prior on image gradient) even tend to favor blurry results [23]. In a majority of situations, the final result requires judicious selection of the prior, its weight, as well as tuning of other parameters. With the advancement in computation and availability of large datasets, deep learning-based deblurring too has come of age. Xu et al. [45] proposed a deep deconvolutional network for non-blind single image deblurring (i.e, the kernel is fixed and known apriori). Schuler et al. [39] came up with a neural architecture that mimics traditional iterative deblurring approaches. Chakrabarti [4] trained a patch-based neural network to estimate the kernel at each patch and employed a traditional non-blind deblurring method in the final step to arrive at the deblurred result. The above-mentioned methods attempt to estimate the blur kernel using a deep network, but finally perform non-blind deblurring outside of the network to get the deblurred result. Any error in the kernel estimate (due to poor edge content, saturation or noise in the image) will impact deblurring quality. Moreover, the final non-blind deblurring step typically assumes a prior (such as sparsity on the gradient of latent image), which again necessitates a careful selection of prior weights; else the deblurred result will be imperfect. Hence, kernel-free approaches are very much desirable. Recent works [30, 31] skips the need for kernel estimation and directly solve for the deblurred frame. But these works are restricted to deblurring and cannot perform resolution improvement.

Blind SR from motion blurred LR images: In situations where the LR frames are affected by motion blur, super-resolution makes little sense without compensating for the effect of the unknown motion blur. Sroubek et al. [41] address the blind SR problem by building a regularized energy function and minimizing it alternately with respect to the original HR image and the camera motion. The method of Ma et al. [26] is based on the premise that the same region is not equally blurred across frames. They propose a temporal region selection scheme to select the least blurred pixels from each frame. The works in [33, 49] perform the joint tasks of alignment, deblurring, and resolution enhancement. It should be noted that the blind SR techniques mentioned above are all multi-frame approaches. Single image blind SR is a much more involved problem and there are at present no traditional approaches to solve it. Very recently, Xu et al. [47] proposed a deep learning algorithm to solve the blind SR problem. They used discriminative image prior based on GAN that semantically favors clear high-resolution images over blurry low-resolution ones and directly regresses for the HR image. In contrast, ours is a sparse coding-based approach and we solve for the HR image by using an invariant feature representation.

6.3   Learning Invariant Features for Faces

Sensory data, including natural images, are sparse in nature and can be described as a superposition of small number of atoms such as edges and surfaces [27]. Dictionary learning methods are built on this very basis. Various image restoration tasks have been attempted with dictionaries (including deblurring and SR). With an added condition that these representations should be invariant to the blur or resolution in the image, dictionary methods have performed these tasks individually by learning coupled dictionaries [43, 48]. However, dictionaries capture only linearities in the data. Blurring process involves non-linearities (high frequencies are suppressed more), hence dictionary methods do not generalize across blurs.

In this chapter, we extend the notion of invariant representations to deep networks that can capture non-linearities in the data. Generalization of dictionary methods using deep networks to capture non-linearities is not new. The work in [44] com-

Stage I

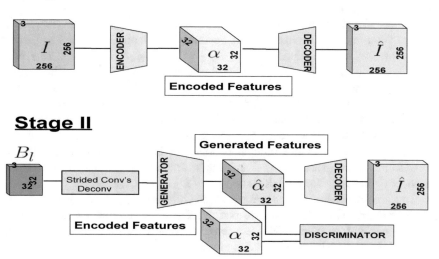

Fig. 6.1 Illustration of proposed architecture

bines sparse coding and denoising encoders for the task of denoising and inpainting. Deep neural networks, in general, have yielded good improvements over conventional methods for various low-level image restoration problems including SR [10], and inpainting and denoising [34, 44]. These networks are learned end-to-end by training with lots of example-data from which the network learns the mapping to undo distortions. We investigate the possibility of such a deep network for the task of single image blind SR. The idea of learning invariant representations is borrowed from our earlier work [31] with the main difference being that the problem we are addressing here is that of a single blind SR rather than just deblurring [31].

Similar to [31], we first require a good feature representation that can capture HR image-domain information. Autoencoders (AE) are apt for this task and have shown great success in unsupervised learning by encoding data to a compact form [15]. Once a good representation is learned for clean HR patches, the next step is to produce an invariant representation (as in [43, 48]) from blurred LR data. We propose to use a GAN for this purpose which involves training of a generator and discriminator that compete with each other. The purpose of the generator is to confuse the discriminator by producing clean features from blurred LR data that are similar to the ones produced by the autoencoder so as to achieve invariance. The discriminator, on the other hand, tries to beat the generator by identifying the clean and blurred features.

A schematic of our proposed architecture is shown in Fig. 6.1. The main difference in architecture vis-a-vis [31] is our generator now has to perform joint SR and deblurring. Since the input LR is of a lower dimension than the HR image, we include fractional strided convolutions in the initial stages of the generator. The number of these layers depends on the SR factor.

Akin to dictionary methods, our encoder-decoder architecture learns a representation in non-linear space. In dictionary approaches, an input HR patch I is sparsely represented with the dictionary atoms D_{HR} as $I = D_{HR}\alpha$. Our encoder-decoder module can be equated to this but in non-linear space. The encoder can be thought of as an inverse dictionary D_{HR}^{-1} that projects the incoming HR data into a sparse representation and decoder (D_{HR}) reconstructs the input from the sparse representation. Generator training can be treated as learning the blur LR dictionary that can project the blurred LR data B_l into the same sparse representation of I, i.e, $\alpha = D_{HR}^{-1} I = D_{b_{LR}}^{-1} B_l$. Once training is done, the input LR blurry image (B_l) is passed through the generator to get an invariant feature which when projected to the decoder yields the deblurred HR result as $\hat{I} = D_{HR}\alpha = D_{HR} D_{b_{LR}}^{-1} B_l$.

Thus, by associating the feature representation learned by the autoencoder with GAN training, our model is able to perform single image blind SR in an end-to-end manner for face dataset. Ours is a kernel-free approach and does away with the tedious task of modeling and selection of prior.

The main contributions of our work are as follows:

- We propose a compact end-to-end regression network that directly estimates the clean HR image from a single blurred LR frame without the need for optimal prior selection and weighting, as well as blur kernel estimation.
- The proposed architecture consists of an autoencoder in conjunction with a generative network for producing blur and resolution invariant features to guide the process.
- The network has shown performance gain in FR surveillance systems and produces good quality face reconstruction from its blurred LR counterpart.

6.4 Network Architecture

Our network consists of an AE that learns the clean HR image domain and a GAN that generates invariant features. We train our network in two stages. We first train an AE to learn the clean image manifold. This is followed by the training of a generator that can produce clean features from a blurred LR image which when fed to the decoder gives the deblurred HR output. Note that instead of combining the task of data representation, SR, and deblurring into a single network, we relegate the task of data-learning to the AE and use this information to guide blind SR. Details of the architecture and the training procedure are explained next.

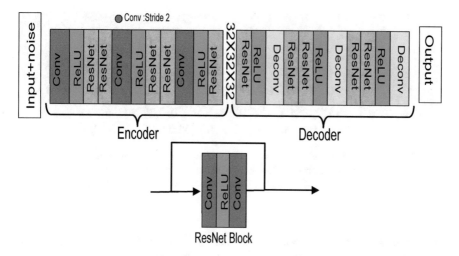

Fig. 6.2 Autoencoder architecture with residual networks

6.4.1   Encoder-Decoder

Autoencoders were proposed for the purpose of unsupervised learning [15] and have since been extended to a variety of applications. AE projects the input data into a low-dimensional space and recovers the input from this representation. When not modeled properly, it is likely that the autoencoder learns to just compress the data without learning any useful representation. Regularization using denoising encoders [42] overcomes this issue by corrupting the data with noise and letting the network undo this effect and get back a clean output. This ensures that the AE learns to correctly represent clean data. Deepak et al. [34] extended this idea from mere data representation to context representation for the task of inpainting. In effect, it learns a meaningful representation that can capture domain information of the data.

We investigated different architectures for AE and observed that including residual blocks (ResNet) [14] helped in achieving faster convergence and in improving the reconstructed output. Residual blocks help by by-passing the higher-level features to the output while avoiding the vanishing gradient problem. The training data was corrupted by noise (30% of the time) to ensure encoder reliability and to avoid learning an identity map. The architecture used in our work along with the ResNet block is shown in Fig. 6.2. A detailed description of the filter and feature map sizes along with the stride values used are as given below.

Encoder: $C^5_{3\to8} \downarrow 2 \to R^{5(2)}_8 \to C^5_{8\to16} \downarrow 2 \to R^{5(2)}_{16} \to C^3_{16\to32} \downarrow 2 \to R^3_{32}$

Decoder: $R^3_{32} \to C^2_{32\to16} \uparrow 2 \to R^{5(2)}_{16} \to C^4_{16\to8} \uparrow 2 \to R^{5(2)}_8 \to C^4_{8\to3} \uparrow 2$

where $C^c_{a\to b} \downarrow d$ represents convolution mapping from a feature dimension of a to b with a stride of d and filter size of c, $\downarrow$ represents down-convolution, $\uparrow$ stands for

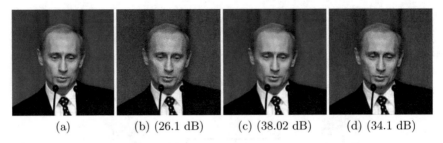

(a)   (b) (26.1 dB)   (c) (38.02 dB)   (d) (34.1 dB)

Fig. 6.3 Effect of ResNet on reconstruction. **a** The target image. **b** Noisy input to the encoder-decoder module. **c** Result of encoder-decoder module of Fig. 6.2. **d** Result obtained by removing ResNet for the same number of iterations. PSNR values are given under the respective figures. (*Enlarge for better viewing*)

up-convolution. $R_a^{b(c)}$ represents the residual block which consists of a convolution and a ReLU block with output feature size a, filter size b, while c represents the number of repetitions of residual blocks.

Figure 6.3 shows the advantage of the ResNet block. Figure 6.3a is the target image and Fig. 6.3c, d are the output of autoencoders with and without ResNet block for the same number of iterations for the input noisy image in Fig. 6.3b. Note that the one with ResNet converges faster and preserves the edges due to skip connections that pass on the information to deeper layers.

6.4.2 GAN for Feature Mapping

The second stage of training constitutes learning a generator that can map from the blurred LR image to clean HR features. For this purpose, we used a generative adversarial network (introduced by Goodfellow [13] in 2014). GANs have since been widely used for various image related tasks. It consists of two models: a Generator ($\mathcal{G}$) and a Discriminator ($\mathcal{D}$) which play a two-player mini-max game. $\mathcal{D}$ tries to discriminate between the samples generated by $\mathcal{G}$ and training data samples, while $\mathcal{G}$ attempts to fool the discriminator by generating samples close to the actual data distribution. The mini-max cost function [13] for training GANs is given by

$$\min_{\mathcal{G}} \max_{\mathcal{D}} C(\mathcal{G}, \mathcal{D}) = E_{x \sim P_{\text{data}}(x)}[\log \mathcal{D}(x)] + E_{z \sim P_z(z)}[\log(1 - \mathcal{D}(\mathcal{G}(z)))]$$

where $\mathcal{D}(x)$ is the probability assigned by the discriminator to the input x for discriminating x as a real sample. P_{data} and P_z are the respective probability distributions of data x and the input random vector z. The main goal of [13] is to generate a class of natural images from z.

Theoretically, GANs are well-defined, but many a time it is difficult to train them. Often there are instability issues that results in artifacts in the generated image. Works exsist that specifically address this issue [37, 38] and try to stabilize the training by introducing new distance metrics [2]. One such work uses conditional GAN (Mirza et al. [29]) which enables GANs to accommodate extra information in the form of conditional input. Training conditional GANs is a lot more stable than unconditional GANs due to the additional guiding input. The inclusion of adversarial cost in the loss function has shown great promise [18, 34]. The modified cost function [18] is given by

$$\min_{\mathcal{G}} \max_{\mathcal{D}} C_{cond}(\mathcal{G}, \mathcal{D}) = E_{x,y \sim P_{\text{data}}(x,y)}[\log \mathcal{D}(x, y)]$$

$$+ E_{x \sim P_{\text{data}}(x), z \sim P_z(z)}[\log(1 - \mathcal{D}(x, \mathcal{G}(x, z)))] \quad (6.1)$$

where y is the clean target feature, x is the conditional image (the blurred input), and z is the input random vector. In conditional GANs, the generator tries to model the distribution of data over the joint probability distribution of x and z. When trained without z for our task, the network learns a mapping for x to a deterministic output y which is the corresponding clean feature.

Following [18] that uses an end-to-end network with generative model to perform image-to-image translation, we initially attempted regressing directly to the clear pixels using off-the-shelf generative networks. However, we observed that this lead to erroneous results. One reason for this could be due to the high dimensionality of data. Hence, we used the apriori-learned features (which are of a lower dimension as compared to image space) of the autoencoder for training GAN. Training a perfect discriminator requires its weights to be updated simultaneously along with the generator such that it is able to discriminate between the generated samples and data samples. This task becomes easy and viable for the discriminator in the feature space for two reasons:

(i) In this space, the distance between blurred LR features and its equivalent clean HR features is higher as compared to the image space. This helps in faster training in the initial stage.

(ii) The dimensionality of the feature space is much lower as compared to that of image space. GANs are known to be quite effective in matching distributions in lower-dimensional spaces [8].

We train the GAN using the normal procedure but instead of asking the discriminator to discern between generated images and clean images, we ask it to discriminate between their corresponding features. The generator ($4 \times$) and the discriminator architectures are as given below.

Generator: $C^5_{3 \to 8} \uparrow 2 \to C^5_{8 \to 8} \to C^5_{8 \to 16} \uparrow 2 \to C^5_{16 \to 16} \downarrow 2 \to R^{5(2)}_{16} \to C^5_{16 \to 32}$
$\downarrow 2 \to R^{5(2)}_{32} \to \hat{C}^3_{32 \to 32} \downarrow 2 \to R^{5(2)}_{32} \to C^3_{32 \to 128} \downarrow 2 \to R^{3(2)}_{128} \to \hat{C}^3_{128 \to 32} \uparrow 2$

Discriminator: $C^5_{32 \to 32} \to C^5_{32 \to 32} \downarrow 2 \to C^5_{32 \to 16} \to C^5_{16 \to 16} \downarrow 2 \to C^5_{16 \to 8} \to$
$C^3_{8 \to 8} \downarrow 2 \to C^3_{8 \to 1}$

Each convolution is followed by a Leaky ReLU and batch-normalization in the discriminator, and ReLU in the generator. The input stage of the generator is a stack of learnable upsampling filters (Deconv layers) and the number of such layers depends on the upsampling factor. Above, we have shown a generator module for $4 \times$ SR factor. $\hat{C}$ indicates a skip connection from that convolution layer till the next $\hat{C}$. Using skip connections help in preserving the finite feature from the lower layers while going deeper helps in reducing the blur. We also tried other models where the generator architecture was similar to that of encoder. Such an architecture helps to preserve details in the final output but residual blur still remains in the output. We observed that going deeper helps in reducing blur at the expense of missing finite details. Hence, we used a generator which goes deeper but at the same time preserves features using skip connections.

Once the second stage is trained, we have a generator module to which we pass the blurred LR input during the test phase. The generator produces features which correspond to clean image features which when passed through the decoder deliver the final deblurred HR result.

6.4.3 Loss Function

Our network is trained in two stages. In the initial phase, the encoder is trained to learn the HR clean feature representation. For this training, we used the widely preferred reconstruction cost. The reconstruction (MSE loss) cost is defined as the l_2 distance between the expected and observed image and is given as

$$\mathcal{L}_{\text{mse}} = \|\mathcal{D}e(\mathcal{E}(I + N)) - I\|_2^2 \qquad (6.2)$$

where $\mathcal{D}e$ is the decoder, $\mathcal{E}$ the encoder, N is noise and I is the target (clean) image. The MSE error captures overall image content but tends to prefer a blurry solution. Hence, training only with MSE loss results in loss of edge details. To overcome this, we used gradient loss ($\mathcal{L}_{\text{grad}}$) as it favors edges as discussed in [28] for video-prediction.

$$\mathcal{L}_{\text{grad}} = \|\nabla \mathcal{D}e(\mathcal{E}(I + N)) - \nabla I\|_2^2 \qquad (6.3)$$

where ∇ is the gradient operator. Adding the gradient loss helps in preserving edges and recovering sharp images as compared to $\mathcal{L}_{\text{mse}}$ alone.

The second phase of training learns the invariant representation using GANs. For training GAN we tried different combinations of cost functions and found that a combined cost function given by $\lambda_{\text{adv}}\mathcal{L}_{\text{adv}} + \lambda_1\mathcal{L}_{\text{abs}} + \lambda_2\mathcal{L}_{\text{mse}}$ in the image and feature space worked for us. Even though l_2 loss is simple and easy to back-propagate, it under-performs on sparse data. Hence, we used l_1 loss for feature back-propagation, i.e.

$$\mathcal{L}_{\text{abs}} = \|\mathcal{G}(B) - \mathcal{E}(I)\|_1 \qquad (6.4)$$

where B is the blurred LR image. The adversarial loss function $\mathcal{L}_{\text{adv}}$ (given in Eq. (6.1)) requires that the samples output by the generator should be indistinguish-

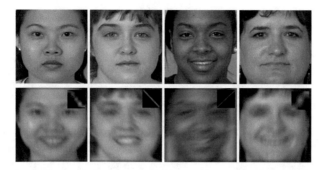

Fig. 6.4 Some example images from gallery (first row) and probe (second row). The kernels used to synthesize the probe images are shown in the inset

able to the discriminator. This is a strong condition and forces the generator to produce samples that are close to the underlying data distribution. As a result, the generator outputs features that are close to the clean HR feature samples. Another advantage of this loss is that it helps in faster training (especially during the initial stages) as it provides strong gradients. Apart from adversarial and l_1 cost on the feature space, we also used MSE cost on the recovered clean image after passing the generated features through the decoder. This helps in fine-tuning the generator to match with the decoder.

6.4.4 Training

We trained the autoencoder using images from the CelebA dataset [25] which consists of around 202,599 face images by resizing them to 256×256. We randomly picked 200K data as training set and rest as test and validation set. The inputs were randomly corrupted with Gaussian noise (standard deviation = 0.2) 30% of the time to ensure learning of useful data representation. We used Adam [20] with an initial learning rate of 0.0002 and momentum 0.9 with batch-size of 16. The training took around 3×10^5 iterations to converge. The gradient cost was scaled by $\lambda = 0.1$ to ensure that the final results are not over-sharpened.

The second stage of training involved learning a blur and resolution invariant representation from blurred LR data. We created blurred face data by synthetically blurring the CelebA dataset with space-invariant parametric blur kernels. We used $\{l, \theta\}$ (l stands for length and θ is the angle) parametrization of the blur and produced blur in the range $l \in \{0, 40\}$ pixels and $\theta \in \{0, 180\}$ degrees. The input clean images were blurred by the parametrized kernels and downsampled by factors of 2, 4, and 8 to generate the training sets for different SR factors. Each set consisted of 4 lakh blurred LR training data. The first stage of the generator was a set of up-convolution learnable filters that scale up the input data to 256. To improve GAN stability, we

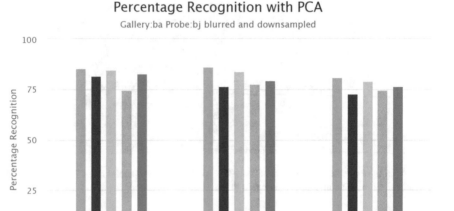

Fig. 6.5 Percentage recognition with a simple PCA FR system. The improvement in accuracy with our blind SR network over other comparative methods can be clearly observed from the figure. Our method performs well with respect to all the matching distance metrics

also used smooth labeling of blur and clean features as discussed in [1]. For around 10^5 iterations, the training was done with feature costs alone with $\lambda_{adv} = 0.001$ and $\lambda_1 = 1$. Fine-tuning of the generator was subsequently done by adding the MSE cost and weighing down the adversarial cost ($\lambda_2 = 1$, $\lambda_1 = 1$ and $\lambda_{adv} = 0.0001$).

6.5 Experiments

We demonstrate the effectiveness of our proposed blind SR network on synthetic as well as real images. We have subdivided the experiments into two sections. In the first section, we quantify performance by analyzing the recognition accuracy of a baseline FR system on the input blurred LR images prior to and after passing through our network. We observed an improvement in accuracy after using our network. The experimental setup for this is as follows. We took the ba and bj folders from the FERET dataset both of which contain 200 subjects (256×256 dimension) with one image per subject. We used ba as our gallery and used bj to produce the probe. The images from bj were subjected to parametric blur and downsampled to get 64×64 probe data. A few examples from the gallery and the probe along with the kernels used to create them are shown in Fig. 6.4. Following this, a basic FR system using PCA was used as the baseline to calculate the percentage recognition rate.

(20.85/0.8289)	(**24.795/0.8886**)	(19.33/0.7620)	(20.2202/0.7817)	
(21.36/0.8093)	(**25.92/0.8792**)	(20.8/0.7926)	(23.12/0.8154)	
(19.9/0.8570)	(**27.64/0.9359**)	(18.74/0.8070)	(23.68/0.8969)	
Input ↑	Our o/p	[32]+[17]	[30]+[22]	HR GT

Fig. 6.6 Results on LFW dataset [16]: The input images were upsampled to [256 × 256]. Results obtained by our blind SR network given in column 2. Results obtained by separately performing deblurring and SR by conventional methods [32]+[17], and deep methods [30]+[22], are given in column 3 and 4, respectively. The ground truth HR image is shown in the last column

The system works by first estimating the PCA basis from the clean HR gallery images. It then projects the probe using the estimated basis and recognizes the subject by matching the features to that of the gallery. We used three distance metrics for matching: Euclidean, Manhattan, and Cosine. Initially, we estimated the recognition rate on the clean probe and found that the recognition was on an average 84.16% only. This was because the probe images had small expression changes from the gallery and our FR system is a simple PCA-based model. Next, we checked the accuracy on the blurred LR probe data and noticed that the accuracy went down from 84.16to 76.8% after blurring and downsampling. We passed these LR probes through our trained network to get a 4 × SR and estimated the accuracy of FR on the output, the accuracy improved to 82.5% using our blind SR model.

Since there are no works on single blind SR for this type of a setting, we performed comparison by independently deblurring the LR frames followed by a single image SR framework. This we did using both conventional methods and deep learning methods. For conventional method, we use a single image deblurring framework of Pan et al. [32] to deblur the LR frames. This is followed by exemplar-based SR as proposed in [17]. The accuracy obtained in this case was quite less (75.5%). The main reason for the reduced accuracy could be due to the artifacts induced by deblurring

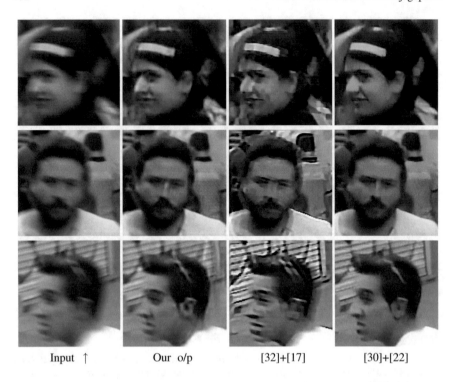

     Input ↑   Our o/p   [32]+[17]   [30]+[22]

Fig. 6.7 Results on Gopro dataset [30]. Faces were cropped from the blurred images provided in the test set. Even though our network was not trained for such a real dataset, it was able to produce comparable results to the work in [30] that was specifically trained on Gopro

which can be attributed to improper selection of prior. The second comparison was with deep learning networks. For this, we used the network in [30] to deblur the probes and these deblurred results were subjected to the SRResnet (proposed in [22]) for 4× upsampling. The accuracy improved to 79.56%. From this experiment, we can conclude that our blind SR network that performs end-to-end simultaneous deblurring and SR can help in improving the recognition accuracy of FR systems. The obtained accuracy using each of the matching methods along with comparisons are provided in Fig. 6.5.

In the second section of our experiments, we show quantitative results on synthetically blurred LR dataset from the LFW dataset [16]. We provide quantification in terms of PSNR (Peak Signal to Noise Ratio) and SSIM (Structure Similarity Index). We also provide qualitative results on a few examples from the real blurred dataset of [36] and Gopro dataset in [30].

For the quantitative experiment in Fig. 6.6, we synthetically blurred the LFW dataset and downsampled it to different scales. For comparison, we tried the existing conventional and deep learning methods as before. Results in the third column of Fig. 6.6 were obtained by deblurring the LR image with the conventional deblurring

Fig. 6.8 Results on real blurred dataset of [36]. Input blurred face and the corresponding result obtained by our network are shown side-by-side

work of Pan et al. [32] which was followed by exemplar-based SR method of [17] for the specified SR factors. Similarly, the results in column four were obtained by the deep learning-based deblurring work of [30] followed by the single image deep learning-based SR work of [22]. The obtained PSNR and SSIM values for each of these examples are provided under each image. Each row corresponds to a different upsampling factor: first row (8 ×), second row (4 ×) and third row (2 ×). Note that the training for each SR factor was done separately in our network, but the encoder training was done only once. From the results, it is evident that a joint approach for deblurring and SR performs much better than individually performing deblurring and SR.

Next, we tested our network on two real blurred datasets provided in [30, 36]. Gopro dataset introduced in [30] was produced by capturing videos using a high frame rate camera and then averaging the frames to produce realistic blurred dataset. We manually cropped faces from their test sets and fed them to our network. Our network was trained with synthetic parametric blur kernels as discussed in Sect. 6.4.4. Even with this training, we obtained results (second column) comparable to that of [30] (fourth column), which was specifically trained on Gopro. The obtained results for visual comparison are provided in Fig. 6.7. A comparison with the traditional method is also provided in the third column of Fig. 6.7. A qualitative result of our method on the real blurred dataset captured by Punnapurath et al. [36] is also provided in Fig. 6.8. Although the blur encountered in the inputs was not high, one can observe an improvement in quality with our network.

It must be mentioned that the work in [47] also addresses blind SR problem for face images. They achieve this by using a direct regression for the HR image from the blurred LR using a generative loss. Our method differs from them in the network architecture. We learn a feature representation with our network that is invariant to the blur and resolution by making use of the generative framework. To compare with the method in [47], we retrained our encoder and generator module on celebA dataset

Input   Our o/p   [47]   Input   Our o/p   [47]

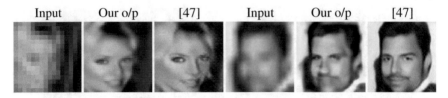

Fig. 6.9 Experimental setup similar to [47]. The input images were of size 64 × 64. These were subjected to blur and downsampling to get LR inputs of size 16 × 16. They trained by cropping faces alone but our training was by resizing. Hence, we had to crop out the face after passing through our network to match their result. The reduction is quality in our result is due to this cropping

for the specified input–output resolution as mentioned in their paper. The inputs (HR) were of 64 × 64 and the LR blurred data were 16 × 16. We modified our architecture to accommodate this input size and learned the invariant features. The input image, the result obtained by our method, and the output of [47] are shown in Fig. 6.9. Our results are comparable to that of [47].

6.6 Conclusions

In this chapter, we proposed an end-to-end deep network for single image blind SR using autoencoder and GAN. Instead of directly regressing for clean pixels, we performed regression over encoder-features to arrive at an invariant representation, which when passed through the decoder produces the desired clean HR output. Our network is kernel-free and does not require any prior modeling. The method shows improvement in FR accuracy even with a baseline FR system. When tested on real datasets, our method showed improved quality when compared to decoupled deblurring and SR.

References

1. Arjovsky M, Bottou L (2017) Towards principled methods for training generative adversarial networks. In: NIPS 2016 workshop on adversarial training, vol 2016
2. Arjovsky M, Chintala S, Bottou L (2017) Wasserstein generative adversarial networks. In: International conference on machine learning, pp 214–223
3. Capel DP (2004) Image mosaicing and superresolution
4. Chakrabarti A (2016) A neural approach to blind motion deblurring. In: ECCV. Springer, pp 221–235
5. Chan TF, Wong C-K (1998) Total variation blind deconvolution. TIP 7(3):370–375
6. Cho S, Lee S (2009) Fast motion deblurring. In: TOG, vol 28. ACM, p 145
7. Delbracio M, Sapiro G (2015) Burst deblurring: removing camera shake through fourier burst accumulation. In: Proceedings of IEEE conference on computer vision and pattern recognition (CVPR), pp 2385–2393

8. Donahue J, Krähenbühl P, Darrell T (2017) Adversarial feature learning. In: ICLR
9. Dong C, Loy CC, He K, Tang X (2014) Learning a deep convolutional network for image super-resolution. In: European conference on computer vision. Springer, pp 184–199
10. Dong C, Loy CC, He K, Tang X (2016) Image super-resolution using deep convolutional networks. TPAMI 38(2):295–307
11. Farsiu S, Robinson MD, Elad M, Milanfar P (2004) Fast and robust multiframe super resolution. IEEE Trans Image Process 13(10):1327–1344
12. Glasner D, Bagon S, Irani M (2009) Super-resolution from a single image. In: 2009 IEEE 12th international conference on computer vision. IEEE, pp 349–356
13. Goodfellow I, Pouget-Abadie J, Mirza M, Xu B, Warde-Farley D, Ozair S, Courville A, Bengio Y (2014) Generative adversarial nets. In: NIPS, pp 2672–2680
14. He K, Zhang X, Ren S, Sun J (2016) Deep residual learning for image recognition. In: CVPR, pp 770–778
15. Hinton GE, Salakhutdinov RR (2006) Reducing the dimensionality of data with neural networks. Science 313(5786):504–507
16. Huang GB, Ramesh M, Berg T, Learned-Miller E (2007) Labeled faces in the wild: a database for studying face recognition in unconstrained environments. Technical Report 07-49, University of Massachusetts, Amherst
17. Huang J-B, Singh A, Ahuja N (2015) Single image super-resolution from transformed self-exemplars. In: Proceedings of the IEEE conference on computer vision and pattern recognition, pp 5197–5206
18. Isola P, Zhu J-Y, Zhou T, Efros AA (2016) Image-to-image translation with conditional adversarial networks. arXiv:1611.07004
19. Ito A, Sankaranarayanan AC, Veeraraghavan A, Baraniuk RG (2014) Blurburst: removing blur due to camera shake using multiple images. ACM Trans Graph Submitt 3(1)
20. Kingma D, Adam JB (2015) A method for stochastic optimisation. In: ICLR. ICLR
21. Krishnan D, Tay T, Fergus R (2011) Blind deconvolution using a normalized sparsity measure. In: CVPR. IEEE, pp 233–240
22. Ledig C, Theis L, Huszar F, Caballero J, Cunningham A, Acosta A, Aitken A, Tejani A, Totz J, Wang Z, Shi W (2017) Photo-realistic single image super-resolution using a generative adversarial network. In: The IEEE conference on computer vision and pattern recognition (CVPR)
23. Levin A, Weiss Y, Durand F, Freeman WT (2011) Understanding blind deconvolution algorithms. TPAMI 33(12):2354–2367
24. Lim B, Son S, Kim H, Nah S, Lee KM (2017) Enhanced deep residual networks for single image super-resolution. In: The IEEE conference on computer vision and pattern recognition (CVPR) workshops
25. Liu Z, Luo P, Wang X, Tang X (2015) Deep learning face attributes in the wild
26. Ma Z, Liao R, Tao X, Xu L, Jia J, Wu E (2015) Handling motion blur in multi-frame super-resolution. In: Proceedings of the CVPR, pp 5224–5232
27. Mairal J, Ponce J, Sapiro G, Zisserman A, Bach FR (2009) Supervised dictionary learning. In: NIPS, pp 1033–1040
28. Mathieu M, Couprie C, LeCun Y (2015) Deep multi-scale video prediction beyond mean square error. arXiv:1511.05440
29. Mirza M, Osindero S (2014) Conditional generative adversarial nets. arXiv:1411.1784
30. Nah S, Kim TH, Lee KM (2017) Deep multi-scale convolutional neural network for dynamic scene deblurring. In: The IEEE conference on computer vision and pattern recognition (CVPR)
31. Nimisha TM, Singh AK, Rajagopalan AN (2017) Blur-invariant deep learning for blind-deblurring. In: The IEEE international conference on computer vision (ICCV)
32. Pan J, Sun D, Pfister H, Yang M-H (2016) Blind image deblurring using dark channel prior. In: CVPR, pp 1628–1636
33. Park H, Lee KM (2017) Joint estimation of camera pose, depth, deblurring, and super-resolution from a blurred image sequence. In: The IEEE international conference on computer vision (ICCV)

34. Pathak D, Krahenbuhl P, Donahue J, Darrell T, Efros AA (2016) Context encoders: Feature learning by inpainting. In: CVPR, pp 2536–2544
35. Perrone D, Favaro P (2014) Total variation blind deconvolution: the devil is in the details. In: CVPR, pp 2909–2916
36. Punnappurath A, Rajagopalan AN, Taheri S, Chellappa R, Seetharaman G (2015) Face recognition across non-uniform motion blur, illumination, and pose. IEEE Trans Image Process 24(7):2067–2082
37. Roth K, Lucchi A, Nowozin S, Hofmann T (2017) Stabilizing training of generative adversarial networks through regularization. arXiv:1705.09367
38. Salimans T, Goodfellow I, Zaremba W, Cheung V, Radford A, Chen X (2016) Improved techniques for training GANS. In: NIPS, pp 2226–2234
39. Schuler CJ, Hirsch M, Harmeling S, Schölkopf B (2014) Learning to deblur. In: NIPS
40. Shan Q, Jia J, Agarwala A (2008) High-quality motion deblurring from a single image. In: TOG, vol 27. ACM, p 73
41. Sroubek F, Cristobal G, Flusser J (2007) A unified approach to superresolution and multichannel blind deconvolution. IEEE Trans Image Process 16(9):2322–2332
42. Vincent P, Larochelle H, Bengio Y, Manzagol P-A (2008) Extracting and composing robust features with denoising autoencoders. In: Proceedings of the 25th iICML. ACM, pp 1096–1103
43. Xiang S, Meng G, Wang Y, Pan C, Zhang C (2015) Image deblurring with coupled dictionary learning. IJCV 114(2–3):248–271
44. Xie J, Xu L, Chen E (2012) Image denoising and inpainting with deep neural networks. In: NIPS, pp 341–349
45. Xu L, Ren JS, Liu C, Jia J (2014) Deep convolutional neural network for image deconvolution. In: NIPS, pp 1790–1798
46. Xu L, Zheng S, Jia J (2013) Unnatural l0 sparse representation for natural image deblurring. In: CVPR, pp 1107–1114
47. Xu X, Sun D, Pan J, Zhang Y, Pfister H, Yang M-H (2017) Learning to super-resolve blurry face and text images. In: The IEEE international conference on computer vision (ICCV)
48. Yang J, Wright J, Huang TS, Ma Y (2010) Image super-resolution via sparse representation. IEEE Trans Image Process 19(11):2861–2873
49. Zhang H, Carin L (2014) Multi-shot imaging: joint alignment, deblurring and resolution-enhancement. In: CVPR, pp 2925–2932
50. Zhu Y, Zhang Y, Yuille AL (2014) Single image super-resolution using deformable patches. In: Proceedings of the IEEE conference on computer vision and pattern recognition, pp 2917–2924

Chapter 7
Hashing A Face

Svebor Karaman and Shih-Fu Chang

Abstract Face recognition methods have made great progress in the recent years. These methods most of the time represent a face image as a high-dimensional real-valued feature, often obtained using a deep network. However, comparisons of this high-dimensional feature can be computationally expensive. Furthermore, when dealing with large face images database this representation can lead to prohibitive storage requirements. Also, in a context where the capture of the face image is performed on a mobile device or in a separate location from the face verification or search process, the amount of data that needs to be transmitted over the network should be minimized.

7.1 Introduction

Face recognition methods have made great progress in the recent years. These methods most of the time represent a face image as a high-dimensional real-valued feature, often obtained using a deep network. However, comparisons of this high-dimensional feature can be computationally expensive. Furthermore, when dealing with large face images database this representation can lead to prohibitive storage requirements. Also, in a context where the capture of the face image is performed on a mobile device or in a separate location from the face verification or search process, the amount of data that needs to be transmitted over the network should be minimized. Thus, the motivations for using hashing for face recognition can be seen as threefold

- Lower the memory consumption: hashing is a way to produce a more compact representation, and thus one could store more face representations on a single machine;

S. Karaman · S.-F. Chang (✉)
Columbia University, New York city, USA
e-mail: shih.fu.chang@columbia.edu

© The Author(s), under exclusive license to Springer Nature Switzerland AG 2021   137
N. K. Ratha et al. (eds.), *Deep Learning-Based Face Analytics*, Advances in Computer
Vision and Pattern Recognition, https://doi.org/10.1007/978-3-030-74697-1_7

- Speed-up the comparison and search time: hamming distance computation can be performed very efficiently thanks to low-level instructions, and storing a hash code as a bucket, hashing can be used as an indexing scheme;
- Reduce the transmission cost: using hashing, very little information needs to be transmitted over the network, this can be, especially advantageous if the recognition system is separated from the encoding system.

In this chapter, we will first discuss the unique challenges of hashing a face. We will then review the state-of-art hashing methods that can be applied to this problem and discuss their strengths and weaknesses. After detailing the face recognition tasks and how they are evaluated, we discuss the datasets on which we trained and evaluated our hashing models. Extensive experiments are conducted to measure how well the different hashing methods explored can solve the face recognition tasks in the compressed binary domain they induce.

There have been comprehensive surveys of compact hashing for large-scale scalable retrieval [29], but without addressing the unique challenges for face recognition. There have been alternative work for efficiently searching large datasets [7] based on quantization, but this approach does not aim to preserve the original features performance. We also focus on hashing approaches due to the unique requirements of compact coding for minimizing transmission cost.

7.2   Unique Challenges of Hashing A Face

Hashing a face representation is especially challenging as the problem of face recognition can be seen as a fine-grained recognition problem, indeed all faces share similar structure and the difference between two identities can be very subtle. Besides, different face images of the same individual can exhibit a very high level of variations due to change of pose or viewing angle, lighting condition, and furthermore can capture changes of the face due to variations of expression, the application of makeup, and also aging. An observation can be a partial observation, with parts of the face being occluded by other persons or objects in the scene. These face images can be captured as a single shot but also as a video, therefore, inducing motion blur and other capture artifacts due to this medium. Also, when dealing with a video, it can be challenging to leverage effectively the multiple frames depicting the same face.

Thus, recent face image representations tend to be highly specialized and discriminative features, often heavily learned and optimized from a large amount of data, and not generic low-level features as what many previous hashing works deal with. Furthermore, in this chapter, our objective is to approximate or maintain as much as possible the performance of the features being hashed, while usually, hashing methods are only evaluated in comparison to other hashing methods.

In this chapter, our goal is to explore how different hashing methods can deal with the challenges of the face recognition tasks in the compact binary space they induce. The large intra-class variations and the small inter-class differences of the

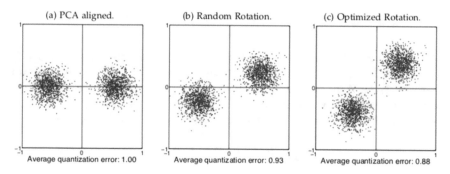

Fig. 7.1 Comparison of **a** PCA aligned, **b** random rotation based, and **c** Optimized Rotation based quantization. Figure from [10]

face recognition problem creates very challenging conditions to learn a hashing model able to produce a compact binary representation that is still effective for the verification and recognition tasks.

We here formalize the hashing task and introduce the notations we will use throughout this chapter. A dataset S is composed of N samples $\{s_i\}_{i=1...N}$. We denote the d-dimensional feature representation of the sample s_i as the row vector $\mathbf{x}_i \in \mathbb{R}^d$, and $\mathbf{X} \in \mathbb{R}^{N \times d}$ is the feature matrix of all samples. A hashing model H maps $\mathbf{x}_i$ to a binary code of length b, i.e., $H(\mathbf{x}_i) = \mathbf{h}_i \in \mathbb{B}^b$ and the full feature matrix to $H(\mathbf{X}) = \mathbf{H} \in \mathbb{B}^{N \times b}$. Note that the binary space $\mathbb{B}$ will be technically either $\{0, 1\}$ or $\{-1, 1\}$ for different methods but there is obviously a strict equivalence between these two choices. We can see the hash model H as a set of b hash functions, i.e., $H = \{f_j\}_{j=1...b}$. The jth bit h_{ij} of $\mathbf{h}_i$ is obtained by applying the jth hash function f_j of H to the feature, i.e., $h_{ij} = f_j(\mathbf{x}_i)$.

In the literature, a large number of hashing methods were proposed to obtain such a hash model. In the next two sections, we review the most important aspects of hashing, and the most recent and effective hashing methods.

7.3 Strategies for Face Hashing

In this section, we briefly review and classify the hashing methods of the literature based on whether they are data-driven in Sect. 7.3.1, linear in Sect. 7.3.2, supervised in Sect. 7.3.3, or dealing with single image input in Sect. 7.3.4. In each of these sections, we will review in more details the set of recent methods that are the most relevant for the task of hashing a face representation and that we will investigate in our experiments. We refer the reader to the survey [29] for a more complete overview of hashing methods in general.

7.3.1 Data-Dependent Versus Data-Independent

One of the first and a very popular hashing method is the Locality-Sensitive Hashing [6, 9, 15, 18] data-independent method. The key idea of Locality-Sensitive Hashing (LSH) is to define a family of hash functions that ensures that "similar" samples are more likely to be mapped to the same hash code or bucket. One common strategy is to sample random projections ($\mathbf{w}$) and thresholds (β) to define the hash functions as

$$f(\mathbf{x}) = sgn(\mathbf{xw} + \beta).$$

LSH is, therefore, a data-independent method with interesting asymptotic theoretical guarantees. This explains its popularity, and that multiple extension of this framework have been proposed [4]. However, in practice, LSH-based approaches need long hash codes and multiple hash tables to produce reasonable retrieval performance. These characteristics are not well suited for our application scenario, where we want to obtain a compact binary representation of a face feature and enable efficient search.

To overcome these limitations, more recent hashing methods learn the hash functions from the data. Principal Component Analysis (PCA) is a well-known method to perform dimensionality reduction, and it has logically been explored as a starting point of several hashing methods [11, 28, 30]. The method proposed in [10], named Iterative Quantization (ITQ), proposed to minimize the quantization error obtained when directly encoding the sign of a PCA-based projection, see Fig. 7.1. To train a hash model generating hash code of b bits, a projection matrix $\mathbf{W} \in \mathbb{R}^{d \times b}$ is obtained by taking the top b eigenvectors of the data covariance matrix $\mathbf{X}^T \mathbf{X}$. The hash code matrix $\mathbf{H}$ could be simply computed as $\mathbf{H} = \text{sgn}(\mathbf{XW})$, but this would correspond to the PCA aligned quantization illustrated in Fig. 7.1a that induces many similar samples to be encoded in different hash codes. One simple solution is to apply a random orthogonal projection $\mathbf{R} \in \mathbb{R}^{b \times b}$ as illustrated in Fig. 7.1b, however, this can be improved with the iterative optimization process the authors proposed, as shown in Fig. 7.1c.

Denoting $\mathbf{V}$ the projected data, i.e, $\mathbf{V} = \mathbf{XW}$, the ITQ optimization objective is to minimize

$$\mathcal{Q}(\mathbf{H}, \mathbf{R}) = ||\mathbf{H} - \mathbf{VR}||_F^2.$$

The optimization is performed in two alternating steps, fixing $\mathbf{R}$ and updating $\mathbf{H}$ (the quantization loss is minimized for $\mathbf{H} = \text{sgn}(\mathbf{VR})$) and fixing $\mathbf{H}$ and updating $\mathbf{R}$ (the quantization loss can be minimized by computing the SVD of the $b \times b$ matrix $\mathbf{H}^T \mathbf{V}$ as $\mathbf{H}^T \mathbf{V} = \mathbf{S} \Omega \hat{\mathbf{S}}^T$ and let $\mathbf{R} = \hat{\mathbf{S}} \mathbf{S}^T$). This two-step procedure is iterated (the authors advise to use about 50 iterations) to obtain the final ITQ hash model. ITQ has been shown to be competitive, especially compared to other PCA based hashing methods.

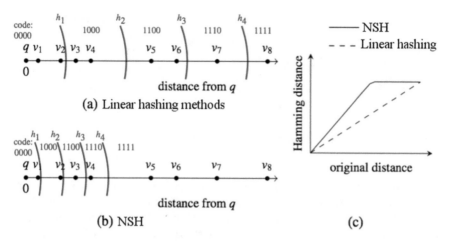

Fig. 7.2 Overview of the NSH method. Figure edited from [23]

7.3.2 Linear Versus Pivots-Based Hashing

Learning the hash functions from the data, as it is done for the ITQ method [10] detailed in the previous section, can enable selecting more effective hash functions. However, any hashing method relying on the sign of linear projections will have hash functions which have limited discriminative ability. Therefore, several hashing methods have been proposed that rely on non-linear hash functions.

Specifically, the common idea of such methods is to sample the feature space to obtain a set of pivot points.[1] Then, similarities or distances with regards to these pivots are exploited to define the hash functions. In this section, we detail two recent methods of that type: the Neighbor-sensitive Hashing [23] and the Spherical Hashing [13] methods.

The Neighbor-sensitive Hashing method (NSH) aims to avoid using hash bits to capture the distances between samples that are far apart. Therefore, the authors propose to place more separators among similar samples. Thus, somewhat counter-intuitively the method aims to increase the distance between similar items in the hamming space. This idea is illustrated in Fig. 7.2. The key element of the NSH approach is what the authors call a Neighbor-Sensitive Transform (NST). They define a function $\hat{f}$ (dubbed a coordinate-transforming function) that is a continuous and monotonic function that for a given distance range $(\eta_{\min}, \eta_{\max})$ produces "larger gaps" in a range $(\eta_{\min}, \eta_{\max})$. That is the difference of distances within the range should be bigger after applying the NST than in the original space.

In practice, the authors rely on multiple NST computed as Radial Basis Functions using the Euclidean distance with regards to a set of pivots. The pivots are obtained by running k-means on the training set. The hash functions are learned as random

---

[1] Sometimes referred to as "anchor points".

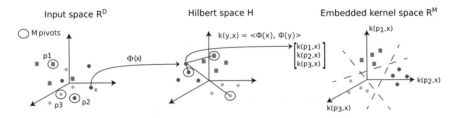

Fig. 7.3 Overview of pivots kernelized-embedding hashing method

orthogonal projections in the embedded space defined by all pivoted transforms in an unsupervised manner which has a complexity linear in the number of samples. NSH is, therefore, a pivots-based kernelized-embedding hashing method, illustrated in Fig. 7.3, similarly to the Kernel Supervised Hashing method we will discuss in 7.3.3. We refer the reader to [23] for additional details.

The Spherical Hashing (SpH) method introduced in [13], proposes to define hashing functions not as hyperplanes but as hyperspheres. More formally, the spherical hashing function $f_i(\mathbf{x})$ is defined by a pivot $\mathbf{p}_i \in \mathbb{R}^d$ and a distance threshold $t_i \in \mathbb{R}^+$ as

$$f_i(x) = \begin{cases} -1 \text{ when } d(\mathbf{p}_i, \mathbf{x}) > t_i; \\ +1 \text{ when } d(\mathbf{p}_i, \mathbf{x}) \le t_i, \end{cases}$$

where $d(\cdot, \cdot)$ denotes a distance function (either Euclidean or Cosine in the experiments we have conducted) between two samples in $\mathbb{R}^d$. The optimization process of the Spherical Hashing method aims to get independent and balanced hashing functions, it iteratively optimizes both the pivots position in space and the threshold values. The threshold values are first estimated to have $N/2$ samples in each hypersphere to satisfy the balancing constraints. Then pairs of pivots are pushed closer together or further away from each other in order to have their overlapping region to contain $N/4$ samples to satisfy the independence constraint. The optimization process is illustrated in Fig. 7.4. The advantage of using hyperspheres for hashing is that we can obtain closed regions with tight distance bounds. The reader is referred to [23] for additional details and experiments on this method.

7.3.3 Unsupervised Versus Supervised Hashing

As we have discussed in the previous section, a data-dependent optimization procedure can help obtain better hash functions. The hash functions can also be optimized to satisfy some supervision provided on the training set. Note that, as pointed out in [26], supervised hashing evaluation only make sense when evaluating the performance on new classes at test time, otherwise hashing a classifier output can easily outperforms state-of-the-art hashing methods. This is exactly what we do in all our

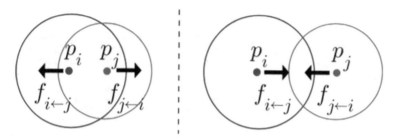

Fig. 7.4 Overview of the SpH repulsive and attractive forces between pivots during the optimization. Figure from [13]

face recognition experiments, as the training set is always disjoint from the test set. Some semi-supervised hashing methods have also been proposed [28], where the objective function combines supervised empirical fitness with unsupervised information theoretic regularization.

The Kernel-based Supervised Hashing (KSH) method, one of the most popular supervised hashing method, was introduced in [20]. This method relies on the kernel-embedding of each sample, estimated with regards to a set of m uniformly sampled points that we refer to as anchors or pivots: $\mathcal{P} = \{\mathbf{p}_j\}_{j=1...m}$. More precisely, a prediction function $\hat{f}_i(\cdot)$ is defined as

$$\hat{f}_i(\mathbf{x}) = \sum_{j=1}^{m} \kappa(\mathbf{p}_j, \mathbf{x})a_{ij} - \beta,$$

where κ is a kernel function and $\mathbf{a_i} = [a_{i1}, \ldots, a_{ij}, \ldots, a_{im}]$ and β are the coefficients and bias respectively. The corresponding hash function is defined as $f_i(\mathbf{x}) = \text{sgn}(\hat{f}_i(\mathbf{x}))$. To obtain a balanced hash function, the bias vector β is computed as the mean of the projected features in the kernel-embedding space, so each of its coordinate β_j is estimated as

$$\beta_j = \sum_{i=1}^{N} \kappa(\mathbf{p}_j, \mathbf{x}_i)/N.$$

The goal of KSH is to learn the coefficient vectors $\mathbf{a}_1, \ldots, \mathbf{a_b}$ using supervised information. Given similar and dissimilar pairs of samples, and leveraging the one-to-one correspondence between hamming distance and hash codes inner product, the authors propose to train hash functions such that hash codes inner product of similar samples are close to 1 and close to -1 for dissimilar samples. This optimization objective is illustrated in Fig. 7.5. In practice, the optimization is performed in a greedy manner, where each hash function parameter (namely the coefficient vector) is learned iteratively. We refer the reader to [20] for the complete optimization procedure description. KSH has been shown to be highly competitive, especially when

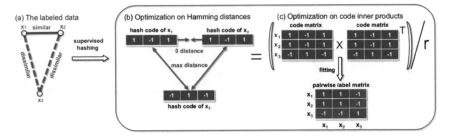

Fig. 7.5 Overview of the KSH optimization procedure. Figure from [20]

dealing with features that are not highly discriminative for the task at hand such as generic hand-crafted features.

7.3.4 Image Versus Set/Video Hashing

Most of the hashing methods consider each sample independently. But in many scenarios involving faces, multiple images of the same person can be provided at once. For example, when extracting face images from a video, a template containing all the appearances of one person in this video can be easily defined. How to leverage efficiently a single hash code representation for that whole set of images has been explored in [8]. The key idea is to build a set representation that is independent of the input size. The set representation captures both local geometric properties (computing the average, variance, minimum, and maximum of the features of all the images in the set) and global distribution features, encoding the image features with the VLAD approach [16]. The image features can be precomputed or can be derived from a deep network, and in that case, the whole model can be trained in an end-to-end fashion.

7.4 Face Recognition Tasks and Evaluation

In this section, we briefly introduce the two main tasks of face recognition: face verification and face search. We also detail how these tasks are usually evaluated.

7.4.1 Face Verification

The verification protocol is the task of comparing two face images (or two sets of face images) and output a score estimating whether or not the two inputs correspond to the same identity. The typical application of that task is access control, where a given person' identity needs to be verified. For example, a building with restricted access may be connected to a database of the face images of the authorized users. To be granted access a user needs to provide an identifier, e.g., a badge with a barcode, that will be used to retrieve the stored face imagery to be compared with a live acquisition of that person. That is also the use case of the "face-unlock" feature of recent smartphones. For this task, the dataset or benchmark usually provides a predefined list of pairs of face images (that are either a genuine match or an impostor comparison) to be evaluated [14, 32].

7.4.2 Face Search

The task of face search is the task of finding the most similar face images in a large database of face images. One of the typical use case would be a watch list situation, where the database holds a list of missing or wanted persons, and queries are face images of persons of interest.

If some probe images do not have a match in the gallery, the task is named an open-set identification. In that case, a good face recognition system should still retrieve the most similar samples from the gallery but they should have a low similarity score. Thus, a filtering approach based on the similarity score should enable flagging the probe as having no match in the gallery.

7.4.3 Evaluation Metrics

To evaluate the face verification protocol, we will report Receiver Operating Characteristic (ROC) curves. At a given threshold T of similarity that would determine if a pair is considered matched, we compute the True Accept Rate (TAR), that is, the fraction of genuine comparisons that have a higher similarity S than the threshold, and thus would correctly be accepted, and the False Accept Rate (FAR), that is the fraction of impostor comparisons that exceed the threshold, and therefore, would incorrectly be seen as matches.

Formally, we denote $\mathcal{G}$ as the set of genuine comparisons and $\mathcal{I}$ as the set of impostors comparisons. Using the previous definitions and the standard notations for the False Positives (FP), True Positives (TP), False Negatives (FN), and True Negatives (TN), we can define the False Accept Rate (FAR) or False Positive Rate as

$$\text{FAR} = \frac{\#\text{impostor with} S > T}{\text{total} \#\text{impostor}} = \frac{|\{x \in \mathcal{I} \mid S(x) > T\}|}{|\mathcal{I}|} = \frac{FP}{FP + TN},$$

while the True Accept Rate (TAR), corresponding to the Recall, is computed as

$$\text{TAR} = \frac{\#\text{genuine with} S > T}{\text{total } \#\text{genuine}} = \frac{|\{x \in \mathcal{G} \mid S(x) > T\}|}{|\mathcal{G}|} = \frac{TP}{TP + FN}.$$

Varying the threshold, we can plot a curve showing the trade-off between these two measures. One can define a target value of one of these measures (e.g., FAR = 10^{-3}) and compare different methods based on the other measure value to select the best approach amongst multiple candidate algorithms at that operating point, i.e., the one with the highest TAR at the target FAR value.

The search performance is evaluated using the Cumulative Matching Curve (CMC). For each probe sample, a ranked list of the most similar samples in the gallery is retrieved. The CMC captures the portion of probe samples for which we have correctly retrieved a matching gallery sample up to a given rank. The retrieval rates up to some specific rank values (e.g., 1, 5, and 10) are also often reported.

7.5 Face Datasets

In this section, we review the different datasets we will use to evaluate our hashing models, as well as the datasets used to train either the hashing models or the deep models which are used to extract the face features we will hash into binary codes.

7.5.1 IJB-A: IARPA Janus Benchmark A

The IJB-A [17] dataset was developed with the goal of pushing the limits of face recognition, especially targeting the unconstrained setting. Most of these face images and videos are captured in a non-collaborative setting thus depicting faces in a wide variety of poses, illuminations, and expressions. This dataset has 5,712 images of 500 different subjects as well as 2,085 videos for these same subjects. The dataset is divided into 10 splits. For each split, 333 randomly selected subjects are placed in the training set. The remaining 167 subjects are used for the test set, with their images randomly sampled into either the probe set or the gallery set.

7.5.2 IJB-B: IARPA Janus Benchmark B

The IJB-B [31] dataset is an extension of the IJB-A dataset. It contains 21,798 images of 1,845 subjects and 7,011 videos. The objectives of this extension are to provide a more uniform geographic distribution of subjects and to evaluate performance on a much larger number of genuine (10,270) and impostor (8,000,000) comparisons when evaluating the verification task, thus producing more meaningful TAR values at low FAR values (0.01 or 0.001%). For the search task, the dataset provides two predefined galleries, as well as three different probe sets grouped based on the media type: either still images only, video frames only or a mix of the two modalities.

One clear distinction between IJB-A and IJB-B is that IJB-B has no pre-defined training set, thus external data must be used to train representation and hash models. We will rely on the UMD faces (see Sect. 7.5.4) to train our hash models for this dataset.

7.5.3 IJB-C: IARPA Janus Benchmark C

The IJB-C [22] dataset is an extension of the IJB-B dataset. It contains 31,334 images of 3,531 subjects and 11,779 videos. The additional data has been collected with the goal of covering a wider set of occupations and geographic origins. This dataset has two gallery sets containing respectively 1,772 and 1,759 subjects, with a single template of about 5 images defined for each subject. There is a single probe set of 19,593 templates, build from both still images and video frames, and one subject may have multiple probe templates. The verification protocol provides 19,557 genuine matches and 15,638,932 impostor matches, allowing a proper evaluation of the performance at low FAR values. As for IJB-B, there is no pre-defined training set and we will rely on the UMD faces to train our hash models for this dataset.

7.5.4 UMD Faces

The UMD Faces dataset [2] is composed of both still images and video frames. The first part containing only still images has 367,888 annotated faces of 8,277 subjects, divided into 3 batches. Human curated bounding boxes for faces and estimated pose (yaw, pitch, and roll), locations of twenty-one keypoints, and gender information generated by a pre-trained neural network are provided. The second part contains 3,735,476 annotated video frames extracted from a total of 22,075 for 3,107 subjects. Similar estimated information is provided. The subjects have been selected to have no overlap with the IJB-A, the IJB-B, and the IJB-C datasets. We will use this dataset to train our hashing models when no other training set is available.

7.5.5 CASIA WebFace Dataset

The CASIA WebFace dataset [33] was the first large-scale face image dataset released. This dataset contains 10,575 subjects (that are famous people such as actors, politicians or other public figures) with a total of 494,414 images collected from the web. There remain 10,548 subjects after removing subjects also present in the IJB-A dataset. This dataset is not used to train hashing models directly but it is used to train some of the deep models used to extract the features (as detailed in the next section) given as input to our hashing models.

7.6 Face Features

In this section, we detail the three different generations of features we use in our experiments.

7.6.1 UMD Features: First Generation

The first generation of UMD features correspond to the Deep Convolutional Neural Networks (DCNN) described in [5]. We refer to the DCNN-S network (described in Table 7.2 of that paper) as the JC feature, while the SW feature correspond to the DCNN-L network (described in Table 7.3 of the paper). The JC feature has 320 dimensions while the SW feature has 512 dimensions. Both networks have been trained on the CASIA WebFace dataset [33] detailed in Sect. 7.5.5. These features will be used for the experiments on the IJB-A dataset.

7.6.2 UMD Features: Second Generation

The second generation of features is also composed of two features that we name JC2 and SW2. The JC2 feature corresponds to the *pool* feature of 320 dimensions of the architecture described in [34]. The SW2 feature is the 512 dimensional identity feature extracted from the "All-in-One" architecture presented in [25]. Each of these features is further embedded into a real-valued 128 dimensional space using the Triplet Probabilistic Embedding approach [27]. We will reference these embedded versions of the JC2 and SW2 as EJC2 and ESW2, respectively. These features will be used for the experiments on the IJB-B dataset.

7.6.3 UMD Features: Third Generation

From the third generation of UMD features, we use the RAG1 feature that is extracted from the bottleneck layer of a resNet-101 network as detailed in [24] This model is trained on a dataset containing over 5 million of images of about 58,000 subjects. These images contain a mixture of about 300,000 still images from the UMD-Faces dataset [3], about 3.7 million still images from the curated version [19] of the MS-Celeb-1M dataset [12] and about 1.8 million video frames from the extension to the UMDFaces dataset [1]. Furthermore, the RAG1 feature is embedded into a real-valued 128 dimensional space using the Triplet Probabilistic Embedding approach [27], we will refer to this embedded version as ERAG1. These features will be used for the experiments on the IJB-C dataset.

7.7 Face Hashing Experiments

In this section, we report multiple experiments conducted on the face verification and face search tasks. We first detail the experimental settings on each of the datasets used. We then report face verification and face search results for the different hashing methods explored. We also report some more detailed analysis based on the type of distance used in pivots-based hashing methods and the type of probe images.

7.7.1 Experimental Settings

In this section, we give the experimental settings used to train the hashing models, to run the experiments, and to evaluate the performance on all datasets.

7.7.1.1 IJB-A

For the IJB-A dataset, we used the JC and SW features of the first generation of UMD features introduced in Sect. 7.6.1. The features are extracted for each face image in a template, l2-normalized and averaged to define the template feature. We train each hash model separately on each feature on the training part of each of the 10 splits of the dataset as detailed in Sect. 7.5.1. The Euclidean distance is used for the pivots embedding hashing methods KSH, NSH, and SpH.

We accumulate the verification statistics and average the retrieval rates over the 10 splits of the dataset to report each method' performance. The baseline performance is obtained by comparing the template feature with the Euclidean distance, while hash codes are compared using the hamming distance.

7.7.1.2 IJB-B

For the IJB-B dataset introduced in Sect. 7.5.2, we used the features of the second generation of UMD features detailed in Sect. 7.6.2. The features are extracted and projected with the TPE approach for each face image in a template, face images coming from different frames of the same video are first averaged into a media-averaged feature, then all media-averaged features and still images features are averaged to define the template feature. As the IJB-B dataset has no training set, we use the UMD

Faces dataset to train the hash models. Specifically, we train a separate model for each hashing method on each of the three batches of the UMD Faces dataset.

We evaluate the verification task using the predefined set of comparisons. For the baseline approach, we compare the template features using the Cosine distance, while hash codes are compared using the hamming distance. For the verification task, we report hashing performance by accumulating the matching statistics for the three different models trained on each UMD faces batch. For the search task, we report the baseline performance averaged over the two galleries. While the hashing performance is averaged over the two galleries and the three different models trained on each UMD faces batch.

7.7.1.3   IJB-C

For the IJB-C dataset introduced in Sect. 7.5.3, we used the features of the third generation of UMD features detailed in Sect. 7.6.3. As for IJB-B, the features are extracted and projected with the TPE approach for each face image in a template, face images coming from different frames of the same video are first averaged into a media-averaged feature, then all media-averaged features and still images features are averaged to define the template feature. The features are not l2-normalized. As the IJB-C dataset has no training set, we use the UMD Faces dataset to train the hash models. We train a single hash model for each feature using all three batches of UMD faces at once. We use the Cosine distance for the pivots embedding hashing methods KSH, NSH, and SpH.

We evaluate the verification task using the predefined set of comparisons. For the baseline approach, we compare the template features using the Cosine distance, while hash codes are compared using the hamming distance. For the search task, we report the performance averaged over the two galleries.

7.7.2   IJB-A

In this section, we report the face verification and search performance of the different hashing methods on the IJB-A dataset.

7.7.2.1   Face Verification Results

We report the verification results obtained on the IJB-A dataset, comparing the original feature performance with hashing results using hash codes of 64, 128, 256 or 512 bits obtained by the ITQ, KSH, NSH, and SpH hashing methods, given as input the JC features in Fig. 7.6, or the SW features in Fig. 7.7.

The JC feature obtains a verification TAR of 0.838, 0.675 and 0.46 at FAR 10^{-2}, 10^{-3} and 10^{-4} respectively. From Fig. 7.6, we can observe that the ITQ method shows

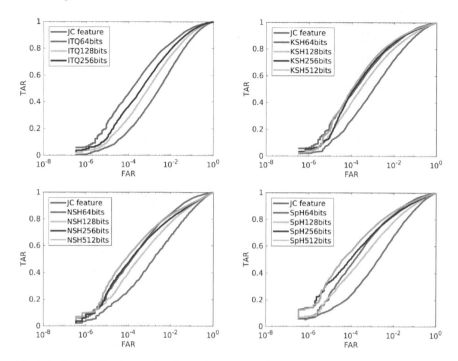

Fig. 7.6 Face verification results obtained with the JC features on the IJB-A dataset using ITQ (top-left), KSH (top-right), NSH (bottom-left) and SpH (bottom-right)

a significant drop in the performance even when using 256 bits (with a TAR of 0.791 and 0.585 at FAR 10^{-2} and 10^{-3}, respectively), note that since ITQ is based on PCA it cannot produce results with 512 bits as the JC feature has only 320 dimensions. The KSH performance when using hash codes of 256 and 512 bits is almost the same as the original features and even slightly better for FAR values higher than 10^{-4}, for example, with 256 bits, KSH obtains a TAR of 0.850 and 0.693 at FAR 10^{-2} and 10^{-3}, respectively. The KSH verification performance is already good with only 128 bits. Similarly, the NSH achieves a similar performance compared to the original JC features when using 256 or 512 bits, with 512 bits NSH obtains a TAR of 0.829 and 0.700 at FAR 10^{-2} and 10^{-3} respectively. The SpH method seems to perform the best for this task and using the JC features, and somewhat surprisingly seems to outperforms the feature performance in the low FAR regime. This may come from the spherical hash function that could better deal with impostors. The SpH hashing with 512 bits obtains a TAR of 0.733 and 0.561 at FAR 10^{-3} and 10^{-4} respectively.

The SW feature obtains a lower verification performance than the JC feature with a TAR of 0.816, 0.639, and 0.433 at FAR 10^{-2}, 10^{-3} and 10^{-4}, respectively. From Fig. 7.7, when using the SW feature, similar conclusions on the hashing methods can be drawn. As the SW feature has 512 dimensions, even ITQ can obtain results with 512 bits. However, the performance with 512 bits is very similar to the 256 bits

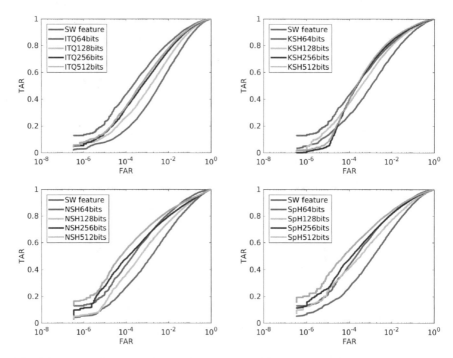

Fig. 7.7 Face verification results obtained with the SW features on the IJB-A dataset using ITQ (top-left), KSH (top-right), NSH (bottom-left), and SpH (bottom-right)

performance, which is specifically of 0.758 and 0.55 TAR at FAR 10^{-2} and 10^{-3}. That can be explained by the fact that the PCA first dimensions are much more informative than the last ones. KSH is able to maintain most of the feature performance with 256 bits (TAR of 0.654 at FAR 10^{-3}) or more, and slightly outperform the original feature performance at high FAR values. The two unsupervised pivots-based hashing methods NSH and SpH slightly outperforms the SW feature, especially at low FAR values. NSH with 512 bits and SpH with 512 bits achieve a TAR value of 0.545 and 0.541, respectively, at a FAR of 10^{-4} which is much higher than the 0.433 of the original feature. Note that as the SW feature has a lower performance than the JC feature, that may give more room for improvement for the hashing methods.

7.7.2.2   Face Search Results

We report the CMC curves obtained on the IJB-A dataset, comparing the original feature performance with hashing results using hash codes of 64, 128, 256 or 512 bits obtained by the ITQ, KSH, NSH, and SpH hashing methods, given as input the JC features in Fig. 7.8, or the SW features in Fig. 7.9.

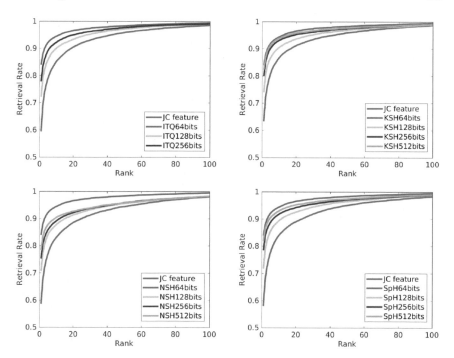

Fig. 7.8 Face search results obtained with the JC features on the IJB-A dataset using ITQ (top-left), KSH (top-right), NSH (bottom-left), and SpH (bottom-right)

The JC features obtain a retrieval rate of 0.839, 0.921, and 0.944 at rank 1, 5, and 10, respectively. From Fig. 7.8, we can observe that KSH is the method that can preserve most of the JC feature search performance, with a limited loss of performance when using 512 bits. KSH achieves a retrieval rate of 0.823 at rank 1 and 0.938 at rank 10. The SpH method exhibits a bit more of performance loss than KSH, but still provides a good representation when using 512 bits, achieving a retrieval rate of 0.819 at rank 1 and 0.93 at rank 10. ITQ achieves reasonable results when using 256 bits, with a retrieval rate of 0.891 at rank 5 and 0.921 at rank 10 but has the lowest of all compared methods rank 1 value with 0.778. Once again this method cannot be trained to produce 512 bits with the JC feature of 320 dimensions. The NSH method is not performing well on this search task, with a significant drop of performance even when using 512 bits with a retrieval rate of 0.783 at rank 1 and 0.904 at rank 10.

The SW features obtain a retrieval rate of 0.832, 0.92, and 0.948 at rank 1, 5, and 10, respectively, which is similar to the JC features performance. From the results reported in Fig. 7.9, when using the SW feature, we can see that all hashing methods exhibit a bigger drop in the performance at rank 1 with the best method SpH with 512 bits reaching a retrieval rate of 0.792. ITQ performs the best at rank 5 and 10 (with retrieval rates of 0.9 and 0.934) followed closely by KSH (with retrieval rates

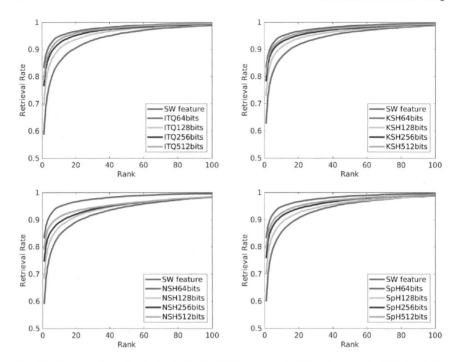

Fig. 7.9 Face search results obtained with the SW features on the IJB-A dataset using ITQ (top-left), KSH (top-right), NSH (bottom-left), and SpH (bottom-right)

of 0.9 and 0.934) and SpH. The NSH method has the lowest performance at higher ranks with the SW feature, reaching a retrieval rate of only 0.909 at rank 10.

7.7.2.3 Discussion

Overall on this IJB-A dataset, the KSH method using kernelized-embedding and supervision is able to maintain each feature verification performance and even slightly outperform them at high FAR values. This method is also able to preserve most of the features' search abilities. The ITQ method is limited by its linear projection scheme and its dependency on PCA. While its performance on the verification task is limited, it does enable reasonable performance on the search task. The two unsupervised pivots-based hashing methods NSH and SpH can maintain or even outperform, especially at low FAR values, the original feature performance on the verification task. The SpH method also performs well on the search task, however, the NSH method exhibits a significant drop of performance.

7.7.3   IJB-B

In this section, we report the face verification and search performance of the different hashing methods on the IJB-B dataset. For this dataset, we use the embedded features EJC2 and ESW2 that were optimized for the Cosine distance. We, therefore, first study the influence of the distance choice (either Euclidean or Cosine) used when performing the pivots-based embedding in the KSH, NSH, and SpH hashing methods. As this dataset also provides three different probe sets based on the type of media (still images, video frames or mixed), we will also explore how this influences the search performance.

7.7.3.1   Distance Choice Analysis

The second generation of UMD features EJC2 and ESW2 are embedded using the TPE approach that is optimizing the Cosine similarity between features. It is, therefore, worth investigating whether using the Euclidean distance (as it was done for the IJB-A dataset) or the Cosine distance in the pivots embedding step would have an influence on the performance of the KSH, NSH, and SpH hashing methods. We have trained models using either the Euclidean distance or the Cosine distance for the pivots embedding step. When using the Euclidean distance, the features are l2-normalized for training. At test time, for the models trained with the Euclidean distance, the features of a template are first l2-normalized before being averaged (and the average is l2-normalized again), while they are not normalized when using the Cosine distance. We will discuss the influence of this choice when reporting results in the following sections.

7.7.3.2   Face Verification Results

We report here the verification results obtained on the IJB-B dataset, comparing the original feature performance with hashing results using hash codes of 128, 256, 512 or 1024 bits obtained by the ITQ, KSH, NSH, and SpH hashing methods. We report, in Tables 7.1 and 7.2 the results obtained with the l2-normalized features and Euclidean distance or with unnormalized features and Cosine distance for the pivots embedding hashing methods, for EJC2 and ESW2 respectively. The best hashing results for each FAR value is reported in bold.

This set of results show that this second generation of features is performing much better, with for the EJC2 feature a TAR of 0.927, 0.852, and 0.753 and for the ESW2 feature a TAR of 0.936, 0.869, and 0.772, at FAR values of 10^{-2}, 10^{-3} and 10^{-4}, respectively. It is, therefore, more challenging to maintain their performance with hashing.

The ITQ method seems to slightly benefit from processing unnormalized features but nevertheless, being limited to the original embedded features dimensions of 128,

Table 7.1 Face verification TAR performance on the IJB-B dataset with the EJC2 feature

Method	Norm.	Pivots Dist.	# bits	FAR = 10^{-1}	10^{-2}	10^{-3}	10^{-4}	10^{-5}	10^{-6}
EJC2	–	–	–	0.973	0.927	0.852	0.753	0.598	0.228
ITQ	l2	–	128	0.950	0.857	0.744	0.588	0.422	0.197
ITQ	–	–	128	0.954	0.865	0.761	0.608	0.448	0.168
KSH	l2	l2	128	0.945	0.852	0.724	0.544	0.366	0.166
KSH	–	cos	128	0.941	0.835	0.677	0.487	0.301	0.130
NSH	l2	l2	128	0.943	0.85	0.703	0.523	0.352	0.158
NSH	–	cos	128	0.947	0.864	0.736	0.59	0.395	0.178
NSH	l2	l2	512	0.95	0.886	0.786	0.648	0.47	0.177
NSH	–	cos	512	0.947	0.892	0.805	0.689	0.526	0.209
NSH	l2	l2	1024	0.953	0.9	0.81	0.682	0.51	0.182
NSH	–	cos	1024	0.953	0.91	0.826	0.711	**0.549**	**0.247**
SpH	l2	l2	128	0.943	0.848	0.713	0.568	0.379	0.167
SpH	–	cos	128	0.947	0.856	0.734	0.598	0.428	0.178
SpH	l2	l2	512	0.965	0.903	0.805	0.673	0.503	0.227
SpH	–	cos	512	0.968	0.912	0.822	0.7	0.513	0.242
SpH	l2	l2	1024	0.968	0.912	0.817	0.687	0.519	0.239
SpH	–	cos	1024	**0.971**	**0.919**	**0.834**	**0.717**	0.543	0.24

is not able to maintain most of the performance at the interesting operating points of FAR 10^{-3} and 10^{-4}, giving TAR values of 0.761 and 0.608 with the EJC2 feature and 0.78 and 0.64 with the ESW2 features. Using the Cosine distance enables a better performance of the pivots embedding hashing methods NSH and SpH. The NSH method with the Cosine distance can preserve most of the performance of each feature with 512 or 1024 bits (e.g., TAR of 0.826 and 0.711 with the EJC2 feature and TAR of 0.847 and 0.735 with the ESW2 feature at FAR 10^{-3} and 10^{-4}) while its performance using the Euclidean distance is lower (TAR of 0.81 and 0.682 with EJC2 and 0.831 and 0.702 with ESW2 at the same FAR values). The SpH method with the Cosine distance can preserve most of the performance of the EJC2 feature, with 1024 bits it obtains a TAR of 0.834 and 0.717 at FAR 10^{-3} and 10^{-4}. It, however, shows a drop in the performance with the ESW2 feature, with 1024 bits obtaining a TAR of 0.815 and 0.66 at FAR 10^{-3} and 10^{-4}, indicating a higher sensitivity to the original feature space than NSH. The SpH method has also lower performance when using the Euclidean distance, e.g., TAR of 0.687 with EJC2 and 0.609 with ESW2 at FAR 10^{-4}. On the contrary, the use of the Cosine distance seems to lower the performance of the KSH method, especially for the ESW2 feature.

7.7.3.3 Face Search Results

We report the retrieval performance obtained on the IJB-B dataset, comparing the original feature performance with hashing results using hash codes of 128, 256, 512 or 1024 bits obtained by the ITQ, KSH, NSH, and SpH hashing methods, given as input the EJC2 features in Table 7.3 or the ESW2 features in Table 7.4. The best hashing results for each rank value are reported in bold. The retrieval rates at rank 1, 5, 10, and 20 are 0.839, 0.909, 0.928, and 0.944 for the EJC2 feature and 0.858, 0.922, 0.94, and 0.956 for the ESW2 feature respectively. For these results, we used the mixed media probes of the IJB-B dataset. We will discuss the influence of the probe type in the next section.

Once again, the performance of the pivots-based hashing methods NSH and SpH is improved when using the Cosine distance. With this configuration the SpH method is able to preserve most of the performance of the EJC2 feature, see Table 7.3, with retrieval rates of 0.822 and 0.899 at rank 1 and 5. It also achieves a good performance for ESW2 in Table 7.4, especially for ranks higher than 10. The NSH method with Cosine distance performs relatively well on the EJC2 features and is the best performing method with the ESW2 features for ranks lower than 5. As for

Table 7.2 Face verification TAR performance on the IJB-B dataset with the ESW2 feature

Method	Norm.	Pivots Dist.	# bits	FAR = 10^{-1}	10^{-2}	10^{-3}	10^{-4}	10^{-5}	10^{-6}
ESW2	–	–	–	0.98	0.936	0.869	0.772	0.589	0.195
ITQ	12	–	128	0.962	0.882	0.757	0.589	0.394	0.137
ITQ	–	–	128	0.966	0.892	0.78	0.64	0.444	0.168
KSH	12	12	128	0.948	0.853	0.706	0.531	0.323	0.094
KSH	–	cos	128	0.947	0.846	0.69	0.469	0.225	0.052
KSH	12	12	512	0.956	0.895	0.8	0.669	0.479	0.151
KSH	–	cos	512	0.963	0.894	0.759	0.545	0.261	0.059
NSH	12	12	128	0.95	0.85	0.714	0.521	0.314	0.105
NSH	–	cos	128	0.95	0.869	0.741	0.593	0.406	0.147
NSH	12	12	512	0.95	0.886	0.786	0.648	0.47	0.177
NSH	–	cos	512	0.96	0.911	0.829	0.719	0.537	**0.214**
NSH	12	12	1024	0.965	0.919	0.831	0.702	0.472	0.171
NSH	–	cos	1024	0.966	**0.925**	**0.847**	**0.735**	**0.542**	0.203
SpH	12	12	128	0.9602	0.868	0.723	0.538	0.306	0.103
SpH	–	cos	128	0.963	0.882	0.763	0.571	0.36	0.129
SpH	12	12	512	0.974	0.903	0.781	0.605	0.319	0.089
SpH	–	cos	512	0.975	0.912	0.799	0.629	0.383	0.109
SpH	12	12	1024	0.977	0.908	0.782	0.609	0.338	0.093
SpH	–	cos	1024	**0.978**	0.922	0.815	0.660	0.417	0.124

Table 7.3 Face search retrieval rates on the IJB-B dataset with the EJC2 feature

Method	Norm.	Pivots Dist.	# bits	Rank 1	Rank 2	Rank 5	Rank 10	Rank 20	Rank 50
EJC2	–	–	–	0.839	0.874	0.909	0.928	0.944	0.964
ITQ	12	–	128	0.711	0.768	0.825	0.860	0.891	0.926
ITQ	–	–	128	0.730	0.784	0.837	0.871	0.899	0.931
KSH	12	12	128	0.696	0.759	0.819	0.855	0.885	0.923
KSH	–	cos	128	0.670	0.730	0.794	0.836	0.873	0.912
NSH	12	12	128	0.687	0.747	0.809	0.848	0.880	0.916
NSH	–	cos	128	0.714	0.772	0.829	0.861	0.891	0.925
NSH	12	12	512	0.769	0.815	0.860	0.885	0.909	0.934
NSH	–	cos	512	0.790	0.831	0.869	0.892	0.912	0.933
NSH	12	12	1024	0.795	0.837	0.876	0.899	0.919	0.941
NSH	–	cos	1024	0.809	0.850	0.885	0.906	0.923	0.942
SpH	12	12	128	0.685	0.745	0.805	0.845	0.881	0.920
SpH	–	cos	128	0.711	0.766	0.823	0.856	0.885	0.922
SpH	12	12	512	0.787	0.833	0.877	0.903	0.925	0.951
SpH	–	cos	512	0.805	0.847	0.887	0.910	0.933	0.955
SpH	12	12	1024	0.802	0.845	0.886	0.911	0.933	0.956
SpH	–	cos	1024	**0.822**	**0.861**	**0.899**	**0.921**	**0.938**	**0.960**

the verification task, the performance of ITQ is slightly better using unnormalized features but is limited due to the ability to only producing hash codes of 128 bits. The KSH performance again decreases when using the Cosine distance. We will, therefore, in the remainder of this chapter report only results using the unnormalized features and the Cosine distance for pivots-based hashing methods SpH and NSH. We will, however, use the normalized and the Euclidean distance for KSH.

7.7.3.4   Probe Type Analysis

Here we further study the influence of the type of media (still images or video frames) on the search performance on the IJB-B dataset which, as detailed in Sect. 7.5.2, has separate probe sets with different media sources. We report the performance with the EJC2 feature for still images probes in Fig. 7.10, and for video frames in Fig. 7.11. The results for mixed media were already reported in Table 7.3. Following the previous section analysis, we use unnormalized features and the Cosine distance for the pivots embedding step of NSH and SpH, but we use 12-normalized features and the Euclidean distance for KSH. We can see that the video frame probes represent a much more difficult task as the search performance is much lower than the still or mixed probes, note the different range on the y-axis between Figs. 7.10 and 7.11. For both probe types, it is the SpH method that is able to better preserve the performance of the original EJC2 features.

7.7.4 IJB-C

In this section, we report face verification and face search results with the ERAG1 feature on the IJB-C dataset.

7.7.4.1 Face Verification Results

We report the verification results on the IJB-C dataset in Fig. 7.12. We can observe that the ERAG1 feature achieves a very high level of face verification performance, with a TAR of 0.979, 0.958, and 0.923 at FAR values of 10^{-2}, 10^{-3}, and 10^{-4}, respectively. This high level of verification performance is very difficult to maintain for ITQ using only 128 bits, the TAR drops significantly to 0.807 and 0.674 at FAR

Table 7.4 Face search retrieval rates on the IJB-B dataset with the ESW2 feature

Method	Norm.	Pivots Dist.	# bits	Rank 1	Rank 2	Rank 5	Rank 10	Rank 20	Rank 50
ESW2	–	–	–	0.858	0.890	0.922	0.940	0.956	0.970
ITQ	12	–	128	0.741	0.795	0.850	0.884	0.911	0.944
ITQ	–	–	128	0.762	0.814	0.864	0.893	0.920	0.950
KSH	12	12	128	0.697	0.755	0.813	0.853	0.888	0.923
KSH	–	cos	128	0.691	0.748	0.807	0.845	0.879	0.917
KSH	12	12	512	0.785	0.827	0.869	0.895	0.916	0.939
KSH	–	cos	512	0.784	0.823	0.866	0.892	0.915	0.943
NSH	12	12	128	0.695	0.754	0.816	0.853	0.888	0.924
NSH	–	cos	128	0.724	0.780	0.833	0.868	0.895	0.927
NSH	12	12	512	0.795	0.836	0.876	0.901	0.922	0.943
NSH	–	cos	512	0.817	0.853	0.887	0.911	0.928	0.948
NSH	12	12	1024	0.817	0.853	0.890	0.915	0.934	0.953
NSH	–	cos	1024	**0.833**	**0.868**	**0.901**	0.922	0.938	0.954
SpH	12	12	128	0.708	0.768	0.829	0.870	0.903	0.937
SpH	–	cos	128	0.741	0.796	0.851	0.883	0.913	0.946
SpH	12	12	512	0.781	0.826	0.873	0.903	0.929	0.956
SpH	–	cos	512	0.811	0.850	0.890	0.917	0.939	0.961
SpH	12	12	1024	0.793	0.835	0.878	0.907	0.933	0.960
SpH	–	cos	1024	0.824	0.863	0.899	**0.923**	**0.944**	**0.964**

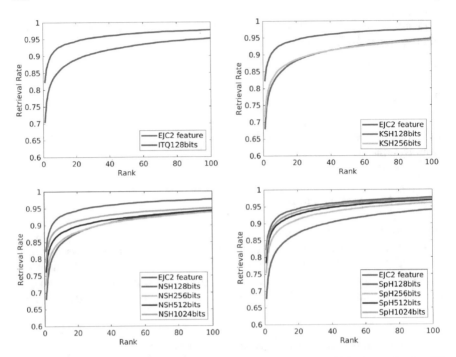

Fig. 7.10 Face search results obtained with the EJC2 features on the IJB-B dataset for the still images probes using ITQ (top-left), KSH (top-right), NSH (bottom-left), and SpH (bottom-right)

values of 10^{-3} and 10^{-4}. The NSH method with 1024 bits is able to preserve most of the original feature verification performance with a TAR of 0.949 and 0.908 at FAR values of 10^{-3} and 10^{-4}. The SpH method saturates its performance at 512 bits, with no or very limited improvement when using 1024 bits.

7.7.4.2 Face Search Results

We report the search results on the IJB-C dataset in Fig. 7.13. We can observe that the ERAG1 feature achieves a very high level of face search performance, with retrieval rates of 0.947 and 0.976 at rank 1 and 10, respectively. For this task, the ITQ hash codes perform relatively well with the best rank 1 retrieval rate of 0.901 when all methods are restricted to using only 128 bits. The NSH methods' performance almost saturates at 512 bits. With 1024 bits, NSH achieves a retrieval rate of 0.934 at rank 1 and 0.961 at rank 5 which is very close to the original features performance of 0.969. The SpH method performance is very good with 512 bits (retrieval rate of 0.933 at rank 1 and 0.971 at rank 10) but surprisingly drops when using 1024 bits. In this experiment, the SpH model is trained using all UMD batches samples at once, when training with many bits this may make the SpH optimization more challenging.

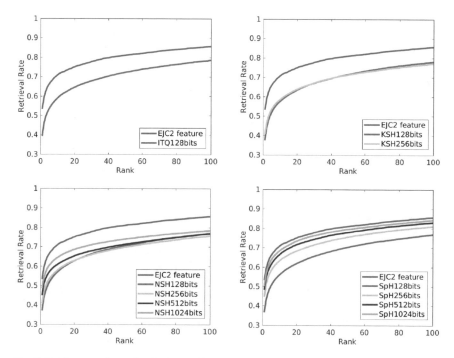

Fig. 7.11 Face search results obtained with the EJC2 features on the IJB-B dataset for the video frames probes using ITQ (top-left), KSH (top-right), NSH (bottom-left), and SpH (bottom-right)

Indeed, the best model parameters selected during the optimization with 1024 bits are obtained after just one iteration, while the models for other number bits are obtained after a few hundred iterations. This also explains why the verification performance was not improved with 1024 bits.

7.8 Open Issues

In this chapter, every hashing method was trained to hash each feature independently. However, for some applications, it may be interesting to combine multiple features to achieve the best performance. Therefore, the question of how to fuse multiple features in a hashing framework is worth investigating. We identify three strategies that could be explored: (i) an early fusion where the features are concatenated and hashed as if they were just one feature; (ii) an intermediate fusion strategy, especially applicable to pivots-based hashing methods, where the distances or similarities between a sample and the pivots in the different feature spaces are aggregated; and (iii) a late fusion strategy where each feature is hashed independently but the hamming distances obtained in each binary space, induced by each hash model, are combined for the final task.

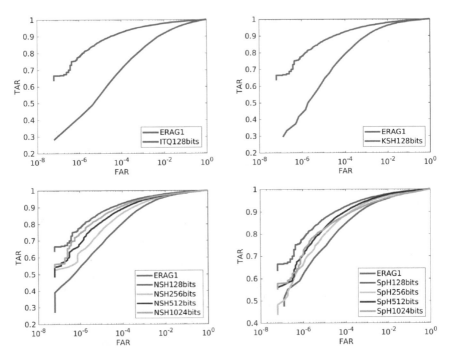

Fig. 7.12 Face verification results obtained with the ERAG1 features on the IJB-C dataset using ITQ (top-left), KSH (top-right), NSH (bottom-left), and SpH (bottom-right)

Recently, graph-based indexing methods [21] have gain popularity due to their high level of performance with close to logarithmic complexity to tackle the approximate nearest neighbors (ANN) search problem. Therefore, it would be interesting to explore how hashing can be combined with such approaches to enable a high level of search performance while enabling storing a large number of database samples thanks to the compression induced by the use of hashing.

7.9 Conclusion

In this chapter, we have addressed the problem of hashing a face, whose goal is to build a compact binary representation of a face image that enables a high level of performance for the face verification and face search tasks.

Extensive experiments conducted with state-of-the-art hashing methods on three challenging face datasets let us draw interesting conclusions. If the features to be hashed are not very discriminative, the supervised hashing method KSH can work better for both the verification and search tasks with a limited number of bits. However, as the features are getting better, the pivots-based but unsupervised methods

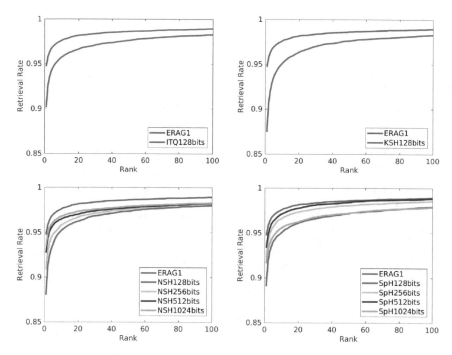

Fig. 7.13 Face search results obtained with the ERAG1 features on the IJB-C dataset for the mixed media probes using ITQ (top-left), KSH (top-right), NSH (bottom-left), and SpH (bottom-right)

tend to outperform the others. We have also shown that the distance used when performing the pivots embedding has to be chosen carefully based on how the features to be hashed have been optimized. The NSH method seems to be the best for the verification task, while the SpH method achieves very compelling results on the search task but has shown sensitivity to the type of features used.

We have shown that on the IJB-A dataset, several hashing methods can preserve the face verification performance using only 256 bits, therefore, allowing a 40 × (for the JC feature) to 64 × (for the SW feature) compression factor without loss of accuracy. As the features get better, more bits are required to preserve the features' performance. Yet, even on the most challenging dataset IJB-C the face verification performance is almost maintained when using 1024 bits. Overall, hashing can be an effective way to compress a face representation while maintaining most of the original features discriminative ability.

Acknowledgements This research is based upon work supported by the Office of the Director of National Intelligence (ODNI), Intelligence Advanced Research Projects Activity (IARPA), via IARPA R&D Contract No. 2014-14071600012. The views and conclusions contained herein are those of the authors and should not be interpreted as necessarily representing the official policies or endorsements, either expressed or implied, of the ODNI, IARPA, or the U.S. Government. The U.S. Government is authorized to reproduce and distribute reprints for Governmental purposes notwithstanding any copyright annotation thereon.

References

1. Bansal A, Castillo CD, Ranjan R, Chellappa R (2017) The do's and don'ts for CNN-based face verification. In: ICCV workshops, pp 2545–2554
2. Bansal A, Nanduri A, Castillo CD, Ranjan R, Chellappa R (2016) Umdfaces: an annotated face dataset for training deep networks, Technical report
3. Bansal A, Nanduri A, Castillo CD, Ranjan R, Chellappa R (2017) Umdfaces: an annotated face dataset for training deep networks. In: 2017 IEEE international joint conference on biometrics (IJCB). IEEE, pp 464–473
4. Bawa M, Condie T, Ganesan P (2005) Lsh forest: self-tuning indexes for similarity search. In: Proceedings of the 14th international conference on World Wide Web. ACM, pp 651–660
5. Chen J-C, Ranjan R, Sankaranarayanan S, Kumar A, Chen C-H, Patel VM, Castillo CD, Chellappa R (2018) Unconstrained still/video-based face verification with deep convolutional neural networks. Int J Comput Vis 126(2–4):272–291
6. Datar M, Immorlica N, Indyk P, Mirrokni VS (2004) Locality-sensitive hashing scheme based on p-stable distributions. In: Proceedings of the twentieth annual symposium on Computational geometry. ACM, pp 253–262
7. Douze M, Sablayrolles A, Jégou H (2018) Link and code: fast indexing with graphs and compact regression codes. In: Proceedings of the IEEE conference on computer vision and pattern recognition, pp 3646–3654
8. Feng J, Karaman S, Chang S-F (2017) Deep image set hashing. In: 2017 IEEE winter conference on applications of computer vision (WACV). IEEE, pp 1241–1250
9. Gionis A, Indyk P, Motwani R, et al (1999) Similarity search in high dimensions via hashing. In: Vldb, vol 99, pp 518–529
10. Gong Y, Lazebnik S, Gordo A, Perronnin F (2013) Iterative quantization: a procrustean approach to learning binary codes for large-scale image retrieval. IEEE Trans Pattern Anal Mach Intell 35(12):2916–2929
11. Gordo A, Perronnin F, Gong Y, Lazebnik S (2014) Asymmetric distances for binary embeddings. IEEE Trans Pattern Anal Mach Intell 36(1):33–47
12. Guo Y, Zhang L, Hu Y, He X, Gao J (2016) Ms-celeb-1m: a dataset and benchmark for large-scale face recognition. In: European conference on computer vision. Springer, pp 87–102
13. Heo J-P, Lee Y, He J, Chang S-F, Yoon S-E (2012) Spherical hashing. In: 2012 IEEE conference on computer vision and pattern recognition (CVPR). IEEE, pp 2957–2964
14. Huang GB, Mattar M, Berg T, Learned-Miller E (2008) Labeled faces in the wild: a database forstudying face recognition in unconstrained environments. In: Workshop on faces in 'Real-Life' images: detection, alignment, and recognition'
15. Indyk P, Motwani R (1998) Approximate nearest neighbors: towards removing the curse of dimensionality. In: Proceedings of the thirtieth annual ACM symposium on theory of computing. ACM, pp 604–613
16. Jégou H, Douze M, Schmid C, Pérez P (2010) Aggregating local descriptors into a compact image representation. In: 2010 IEEE conference on computer vision and pattern recognition (CVPR). IEEE, pp 3304–3311
17. Klare BF, Klein B, Taborsky E, Blanton A, Cheney J, Allen K, Grother P, Mah A, Jain AK (2015) Pushing the frontiers of unconstrained face detection and recognition: Iarpa janus benchmark a. In: Proceedings of the IEEE conference on computer vision and pattern recognition, pp 1931–1939
18. Kulis B, Grauman K (2009) Kernelized locality-sensitive hashing for scalable image search. In: 2009 IEEE 12th international conference on computer vision. IEEE, pp 2130–2137
19. Lin W-A, Chen J-C, Ranjan R, Bansal A, Sankaranarayanan S, Castillo CD, Chellappa R (2018) Proximity-aware hierarchical clustering of unconstrained faces. Image Vis Comput 77:33–44
20. Liu W, Wang J, Ji R, Jiang Y-G, Chang S-F (2012) Supervised hashing with kernels, In: 2012 IEEE conference on computer vision and pattern recognition (CVPR). IEEE, pp 2074–2081
21. Malkov YA, Yashunin DA (2016) Efficient and robust approximate nearest neighbor search using hierarchical navigable small world graphs. arXiv:1603.09320

22. Maze B, Adams J, Duncan JA, Kalka N, Miller T, Otto C, Jain AK, Niggel WT, Anderson J, Cheney J, et al (2018) Iarpa janus benchmark–c: face dataset and protocol. In: 11th IAPR international conference on biometrics
23. Park Y, Cafarella M, Mozafari B (2015) Neighbor-sensitive hashing. Proc VLDB Endow 9(3):144–155
24. Ranjan R, Bansal A, Xu H, Sankaranarayanan S, Chen J-C, Castillo CD, Chellappa R (2018) Crystal loss and quality pooling for unconstrained face verification and recognition. arXiv:1804.01159
25. Ranjan R, Sankaranarayanan S, Castillo CD, Chellappa R (2017) An all-in-one convolutional neural network for face analysis. In: 2017 12th IEEE international conference on automatic face & gesture recognition (FG 2017). IEEE, pp 17–24
26. Sablayrolles A, Douze M, Usunier N, Jégou H (2017) How should we evaluate supervised hashing?. In: 2017 IEEE international conference on acoustics, speech and signal processing (ICASSP). IEEE, pp 1732–1736
27. Sankaranarayanan S, Alavi A, Castillo CD, Chellappa R (2016) Triplet probabilistic embedding for face verification and clustering. In: 2016 IEEE 8th international conference on biometrics theory, applications and systems (BTAS). IEEE, pp 1–8
28. Wang J, Kumar S, Chang S-F (2012) Semi-supervised hashing for large-scale search. IEEE Trans Pattern Anal Mach Intell 12:2393–2406
29. Wang J, Liu W, Kumar S, Chang S-F (2016) Learning to hash for indexing big data-a survey. Proc IEEE 104(1):34–57
30. Weiss Y, Torralba A, Fergus R (2009) Spectral hashing. In: Advances in neural information processing systems, pp 1753–1760
31. Whitelam C, Taborsky E, Blanton A, Maze B, Adams J, Miller T, Kalka N, Jain AK, Duncan JA, Allen K, et al (2017) Iarpa janus benchmark-b face dataset. In: CVPR workshop on biometrics, vol 1
32. Wolf L, Hassner T, Maoz I (2011) Face recognition in unconstrained videos with matched background similarity. In: 2011 IEEE conference on computer vision and pattern recognition (CVPR). IEEE, pp 529–534
33. Yi D, Lei Z, Liao S, Li SZ (2014) Learning face representation from scratch. arXiv:1411.7923
34. Zheng J, Chen J-C, Bodla N, Patel VM, Chellappa R (2016) Vlad encoded deep convolutional features for unconstrained face verification. In: ICPR, pp 4101–4106

Chapter 8
Evolution of Newborn Face Recognition

Pavani Tripathi, Rohit Keshari, Mayank Vatsa, and Richa Singh

Abstract Accidental new born swapping, health-care tracking, and child-abduction cases are some of the scenarios where new born face recognition can prove to be extremely useful. With the help of the right biometric system in place, cases of swapping, for instance, can be evaluated much faster. In this chapter, we first discuss the various biometric modalities along with their advantages and limitations. We next discuss the face biometrics in detail and present all the datasets available and existing hand-crafted, learning-based, as well as deep-learning-based techniques which have been proposed for new born face recognition. Finally, we evaluate and compare these techniques. Our comparative analysis shows that the state-of-the-art SSF-CNN technique achieves an average of rank-1 new born accuracy of 82.075%.

8.1 Introduction

In a recent incident in India, two children were swapped just after birth. The parents had their doubts; however, due to lack of biometric records, their babies could not be identified. It was only after two years and nine months, post-birth, the DNA test proved that the children had been exchanged at the hospital [1]. Similarly, there are cases in Russia, where the knowledge of accidental newborn swapping came to light almost after 28 years [2]. Many such incidents are reported all around the world [3]. Had there been systems which could evaluate biometric modalities such as ear, fingerprint or face, the case of swapping could have been evaluated much faster. Therefore, the development of biometric systems, which can enroll newborns, is an imperative step towards the safety of children.

P. Tripathi · R. Keshari
IIIT Delhi, New Delhi, India

M. Vatsa · R. Singh (✉)
Indian Institute of Technology Jodhpur, Jodhpur, India
e-mail: richa@iitj.ac.in

© The Author(s), under exclusive license to Springer Nature Switzerland AG 2021   167
N. K. Ratha et al. (eds.), *Deep Learning-Based Face Analytics*, Advances in Computer
Vision and Pattern Recognition, https://doi.org/10.1007/978-3-030-74697-1_8

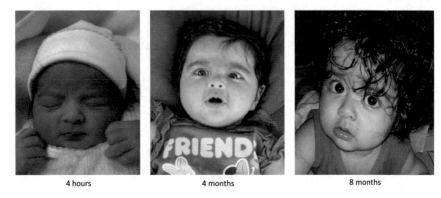

4 hours   4 months   8 months

Fig. 8.1 Illustrates the rapid development of craniofacial features within a few months

Apart from security applications, many other domains, viz. vaccination, health-care tracking, civil-ID programs, and child abduction cases too call for designing techniques which can perform authentic identification of newborns. For instance, India's Aadhaar program does not record any biometric information of children below the age of 5 years [4]. Every infant is registered for Aadhaar using one of the parents' biometric modalities. The details need to be updated once the child is 5 years old and again when the child turns 15. Since, the average birth rate in India is 21.84% per year per 1000 population [5], a large portion of the population does not get enrolled due to unavailability of technology and biometric standards pertaining to newborns, toddlers, and young kids. Consequently, many social programs use not-so-accurate non-biometric approaches.

In this chapter, approaches which can perform newborn recognition using bio-metrics, especially face, are presented. We first present a review of different bio-metric modalities, viz. fingerprint, palm print, ear, face, and iris, that have been explored for the identification of newborns. Face being the most non-intrusive bio-metric modality has been proven to be the most accurate and reliable modality for recognition of newborns. However, as shown in Fig. 8.1, there is a rapid develop-ment in the features of the baby within a few months. It is evident from the series of images that the craniofacial features of the newborn changes a lot within 1–2 years. Hence, the chapter discusses the characteristics of newborn faces followed by challenges involved in newborn face recognition. Section 8.2 then discusses face databases developed for promoting research in newborn recognition. Section 8.3 dis-cusses existing hand crafted, learning-based, and deep learning-based approaches, that have been developed for newborn face recognition. Finally, in Sect. 8.1.2, we show results and analysis of some of these existing techniques.

8.1.1 Biometric Modalities for Newborns

One of the major challenges is identifying the best biometric modality for the age group of 0–4 years. The research community has studied different biometric modalities, viz. footprints, palm prints, fingerprints, iris, face, and fusion of some of these modalities. Table 8.1 presents an account of various biometric modalities and the respective techniques that have been developed for newborn identification.

In early days, hospitals used to capture footprints of a newborn using a predefined process. However, scientific studies later showed that footprint cannot be used for recognition of newborns. For instance, Shepard et al. [6] examined footprints of 51 newborns at California State's Department of Justice. Only 10 babies were identified by the experts using only the footprints. Similarly, Pela et al. [7] examined 1917 footprints collected by trained personnel from a hospital in Brazil. It was reported that none of the images were sufficient for accurate identification. Based on these studies, it was concluded that with state-of-the-art capturing techniques, it is difficult to use footprint for identification of newborns. Thus, the American Academy of Pediatrics and the American College of Obstetricians and Gynecologists stated that *individual hospitals may want to continue the practice of footprinting or fingerprinting, but universal use of this practice is no longer recommended.*

Research community also explored several other biometric modalities such as fingerprint, palm print, and ear for newborn identification. Even though palm prints and fingerprints give promising results in adult identification, they have not achieved success in identifying newborns. Weingaertner et al. [8] and Lemes et al. [9] carefully captured footprints and palm prints of 106 newborns at high resolution (≥ 1500 ppi) using specialized sensors to improve useful biometric information content. Images captured from only 20 newborns (about 5% of the entire captured dataset) were deemed to be of useful quality. For 106 newborns, two images were collected at an interval of 24 hr just after the birth, using which the experts reported 67.7% and 83% identification accuracy using footprints and palm prints, respectively. Since many of the biometric modalities are, by their nature, contact based, therefore, they introduce capture challenges, such as, it is difficult to hold the hands and legs of a newborn to make them stay still. Further, sensors are not designed to capture small fingerprints and footprints. Fields et al. [10] and Tiwari et al. [11] studied the feasibility of ear for newborns. Fields et al. [10] manually analyzed the samples and concluded that, visually, ears can be used to differentiate between two children. Tiwari et al. [11] prepared a database for newborn ear to understand the effectiveness of ear recognition for newborns. Similar to other modalities, capturing ear images require user cooperation, and at times, due to *unintentional uncooperative* nature of newborns makes acquiring ear images challenging.

Iris has shown to be very successful for adult identification [20]. However, only one study has been conducted on iris recognition for children, that too only for toddlers and pre-school children. Basak et al. [16] prepared a multimodal biometric database for toddlers and pre-school children and showed the effectiveness of iris recognition

Table 8.1 Summarizes the techniques developed for the use of different biometric modalities for identification of newborns

Modality	Authors	Comments
Footprint	Shepard et al. [6]	Concluded that with state-of-the-art capturing techniques, it is difficult to use footprint for identification of newborns
	Pela et al. [7]	
Footprint & palm print	Weingaertner et al. [8]	It was concluded that even best capturing techniques are insufficient to perform accuratenewborn identification because it is difficult to hold their hands and legs still
	Lemes et al. [9]	
Ear	Fields et al. [10]	Showed that there is adequate discriminability in the ear to distinguish identities; however, capture, segmentation and orientation correction is challenging
	Tiwari et al. [11]	
Fingerprint	Jain et al. [12]	Fused predictions of input fingerprints from various commercial SDKs
Face	Bharadwaj et al. [13]	Non-deep learning as well as deep learning- based methods have shown that newborn face recognition can be accurate, friendly and cost-effective
	Bharadwaj et al. [14]	
	Jain et al. [15]	
	Basak et al. [16]	
	Siddiqui et al. [17]	
	Keshari et al. [18]	
Fusion	Tiwari et al. [19]	Fused ear and soft-biometrics for newborn recognition

for such potential users. Since it is very difficult to capture iris for newborns, no such study has been conducted for using iris to identify newborns.

Face recognition is one of the most popular non-contact biometric modality and existing algorithms show high performance on identifying adult faces, particularly in a controlled and semi-controlled environment. Nonetheless, face recognition for newborns is not well explored and is considered an open problem due to various challenges such as pose and emotion variation, excessive blur, and rapid development of the facial structure. Figure 8.1 shows how the facial features change within a few months of birth. Such growth makes the problem arduous and interesting. In this chapter, we discuss the use of face biometric for identification of newborns, along with characteristics of newborn faces and challenges involved in newborn face recognition. We present a review of existing databases and algorithms focusing on this research problem. Lastly, we present an experimental analysis of existing algorithms and discuss that face recognition of newborns can be "friendly" and "cost-effective".

8.1.2   Characteristics and Challenges of Newborn Face Recognition

As discussed by Bharadwaj et al. [14], newborn faces are structurally different from the adult faces and they cannot be viewed as miniature version of adult faces. Following studies/observations provide some characteristics of newborn faces:

1. It is believed that for adults, newborns are difficult to differentiate. Kuefner et al. [21] reported that adults have difficulty in differentiating between child faces compared to adult faces. They selected 31 adults who had no extensive interaction with infants and collected data based on a questionnaire. They statistically showed that there was a significant decrease when asked to recognize infants compared to adults. On the other hand, the infant recognition accuracy drastically increased when the same experiments were conducted on 18 pre-school teachers who were in contact with children for at least 30 hours per week. The analysis suggests that adults maybe unable to extract biometric features from newborn faces due to the *other age effect*. In another study conducted by Anastasi and Rhodes [22], it was reported that both children and adults are good at recognizing own age faces compared to other age faces.
2. Every face possess some unique facial traits and subtle differences in shape, proportions of hard and soft tissues, and topographical contours. To understand the applicability of face recognition for newborns, it is important to identify those facial characteristics that lead to unique and discriminative features. For instance, in newborns, the growth of the mandible (lower jaw) and chin is slower and continues longer than mid-facial development, thus resulting in newborns often being characterized by large foreheads. Hence, local facial regions of newborn face provides evidence of identity with varying levels of confidence.
3. The nasal region of a face is generally an important point in the facial architecture as the surrounding arches rely on it for support. The region is extensively studied to improve the aesthetics of plastic surgery procedures. Bharadwaj et al. [14] observed that the nasal region of newborns is shallow as compared to adults. They also noticed that infant faces do not possess bold topographical features as compared to adults, as shown in Fig. 8.2. Further, the craniofacial structure of newborn faces is characterized by prominent eyes, small jaws, puffy cheeks, and a high forehead. Moreover, there are differences in the eyebrow ridge and

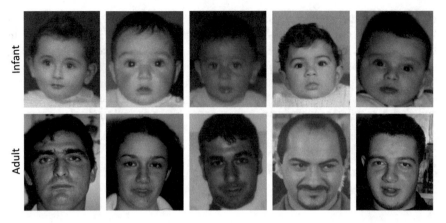

Fig. 8.2 Sample images from FG-Net database showing that the craniofacial characteristics of an infant face are not proportionally equivalent to a miniature adult face. However, several studies have been performed which show that faces of newborns do possess unique characteristics that could be utilized in face recognition of newborns

the overall proportions as well. These observations indicated that the shape and structure of infants are not miniature adult faces.

4. While planning facial reconstruction of infant patients (particularly cleft lip patients) [23], studies have shown that race, gender, and age significantly influence the planning of the re-constructive surgery. These observations suggest that *newborn faces also posses distinguishing and characteristic facial features, that can be utilized for newborn face recognition.*

Though studies have shown that face recognition for newborns is feasible, it is challenging to create a robust system for the same. The primary challenges that make face recognition of newborns difficult are discussed below:

1. Bharadwaj et al. [14] refer to the newborns as *unintentionally non-cooperative users* of face recognition. This is so because, while capturing the images, it is not possible to ensure that the subject has neutral expression and shows the frontal face due to their unpredictable behavior and elastic faces. Further, motion blurriness might be caused due to excessive movement. Figure 8.3 presents a set of images showcasing the variation in expressions and motion blur.

2. Another challenge that is faced by newborn face recognition is the recognition of identical twins. In later stages of life, the twins might acquire some distinct features, but at the time of the birth it is very challenging to distinguish one from the other. In Fig. 8.4, one can see that it is difficult to differentiate between twins since they do not possess the distinguishing characteristics that adult twins may possess. In certain cases, the face of the newborn maybe covered with soft hair, known as *lanugo*, especially on cheek and forehead. They shed off as the baby ages resulting in variations in image texture.

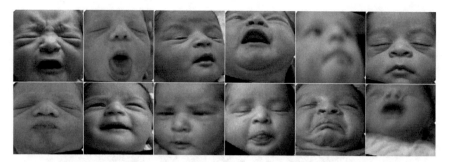

Fig. 8.3 Illustrates the variation in expressions and pose that occur while capturing images along with motion blur in some images [14]

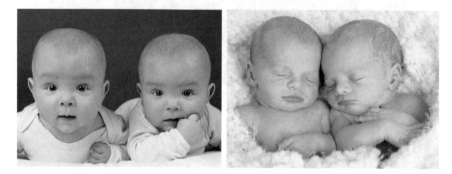

Fig. 8.4 It can be seen visually that it is difficult to differentiate between identical twins. (http://easydoesit.org/category/announcements/, http://www.weebeedreaming.com/my-blog/top-10-twin-sleep-tips)

3. Figure 8.1 illustrates the rapid rate at which the facial features of newborns change. It can be visually observed that the craniofacial features vary a lot even in a short time span of four months. This makes newborn face recognition a challenging problem since the technique designed for newborn face recognition must be able to understand and adapt itself to the structural differences in the faces of newborns.

8.2 Datasets for Newborn Face Recognition

The research in this domain is mainly restricted due to the lack of publicly available databases. Even the ones which are available to the research community are small sample size databases. In the following subsections, we discuss the databases that can be used for conducting face recognition research on newborn, toddlers, and pre-school children.

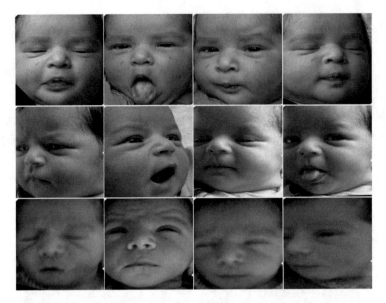

Fig. 8.5 Sample images from the newborns face database [14]

8.2.1 Newborns Face Database

Bharadwaj et al. [14] proposed the Newborns Face Database that consists of more than 1200 images pertaining to over 450 newborns from various hospitals. Each subject has 1–10 images. Among these, images for 96 babies have been acquired in multiple sessions ranging from one hour to a few weeks after birth. No constraint was put on the babies, however, the authors tried to capture near frontal face images. Figure 8.5 presents sample images from the database. It is evident from the images that there is large variation in terms of pose and expression.

The database was further extended to include 1185 images pertaining to 204 newborns, each having 1–17 images, collected from various hospitals. The time of capture varies from one hour to a few weeks after birth. Siddiqui et al. [17] performed experiments on this extended newborn face dataset.

8.2.2 Newborns, Infants, and Toddler Longitudinal Face Database

Jain et al. [24] proposed face image database for newborns, infants and toddlers. The database is called *longitudinal* because it contains images of a subject captured during four different sessions over the course of one year. There are a total of 161

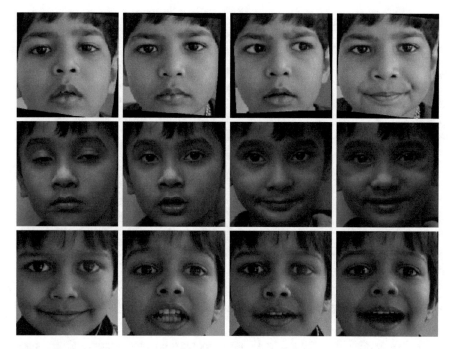

Fig. 8.6 Presents sample face images of toddlers from the children multimodal biometric database [16]

subjects with face images from all four sessions and ages ranging from 0 to 4 years. However, this database is not publicly available for research.

8.2.3   Children Multimodal Biometric Database (CMDB)

Basak et al. [16] proposed a database containing 2590 face images of 141 toddlers and pre-school children (age range of 18 months to 4 years), each having 10–20 images. Along with face images, the database also consists of iris and fingerprint images of 100 toddlers. This database has been collected in two sessions which are months apart. Figure 8.6 presents sample images from the database.

8.3   Existing Techniques for Newborn Face Recognition

In this section, we discuss existing face recognition techniques which have yielded promising results. Table 8.2 presents a brief summary of the existing techniques.

Table 8.2 Summarizes the research findings for newborn face recognition

Method	Comment
Bharadwaj et al. [13]	Articulated the challenges of newborn face recognition
Bharadwaj et al. [14]	Proposed a two-stage domain-specific learning for newborn face recognition
Jain et al. [24]	Proposed a newborn face database and demonstrated the results of COTS face matcher
Basak et al. [16]	Proposed a newborn and toddler face database and evaluated existing tools and algorithms
Siddiqui et al. [17]	Proposed a deep learning technique which applies class-based penalties while learning the filters of the convolutional network
Keshari et al. [18]	Proposed SSF-CNN which focuses on learning the "structure" and the "strength" of the convolutional filters

8.3.1 Handcrafted Feature Extraction Methods

Bharadwaj et al. [13] were the first ones to explore face recognition of newborns. They conducted a preliminary study on 34 newborns and concluded that automatic face recognition of newborns is feasible. As discussed in Sect. 8.4, newborn face recognition faces challenges due to the pose variation and blurriness in the images. To reduce the effect of blurriness in the images, Bharadwaj et al. applied Gaussian smoothing on the images to filter the excessive wrinkle information, while preserving the discriminating texture information. After that, SURF [25] and LBP [26] were used to extract facial features from the original face image and the low frequency images, respectively. They further used χ^2 distance to measure the dissimilarity between corresponding levels of Gaussian pyramid. Weighted sum rule was applied to fuse the three match scores.

The results of the proposed algorithm were compared with other appearance based algorithms, namely, Principal Component Analysis (PCA) [27], Linear Discriminant Analysis (LDA) [27] and Independent Component Analysis (ICA) [28]. The results negated the human (adult) perception that children's faces look alike and cannot be accurately distinguished.

8.3.2 Autoencoder Learning-Based Method

Newborn face recognition faces several challenges that arise due to their uncooperative nature. In these scenarios, handcrafted feature extraction and matching techniques may not be optimal for newborn face recognition. In 2016, Bharadwaj et al.

[14] proposed a learning based feature extraction technique along with a matching technique that explicitly encode the properties of the feature space to improve the recognition performance.

They observed that the asymmetric development of faces of newborns results in a unique craniofacial structure, however, the structure is not proportionally equivalent to the miniature adult face. Further, they also observed that since newborns are unconstrained users of face recognition, the acquired images suffer from large variations in pose and expression. Based on these observations, they proposed a two-stage learning algorithm that first learns a *domain-specific* representation of the face with variable expressions using a stacked denoising autoencoder (SDAE) [29], and then learns the representation with a *problem-specific* learned distance metric that benefits from the availability of additional unlabeled problem-specific face images of newborns to reduce the semantic gap during matching thereby improving newborn face recognition performance.

In the first stage, a domain-specific representation of the human face with a deep learning architecture is learned. Firstly, the input image is divided into nine overlapping patches. Then, separate sets of multi-layer encoders are learned for each overlapping patch of a face image, that help enhance the depth of the encoders. The patch encoding ensures spatial coherence of the resultant representation. Each encoder provides the representation of a component of the face image, such as the forehead, periocular, mouth, and chin regions. The representations thus obtained are concatenated into a single feature vector. SDAE was trained on a large number of domain-specific samples, i.e., face images with varying illumination and expression.

The second stage consisted of learning a problem-specific distance metric via one-shot similarity with 1-Class-Online-SVM. While performing matching for newborn faces, low-level features suffer from *semantic gap effect*, adult to newborn faces. Since the traditional distance metrics χ^2 and Euclidean distance may not account for the semantic gap effect, they proposed a distance metric learning technique which can reduce the semantic gap effect in newborn face representations. Since the intra-class variations in newborn face images are significantly high, rather than viewing the feature manifold as a linear space, they proposed to project the samples in the kernelized feature space which ensured a robust, non-linear and high-dimensional representation of the distance space and provided a generalizable solution.

8.3.3 Class-Based Penalty in CNN Filter Learning

Siddiqui et al. [17] proposed a deep learning-based face recognition technique for newborns, toddlers, and pre-school children. In their research, they presented a modification in the convolutional filter learning and imposed a class-based penalty on the weights of the filters. They showed that class-based penalty helps in learning class-specific discriminative features and avoids overfitting. Further, they trained the network with residual skip connections [30] to learn higher level features with increasing number of layers.

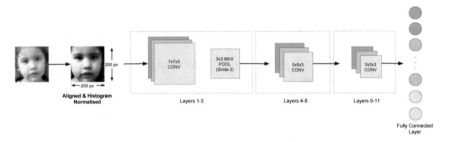

Fig. 8.7 Steps involved in the proposed CNN feature extraction with class-based penalty for filter learning. For matching, the classifier is added separately depending on the identification or verification experiments [17]

They designed their own 11-layer network using the above-mentioned approach. In the first three layers of CNN, three filters of 7×7 were used, followed by max pool with size 3×3 and stride size of 2. In the upcoming five layers, three filters of size 5×5 were used followed by three convolution layers with filter size 3×3. Finally, for classification, a fully connected layer followed by a softmax layer were appended at the end of the network.

As discussed in the database section, newborn face recognition is a small sample size problem. Thus, the proposed CNN was first pretrained using the CMU-MultiPIE [31] database. It was then finetuned by augmenting the Newborn Face Database and the Children Multimodal Biometric Database. The network was trained in verification mode and for identification, N-way verification was performed. Figure 8.7 presents the proposed 11-layer network architecture, used for newborn face recognition.

8.3.4   Learning Structure and Strength of CNN Filters

Collecting large databases which contain biometric modalities, especially the face of newborns, is a challenging task. Hence, all existing newborn databases contain a small number of samples. While convolutional neural networks have achieved high accuracies in several computer vision tasks, due to a large number of parameters, they require large number of training samples. To overcome this limitation, Keshari et al. [18] proposed a deep learning-based technique for newborn face recognition, which can attain best results with small sample size databases.

The proposed Structure and Strength Filtered CNN, termed as SSF-CNN, focuses on learning the "strength" and "structure" of the convolutional filters of the network. Keshari et al. [18] hypothesized that domain-specific large databases or other representation learning paradigms that require less training data such as dictionary learning [32, 33], can help in learning the structure of the filter. Dictionary learning or matrix factorization can help in learning the dictionary that helps to encode the representative features. Thus, CNN filters can be represented by dictionary to

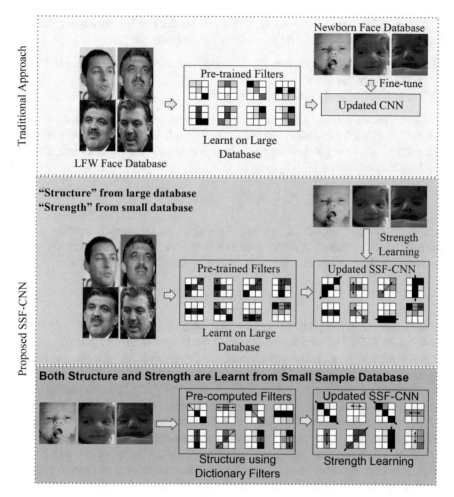

Fig. 8.8 Face recognition models trained on adult faces may not give good results on face recognition of newborns. SSF-CNN learns strength and structure of the filters for improving the classification performance for small sample databases [18]

learn the "structure" of the domain. However, this structure may not be optimal for problem-specific features. Hence, the strength of the filter is learned to adapt the weights of these filters according to the problem-specific data characteristics. The proposed architecture is presented in Fig. 8.8.

To initialize the CNN first the hierarchical dictionary filters are learned, followed by learning the strength parameter to train the CNN model. The strength parameter is **'t'** for the CNN filters **'W'**, which allows the network to assign *weight* for each filter based on its structural importance. We, next, describe the approach to hierarchically learn **W**, filters of the CNN model using dictionary learning followed by learning approach for the strength parameter **t**.

Learning Structure of Filters: In SSF-CNN, dictionary learning is proposed to learn the structure of the filters. This step is divided into two parts:

1. Learn hierarchical dictionary filters and utilize trained dictionary filters for initialization of the CNN.
2. Train CNN with dictionary initialized filters.

Hierarchical Dictionary Filter Learning: Through dictionary learning, sparse representation of the input data is learned in the form of a linear combination of basic elements. A dictionary $\mathbf{D}$ and the coefficient α is learned for a given input $\mathbf{Y}$ using the following equation:

$$\min_{\mathbf{D},\alpha} \|\mathbf{Y} - \mathbf{D}\alpha\|_F^2, \; such \; that \; \|\alpha\|_0 \leq \tau \tag{8.1}$$

where, the ℓ_0-norm imposes a constraint of sparsity on the learned coefficients and the maximum number of non-zero elements is represented by τ. Many at times, the ℓ_0-norm is relaxed, thus, changing the dictionary learning formulation to:

$$\min_{\mathbf{D},\alpha} \|\mathbf{Y} - \mathbf{D}\alpha\|_F^2 + \lambda \|\alpha\|_1 \tag{8.2}$$

where λ represents a regularization parameter which controls the sparsity promoting ℓ_1-norm. In SSF-CNN dictionary learning is used to pre-train the filters of the CNN model in a hierarchical manner. A hierarchical dictionary learning technique is used to initialize the CNN model. The trained dictionary atoms are used to convolve over the input image. After convolution, feature maps are normalized according to the activation function (e.g., ReLU) used in CNN models. The extracted feature map is the input for the next level of the hierarchical dictionary. In this manner, the number of dictionary layers is the same as the number of convolutional layers in CNN models. The trained dictionary is organized in the two-dimensional array, where each filter is arranged in one column. These learned filters are reshaped and convolved over the input image to produce the feature maps for the next level of the dictionary.

Training CNN with Dictionary Initialized Filters: CNN models have multiple covolutional and pooling layers stacked on top of each other. It typically has multiple convolutional layers, with each layer having multiple filters. These filters are trained using optimisers such as stochastic gradient descent [34]. Given an input image $\mathbf{X}$ and a convolutional filter $\mathbf{W}$, the convolution operation can be expressed as

$$f(\mathbf{X}, \mathbf{W}, b) = \mathbf{X} * \mathbf{W} + b \tag{8.3}$$

where $*$ and b represent the convolution operation and bias respectively. Two passes namely, forward pass and backward pass are part of the training process. In the former, the network propagates the input signal to the last classification layer. On the other hand, in the latter, the error δ_j^l for each layer l on node j is computed with respect to the total cost and the weights of the CNN filters are updated accordingly.

In traditional CNN, the weights are initialized using Xavier [35], or MSRA [36] approach and even randomly. In the proposed method, Keshari et al. initialized the

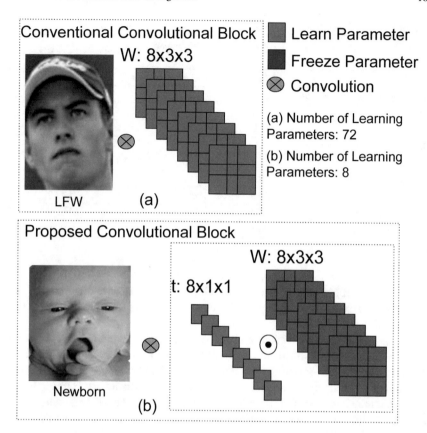

Fig. 8.9 Visually illustrates the concept of learning the "strength" of the filter which in turn reduces the number of training parameters [18]

CNN filters using dictionary learned filters. They have shown that initialization with dictionary learned filters have more "structure" compared to traditional initialization techniques, especially for small sample size database. Though this type of initialization helps in finding improved filters, updating the filters in a traditional manner still requires training of a large number of parameters. To avoid this, only a "strength" parameter for each filter is proposed to learn, thus reducing the number of parameters significantly.

Learning Filter Strength: Figure 8.9 visually presents the concept of learning the strength of the filter. The weights learned from dictionary learning are frozen and only the strength parameter is learned while training. For lth layer, the strength parameter '$\mathbf{t}^l$' is learned using stochastic gradient descent method; i.e., a scalar value is learned rather than learning the complete filter. The proposed process can be written as

$$f(\mathbf{X}, \mathbf{W}, b, \mathbf{t}) = \mathbf{X} * (\mathbf{t} \odot \mathbf{W}) + b \tag{8.4}$$

Table 8.3 Rank-I identification accuracies (%) on Newborn Face Database

Algorithm	Number of gallery images			
	1	2	3	4
COTS	41.0	53.6	60.1	4.6
Handcrafted feature techniques				
LBP + χ^2 [26]	21.1	32.5	39.5	44.7
DSIFT + χ^2 [25]	31.3	41.9	48.0	53.4
Deep learning techniques				
Bharadwaj et al. [14]	51.1	66.0	73.1	78.5
Siddiqui et al. [17]	58.3	63.7	74.8	80.8
Vinyals et al. [43]	59.4	63.7	68.6	72.2
Hariharan et al. [44]	65.4	73.3	79.3	85.4
Keshari et al. [18]	70.4	81.4	86.5	90.0

where, $(\mathbf{t} \odot \mathbf{W})$ represents element-wise multiplication. $\mathbf{t}$ can be learnt using stochastic gradient descent. Since $|\mathbf{W}| >> |\mathbf{t}|$, even small training data can be used to train the network.

Learning SSF-CNN for Newborn Face Recognition: The predefined experimental protocol limited the authors to only use images pertaining to 10 newborns for training and rest 86 newborns for testing from the Newborn Face Database. The performance was computed on ResNet architecture by employing the dictionary learning initialization technique to learn the structure of the filters from 10 newborns, followed by learning of strength parameter by attuning the filters. They also reported that if the CNN model is trained from scratch, the accuracies are extremely low.

Apart from learning the "structure" of the database using dictionary learning, the authors reported that the "structure" can also be learned from large domain-specific databases, followed by attuning the strength parameters for the problem-specific database. Hence, the experiments are performed with pre-trained networks (pre-trained filters are obtained after learning from either ImageNet or Labeled Faces in the Wild dataset (LFW) [37] and YouTube Faces (YTF) [38] databases) and use strength parameter to attune it for newborn face recognition based on training data of 10 newborns. For this experiment, they have used variants of ResNet [30], VGG [39], VGGFace [40], LightCNN [41], and DenseNet [42] architectures.

In [18], it is observed that learning strength of the filters improves the performance of CNN models compared to conventional fine-tuning approach. With single gallery image per subject, the best rank-1 accuracy of over 70% is obtained when the proposed strength parameter is used with pre-trained VGG-Face [40], which is at least 10% better than the conventional fine-tuning based approach. Thus, showcasing that in real-world applications, the concept of learning structure and strength helps in achieving improved performance.

8.4   Results and Analysis of Existing Newborn Face Recognition Techniques

In this section, we perform a comparative analysis of existing techniques using Newborn Face Database and experimental protocol as defined in [14]. Two handcrafted features, LBP [26] and DSIFT [25], and five feature learning-based algorithms are used for evaluation. We have also used a commercial face recognition system, referred as COTS, for evaluation and comparison. Table 8.3 shows the comparison between these techniques discussed in this chapter for newborn face recognition. It should be noted that some results have been reported directly from the respective papers. Our observations are discussed below:

1. **COTS versus Handcrafted Techniques**: As can be observed from Table 8.3, handcrafted features such as LBP [26] and DSIFT [25] achieve least accuracies due to the limited representation capacity for newborn face recognition. Figure 8.10 illustrates that LBP [26] generates feature space based on the local information irrespective of discriminative information which is required for newborn face identification. Similarly, DSIFT [25] finds keypoints in a dense manner irrespective of how much discriminative information they provide for identification. On the other hand, COTS performs better than handcrafted features, since the system is trained for face recognition. However, the accuracy achieved is still low compared to other existing techniques. Perhaps the system is trained on adult faces, and hence it is unable to accurately adapt to newborn faces.

2. **Efficacy of Learning-Based Methods**: Bharadwaj et al. [14] proposed an autoencoder-based feature representation followed by newborn face recognition specific distance metric learning. The latter is done via one-shot similarity with one class-online support vector machine for newborn face recognition. The proposed method achieved a rank-1 identification accuracy of 51.1% with a single gallery, a 10.1% improvement compared to COTS. This shows that learning-based frameworks are more effective for newborn face recognition compared to handcrafted techniques or COTS. This happens because learning-based methods allow the algorithm to capture the semantic understanding of the encoding scheme. Siddiqui et al. [17] proposed a deep learning-based technique for newborn face recognition. They modified the learning of the convolution filter and imposed a class-based penalty on the weights of the filters. The class-based penalty helped in learning class-specific discriminative features. Based on the results, it can be concluded that learning-based methods can be developed which can predict the most discriminative features required for accurate newborn face recognition.

3. **Deep Learning Methods for Small Sample Size Problems**: Due to a large number of parameters, deep learning methods require a large number of sample images to train the network. With Newborn Face Database being a small sample size database, we compare algorithms proposed by Vinyals et al. [43], Hariharan et al. [44] and Keshari et al. [18], which are specifically designed for small sample size databases. The results show that such methods outperform the deep learning

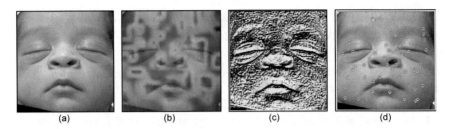

Fig. 8.10 Illustrates the visualizations of deep learning as well as handcrafted features. **a** Input image. **b** Heat map showing the discriminative regions considered by method proposed by Keshari et al. [18]. **c** LBP feature representation. **d** DSIFT feature representation (Best viewed in color)

method proposed by Siddiqui et al. [17]. Figure 8.10 visually illustrates that deep learning-based methods focus on the most discriminative regions, and hence achieve the highest accuracy. Therefore, one can say that by learning the features focusing on the periocular, nose, and mouth region, high recognition accuracy can be achieved. It is our assertion that zero-shot or one-shot learning techniques, which are specially designed for newborn face recognition, can further boost the performance since such methods exploit deep learning models for learning the optimal features even in cases when there is one or no image of a particular subject.

4. **Effect of the gallery size on recognition accuracy**: It can be noted that the accuracy for any particular algorithm increases when the number of gallery images increases from one to four. Figure 8.11 presents cases when state-of-the-art method proposed by Keshari et al. [18], failed to accurately match probe images with the gallery images. It can be observed from the same figure that this happens because the newborn face images suffer from expression variation or excessive blur due to unintentional uncooperative behavior of newborns. Based on this observation, it can be concluded that increasing the number of gallery images improves the newborn face recognition performance.

8.5 Conclusion

Designing an automatic verification/identification of newborns has become imperative due to its application in various domains, such as health-care, child security, and civil-ID programs. However, due to the unintentional uncooperative behavior of newborns, the acquired images consist of varied expressions, while some images suffer from excessive blur. These covariates make matching images of the same subject a challenging task. Further, the asymmetrical development of the craniofacial structure of the newborns is not similar to the miniature adult face. Hence, it is imperative to design algorithms specifically for newborn face recognition. In this chapter, we

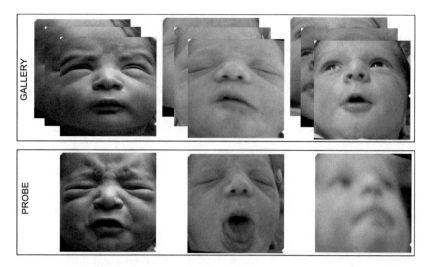

Fig. 8.11 Keshari et al. [18] achieved the state-of-the-art results on the Newborn Face Database, however, the above images suffer from excess blur and expression variation. Hence, these images were incorrectly identified

discussed various databases of newborn faces and techniques which have shown to perform well in newborn face recognition.

It is observed that learning-based techniques perform better than algorithms dependent on handcrafted feature representations or existing distance metrics. However, to successfully learn all the discriminative features, it is imperative to have large databases which have images with varied expressions, pose and illumination. In future, data fine-tuning [45] and guided-dropout [46] can be used to make the learning algorithm more robust to variations in the test images. Further, the synthetic images of the newborn faces can also be generated through GANs.

Moving forward, video-based face recognition for recognizing face images pertaining to newborns can be designed. Videos would allow to capture expression variations on various time steps and hence enable the development of techniques robust to expression and pose variations. Algorithms such as frame selection in videos [47], joint-feature learning [48], and fusion [49] can help improve the performance of face recognition in realistic conditions.

References

1. Pandey G. https://www.bbc.com/news/world-asia-india-42743982. Accessed 15 Nov 2018
2. Stewart W. https://www.dailymail.co.uk/news/article-6014157/Two-women-discover-raised-wrong-families-hospital-blunder-Russia.html. Accessed 15 Nov 2018
3. Sharma A. https://indianexpress.com/article/india/india-news-india/shimla-five-months-after-swap-babies-will-return-to-their-original-parents-3095974/. Accessed 15 Nov 2018

4. Nambiar N. https://timesofindia.indiatimes.com/city/pune/aadhaar-for-newborns-at-500-hospitals-from-august/articleshow/64981490.cms. Accessed 15 Nov 2018
5. https://www.indexmundi.com/g/g.aspx?c=in&v=25. Accessed 15 Nov 2018
6. Shepard KS, Erickson T, Fromm H (1966) Limitations of footprinting as a means of infant identification. Pediatrics 37(1):107–108
7. Pelá NTR, Mamede MV, Tavares MSG (1976) Analise critica de impressoes plantares de recem-nascidos. Revista Brasileira de Enfermagem 29(4):100–105
8. Weingaertner D, Bellon ORP, Silva L, Cat MN (2008) Newborn's biometric identification: can it be done? In: International conference on computer vision theory and applications, pp 200–205
9. Lemes RP, Bellon OR, Silva L, Jain AK (2011) Biometric recognition of newborns: identification using palmprints. In: IEEE international joint conference on biometrics, pp 1–6
10. Fields C, Falls HC, Warren CP, Zimberoff M (1960) The ear of the newborn as an identification constant. Obstet Gynecol 16(1):98–102
11. Tiwari S, Singh A, Singh SK (2011) Newborn's ear recognition: can it be done? In: IEEE international conference on image information processing, pp 1–6
12. Jain AK, Cao K, Arora SS (2014) Recognizing infants and toddlers using fingerprints: increasing the vaccination coverage. In: IEEE international joint conference on biometrics, pp 1–8
13. Bharadwaj S, Bhatt HS, Singh R, Vatsa M, Singh SK (2010) Face recognition for newborns: a preliminary study. In: Fourth IEEE international conference on biometrics: theory applications and systems, pp 1–6
14. Bharadwaj S, Bhatt HS, Vatsa M, Singh R (2016) Domain specific learning for newborn face recognition. IEEE Trans Inf Forensics Secur 11(7):1630–1641
15. Best-Rowden L, Jain AK (2018) Longitudinal study of automatic face recognition. IEEE Trans Pattern Anal Mach Intell 40(1):148–162
16. Basak P, De S, Agarwal M, Malhotra A, Vatsa M, Singh R (2017) Multimodal biometric recognition for toddlers and pre-school children. In: IEEE international joint conference on biometrics, pp 627–633
17. Siddiqui S, Vatsa M, Singh R (2018) Toddlers and preschool children: a deep learning approach. In: International conference on pattern recognition
18. Keshari R, Vatsa M, Singh R, Noore A (2018) Learning structure and strength of CNN filters for small sample size training, *Conference on Computer Vision and Pattern Recognition* pp 9349–9358
19. Tiwari S, Singh A, Singh SK (2012) Fusion of ear and soft-biometrics for recognition of newborn. Signal Image Process: Int J 3(3):103–116
20. Bowyer KW, Burge MJ (2016) Handbook of iris recognition. Springer
21. Kuefner D, Cassia VM, Picozzi M, Bricolo E (2018) Do all kids look alike? evidence for an other-age effect in adults. J Exp Psychol: Hum Percept Perform 34(4):811
22. Anastasi JS, Rhodes MG (2005) An own-age bias in face recognition for children and older adults. Psychon Bull Rev 12(6):1043–1047
23. Radhakrishnan V, Sabarinath V, Thombare P, Hazarey P, Bonde R, Sheorain A (2010) Presurgical nasoalveolar molding assisted primary reconstruction in complete unilateral cleft lip palate infants. J Clin Pediatric Dent 34(3):267–274
24. Best-Rowden L, Hoole Y, Jain A (2016) Automatic face recognition of newborns, infants, and toddlers: a longitudinal evaluation. In: IEEE international conference of the biometrics special interest group, pp 1–8
25. Bay H, Tuytelaars T, Van Gool L (2006) Surf: speeded up robust features. In: European conference on computer vision. Springer, pp 404–417
26. Ahonen T, Hadid A, Pietikainen M (2006) Face description with local binary patterns: application to face recognition. IEEE Trans Pattern Anal Mach Intell 12:2037–2041
27. Belhumeur PN, Hespanha JP, Kriegman DJ (1997) Eigenfaces versus fisherfaces: recognition using class specific linear projection. IEEE Trans Pattern Anal Mach Intell 19(7): July 1997
28. Bartlett MS, Movellan JR, Sejnowski TJ (2002) Face recognition by independent component analysis. IEEE Trans Neural Netw 13(6):1450

29. Vincent P, Larochelle H, Bengio Y, Manzagol P-A (2018) Extracting and composing robust features with denoising autoencoders. In: International conference on machine learning, pp 1096–1103
30. He K, Zhang X, Ren S, Sun J (2016) Deep residual learning for image recognition. In: IEEE conference on computer vision and pattern recognition, pp 770–778
31. Gross R, Matthews I, Cohn J, Kanade T, Baker S (2010) Multi-PIE. Image Vis Comput 28(5):807–813
32. Tariyal S, Majumdar A, Singh R, Vatsa M (2016) Deep dictionary learning. IEEE Access 4:10096–10109
33. Tosic I, Frossard P (2011) Dictionary learning. IEEE Signal Process Mag 28(2):27–38
34. LeCun Y, Touresky D, Hinton G, Sejnowski T (1988) A theoretical framework for back-propagation. In: Connectionist models summer school, vol 1. Morgan Kaufmann, CMU, Pittsburgh, Pa, pp 21–28
35. Glorot X, Bengio Y (2010) Understanding the difficulty of training deep feedforward neural networks. In: International conference on artificial intelligence and statistics, pp 249–256
36. He K, Zhang X, Ren S, Sun J (2015) Delving deep into rectifiers: surpassing human-level performance on ImageNet classification. In: IEEE international conference on computer vision, pp 1026–1034
37. Huang GB, Ramesh M, Berg T, Learned-Miller E (2007) Labeled faces in the wild: a database for studying face recognition in unconstrained environments. University of Massachusetts, Amherst, Tech Rep 07–49
38. Wolf L, Hassner T, Maoz I (2011) Face recognition in unconstrained videos with matched background similarity. In: IEEE conference on computer vision and pattern recognition, pp 529–534
39. Simonyan K, Zisserman A (2014) Very deep convolutional networks for large-scale image recognition. arXiv:1409.1556
40. Parkhi OM, Vedaldi A, Zisserman A (2015) Deep face recognition. In: british machine vision conference, vol 1, no 3, p 6
41. Wu X, He R, Sun Z, Tan T (2018) A light CNN for deep face representation with noisy labels. IEEE Trans Inf Forensics Secur 13(11):2884–2896
42. Huang G, Liu Z, van der Maaten L, Weinberger KQ (2017) Densely connected convolutional networks. In: IEEE conference on computer vision and pattern recognition, vol 1, no 2, p 3
43. Vinyals O, Blundell C, Lillicrap T, Wierstra D, et al (2016) Matching networks for one shot learning. In: Advances in neural information processing systems, pp 3630–3638
44. Hariharan B, Girshick RB (2017) Low-shot visual recognition by shrinking and hallucinating features. In: IEEE international conference on computer vision, pp 3037–3046
45. Chhabra S, Majumdar P, Singh R, Vatsa M (2019) Data finetuning. In: AAAI conference on artificial intelligence
46. Keshari R, Vatsa M, Singh R (2019) Guided dropout. In: AAAI conference on artificial intelligence
47. Goswami G, Bhardwaj R, Singh R, Vatsa M (2014) Mdlface: memorability augmented deep learning for video face recognition. In: IEEE international joint conference on biometrics, pp 1–7
48. Goswami G, Vatsa M, Singh R (2017) Face verification via learned representation on feature-rich video frames. IEEE Trans Inf Forensics Secur 12(7):1686–1698
49. Bhatt HS, Singh R, Vatsa M (2014) On recognizing faces in videos using clustering-based re-ranking and fusion. IEEE Trans Inf Forensics Secur 9(7):1056–1068

Chapter 9
Deep Feature Fusion for Face Analytics

Nishant Sankaran, Deen Dayal Mohan, Sergey Tulyakov, Srirangaraj Setlur, and Venu Govindaraju

9.1 Introduction

Data produced from a particular source often exhibit correlations with those arising from other sources. Common data sources include (i) *sensors*—that gather raw data, (ii) *feature extractors*—which process the raw data from sensors to generate features representing the original data, and (iii) *evaluators*—which produce a score or a measure that conveys the likelihood of the provided features belonging to an application specific hypothesis. Fusion methods that can effectively capture the said correlations provide the capability to arrive at more informed decisions. Feature fusion methods, specifically, work with data produced from feature extractors and attempt to leverage the shared information contained within various features by generating a fused representation that is more efficient at a particular task. Depending on the number of sensors involved, there can be two feature fusion approaches: unimodal feature aggregation, multi-modal feature fusion.

Unimodal feature aggregation consolidates data produced from a single source (modality) into a robust representation which summarizes the important features of the data (sensor) signals. It is typically used to remove noise inherently present in the features and refine the information contained in it. A naive approach here would be to simply average or compute the mean of the feature vectors which would result in the desired de-noising effect. However, more nuanced approaches exist which adaptively elicit more information from the feature vectors. Multi-modal feature fusion captures the correlations among the multiple data sources (modalities) and exploits it to improve performance in the target optimization task. There are primarily two methods of multi-modal representation learning: *joint representation* learning that combines unimodal features into a common space, and *coordinated representation* learning that discovers modality specific representations but enforces similarity

N. Sankaran · D. D. Mohan · S. Tulyakov · S. Setlur (✉) · V. Govindaraju
Department of Computer Science Engineering, University at Buffalo, Buffalo, NY, USA
e-mail: setlur@buffalo.edu

constraints on them. Typical joint representation learning approaches stack or concatenate unimodal data representations and apply a logistic regression based model or, more recently, deep neural networks that learn to fuse them into a unified transformed representation that embodies characteristics of all the modalities involved. In this chapter, we explore a unimodal feature aggregation scheme applied to the task of face recognition (Sect. 9.2) and a multi-modal feature fusion method that has elements of both joint representation and coordinated representation learning applied to facial action unit recognition (Sect. 9.3).

9.2 Feature Aggregation for Face Recognition

Face recognition is the problem of classifying faces to particular identities or verifying the possibility that two given faces are of a common identity or not. Over the past few years, face recognition has seen tremendous advances in pushing the state-of-the-art performances to near human [18, 26] and sometimes even surpassing human capabilities [17, 22]. Though these systems have demonstrated exemplary performances leading to the community considering constrained face recognition as generally a solved problem, unconstrained face recognition, however, presents a different challenge.

Unconstrained face recognition attempts to address the fact that many face recognition systems are deployed in settings where there is no control over the conditions under which faces are captured with the possibility of uncooperative subjects. In the unconstrained setting (e.g., video surveillance), the goal of face recognition systems is to identify subjects (referred to as probe) from a media collection (referred to as gallery) that may have been compiled previously. The probes and galleries are stored as templates—each of which can constitute one or more face images corresponding to a specific identity. These face images are typically generated through a pipeline of face detection [11, 31], landmark identification [19], and finally alignment. The aligned faces are then transformed into a discriminative representation (such as CNN based features [24, 26]) that is compared with similar representations of other face images to determine if the identities present in the images are the same. Several metrics are employed for the purpose of estimating the similarity of face representations such as the euclidean distance, cosine proximity, and even metric learning methods [1, 20].

Matching face templates which are comprised of only single images for the probe and gallery each is relatively straight forward with the use of the above mentioned similarity functions, the most common one being the cosine similarity. However, in the unconstrained datasets like IJB-A [12] and YTF [27], face templates contain multiple images, and therefore, poses a new challenge of determining how to fuse/pool the face features to a single feature vector representative of the template. Typically, the simplest solution is employed, i.e., naive average/max pooling [3, 4, 18] of the features to yield the template representation. In recent works, more intelligent solutions using weighted averaging have been proposed [15, 29], where the weights are determined by analyzing the features and evaluating their representativeness.

Here, we present a new approach for pooling features of a template trained in the context of a face verification task. We use metadata accompanying the face images in the template for the purpose of evaluating the importance of each feature in the aggregation process. Metadata for face images include, but are not limited to, the yaw, pitch, and roll of the face in the image, as well as other external details such as the size of the face crop, positions of the landmarks, etc. The motivation behind our approach stems from the fact that all previous approaches [15, 29] only consider the features for determining the aggregation weights. Generally speaking, the features are generated by a CNN or other embedding system whose optimization criteria is to map all the face images of an identity to a single distinct cluster with minimal within-class variances (to enhance discriminability) and maximal inter-class variances (to enhance separability). But it becomes evident that, in doing so, this very optimization function restricts the ability of a system to exploit the variances among the features to determine optimal relative weights for pooling. Hence, we conjecture using additional data/metadata which is unperturbed by the optimization process for generating discriminative features would lead to discovering better aggregation weights.

We use CNNs described in [2, 19] to obtain the metadata and features used in our approach. We design a Metadata-based Feature Aggregator Network (M-FAN) which takes as input, features, metadata, and an extra parameter called seed weights to produce a weighted feature representation for the template. The seed weights are simply initial weight estimates provided to the network intended as a starting point for the optimization process and the network is trained to transform these seed weights based on the corresponding metadata. This parameter presents the possibility of providing the network with previously handcrafted weights, which it can then fine-tune according to the metadata, thereby boosting the performance as compared to using the handcrafted aggregation weights. We experiment the model on IJB-A and Janus CS4 datasets and obtain compelling improvements over the previous state-of-the-art approaches and show that the M-FAN model improves the performance of face recognition systems traditionally using naive pooling strategies.

9.2.1 Metadata-Based Feature Aggregator Network (M-FAN)

The entire objective of the M-FAN model is to function as a feature quality evaluator and produce weights corresponding to the "worthiness" of the feature vector as being a part of the template. Let f_i and m_i be the ith feature vector and corresponding metadata vector in a template. We define an evaluator function h_θ to be a function of the metadata vector, parametrized by θ, producing a weight that qualifies the provided metadata. If T denotes the template vector or the pooled features for the template, we have

$$T = \sum_i h_\theta(m_i) f_i \tag{9.1}$$

Here, h_θ could be realized as any function approximator, and in our case, it is represented by a Fully Connected Network (FCN). The above formulation ensures that the M-FAN network does not rely on the features to make its predictions, which is crucial to the performance of our model based on the following reasoning. The feature vectors are typically generated by a face recognizer whose task is to map any and all variations of face images for a particular subject to a single tightly bound cluster in the feature space. It would, therefore, imply that the feature vectors corresponding to the set of face images for a subject would have minimal variations so as to maximize discriminability for the concerned subject. Now this presents a problem for any aggregation system that attempts to evaluate the relative "richness" of the feature vectors in a template since they would all be extremely similar. This motivates the intuition why the same system would need orthogonal information such as metadata, which is not affected by the feature generation process, to yield context that can help to discriminate between face images of a subject.

Given the template vector construction, the objective of our system then becomes to determine the optimal set of parameters θ that minimizes our cost function defined as

$$E_{pg} = \left[\frac{T_p . T_g}{\|T_p\| \|T_g\|} - Y_{pg} \right]^2 \tag{9.2}$$

$$Cost = \sum_p \sum_g E_{pg} \tag{9.3}$$

where T_p and T_g are the probe and gallery template vectors obtained using (9.1), $Y_{pg} \in [0, 1]$ is the match label for the given probe and gallery templates, E_{pg} is the error in match score prediction, and as it is evident, the similarity between the two templates is obtained using the cosine similarity. With these goals in mind, Sect. 9.2.2 presents the design of the M-FAN structure.

9.2.2 Architecture

The setup of the M-FAN architecture is illustrated in Fig. 9.1. The essence of the model is the Fully Connected Network (FCN) that assesses the metadata and outputs a weight for the corresponding feature vector. In practice, the network is also provided a set of seed weights w_i (for example, setting $w_i = \frac{1}{n}$, n being the number of images in the template) which it can use as an origin to begin the optimization process. Consequently, the FCN block does not explicitly produce weight predictions as output, rather, produces parameters used to transform the seed weights. Providing

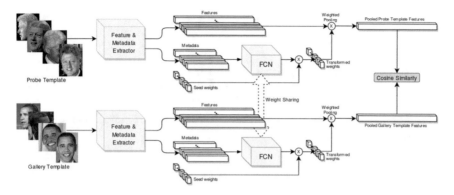

Fig. 9.1 M-FAN architecture. This figure shows the training setup with the M-FAN model deployed as a siamese network. The Feature and Metadata extractor are the networks described in [2, 19]. The FCN is the only trainable block in the structure

seed weights produced by elaborate handcrafted functions generally improves the ability of the model to converge on better weight predictions.

For training, the M-FAN network is built as a siamese network. The network is provided features, metadata and seed weights for the probe template as the left input and the same for the gallery template as the right input. The features and metadata are obtained from the networks described in [2, 19]. The FCN block that these inputs go through use shared weights as is typical in siamese architectures. The output of the FCN block transforms the seed weights w_i producing w_i' such that $\sum w_i' = 1$, which is then used in conjunction with the corresponding features f_i to create the template vector for each probe and gallery template. The cosine similarity loss layer computes the distance between the templates and is optimized against the match label. During testing, we don't use the siamese setup, and instead, present all the inputs for a template to the M-FAN model which produces the aggregated feature vector.

9.2.3   Gradient Backpropagation

The error E_{pg} defined in (9.2) can be used to derive the gradients for the parameters θ in the FCN being optimized. The gradient for an individual probe gallery template match is computed as

$$\frac{\partial E_{pg}}{\partial \theta} = 2\sqrt{E_{pg}} \left[\frac{\|T_p\| \, \|T_g\| \, \frac{\partial T_p.T_g}{\partial \theta} - T_p.T_g \frac{\partial \|T_p\| \|T_g\|}{\partial \theta}}{\|T_p\|^2 \, \|T_g\|^2} \right] \tag{9.4}$$

where

$$\frac{\partial T_p.T_g}{\partial \theta} = T_p.T_g' + T_g.T_p' \tag{9.5}$$

and

$$\frac{\partial \|T_p\| \|T_g\|}{\partial \theta} = \|T_p\| \cdot \|T_g\|' + \|T_g\| \cdot \|T_p\|' \tag{9.6}$$

follows the product rule of calculus. Similarly, we obtain the gradients of the template vector T and its norm as

$$T' = \sum_i h_\theta'(m_i) f_i \tag{9.7}$$

$$\|T\|' = \frac{T}{\|T\|}.T' \tag{9.8}$$

The interesting thing to note here is that the gradients for the parameters θ are also a function of the feature vectors f_i. This has a nice effect on the overall training procedure in that even though the FCN block never sees the feature vector for making its predictions, its parameter updates are influenced by f_i, thereby forcing it to learn the implicit correlations between the metadata and features. Moreover, with the presence of only a few dimensions in the input space, as compared to 100s or 1000s when taking the face feature vector also as input, the training algorithm is able to converge faster using fewer network parameters.

9.2.4 Batch Training

During the design of the training setup, it became clear that the network would only be able to train on a single pair of probe and gallery templates at each iteration. This was owing to the fact that each template may be comprised of a variable number of face images, which implies that making batches of probe and gallery templates would be difficult. However, as mentioned in [14], networks generally converge faster and to better minima with batch sizes >1. In order to work around this problem, we introduced an additional input—indices k_i which held the template indices in the batch that each face image (and the corresponding features and metadata) would be mapped to. This enabled us to group multiple sets of templates as a batch (Fig. 9.2 and to compute the aggregated template vectors using only the corresponding feature vectors as indexed by k_i.

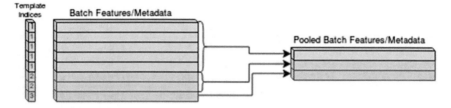

Fig. 9.2 Batch Processing of Templates. The template indices provided indicate which features/metadata comprise a template

9.2.5 Experiment Setup

The CNN described in [2] produces a 128 dimensional feature vector for each face image. Alongside that, the CNN detailed in [19] produces multiple metadata outputs of which we use yaw, pitch, roll, face bounding box area, gender classification confidence, and the face detection score for our experiments. Our M-FAN model is created with a 4 layer FCN having ReLU as activations. We group the subjects in the datasets into 3 sets—60% for training (of which 20% is for validation) and the rest for testing. We then used the provided template protocols to generate probe vs gallery matches for the three sets. We train our model on the training set for a maximum of 100 epochs with a learning rate of 0.15 with a decay rate of 0.99 every epoch. We report results on the test set with the model that had the best performance on the validation set. Since the network performance would be influenced by the seed weights provided to it, we conducted 2 sets of experiments—one with the naive average weights and the other by grouping images by their media source and weighting it by the face detection score on each image. For the latter, we first pool all the images corresponding to a particular media source weighted by their face detection scores s_i, i.e., $f_m = \sum_i \frac{exp(s_i)}{\sum_j exp(s_j)} f_i$. The weighted face detection score for the media-pooled images is $s_m = \sum_i \frac{exp(s_i)}{\sum_j exp(s_j)} s_i$. In a similar manner, we aggregate all the media-pooled features f_m with their respective scores s_m to get the template vector T. We record the final weights computed for each image via this method and use them as the seed weights for this experiment which we'll refer to as "media average weights". The models we train on both these experiments are referred to as M-FAN (naive) and M-FAN (media) respectively.

9.2.6 Results on IJB-A

Here we present the results of the M-FAN model on the IJB-A verification protocol. IJB-A contains 5712 images and 2085 videos of 500 subjects, for an average of 11.4 images and 4.2 videos per subject. We divide the subjects in the provided training split into training and validation (80:20 splits) and evaluate the trained model on the

Table 9.1 IJB-A 1:1 Verification TAR (%)

Method	10^{-1} FAR	10^{-2} FAR	10^{-3} FAR
TE	96.40 ± 0.5	90.00 ± 1.0	81.30 ± 2.0
TA	97.90 ± 0.4	93.90 ± 1.3	83.60 ± 2.7
NAN	97.80 ± 0.3	94.10 ± 0.8	88.10 ± 1.1
M-FAN[†]	97.97 ± 0.3	96.34 ± 0.3	94.10 ± 0.7
M-FAN[‡]	**98.00** ± 0.3	**96.56** ± 0.4	**94.44** ± 0.5

TE: Template Embedding [20], TA: Template Adaptation [5], NAN: Neural Aggregation Network [29], [†]M-FAN (naive), [‡]M-FAN (media)

protocol provided in the test splits. The 1:1 verification results are evaluated using the ROC curve and the TAR (True Accept Rate) performance is reported for different FAR (False Accept Rate) values. We present the results reported by the previous state-of-the-art approaches for IJB-A and compare them to our system. The results shown in Table 9.1 clearly indicate the ability of M-FAN to capture the correlations of the metadata and the features constituting a template and proves its utility as an intelligent aggregation unit.

9.2.7 Results on Janus CS4

We conduct our experiments on the IARPA Janus Challenge Set 4 (CS4) dataset, which is a superset of the IJB-A dataset [12]; the comparison between CS4 and IJB-A sets is given in [2]. A sample of the weights predicted by M-FAN is shown in Fig. 9.3. Table 9.2 shows the improvements in performances while using the M-FAN model seeded with naive average weights. It is interesting to note that even when M-FAN is provided the naive average seed weights, it is able to perform better than the handcrafted media average weights. When it is provided the more complex media average weights, it can improve upon its earlier performance by over 1.5%. We also analyzed the weights predicted (during test phase) by the M-FAN model with respect to the various metadata provided to it and plotted the results shown in Fig. 9.4. We can see that the network has learnt to predict low aggregation weights for any orientation that strays far from the frontal pose. Figure 9.4d depicts the weights for various face detection (FD) scores and here too we see it assign higher confidence to images having higher FD scores.

Fig. 9.3 M-FAN predictions on Janus CS4. The pooling weights are influenced by the orientation of the face, source image size, etc

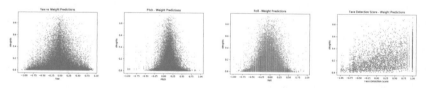

(a) Yaw vs Weights (b) Pitch vs Weights (c) Roll vs Weights (d) FDS vs Weights

Fig. 9.4 Plots showing aggregation weight predictions of M-FAN against various metadata features. FDS indicates face detection scores

Table 9.2 Janus CS4 1:1 Verification TAR(%)

Pooling method	10^{-2} FAR	10^{-3} FAR	10^{-4} FAR
Naive average	95.27	90.40	86.54
Media average	95.38	90.85	86.88
M-FAN (naive)	95.65	90.99	87.35
M-FAN (media)	95.98	92.19	88.63

9.3 Feature Enhancement for Facial Action Unit Recognition

In recent years, methods that combine features from multiple data sources have been gaining popularity. This can be primarily attributed to the large amount of data produced by different multi-modal sensors. Most of the methods that try to combine the multi-modal data, broadly fall under two major categories. Multi-view learning methods [28], which look at finding a subspace or a shared space between the data of multiple modalities and employ that as a unified representation. These methods generally try to enforce a constraint that increases the similarity of features learned by the views. On the other hand, multi-modal fusion methods [13], try to combine features of different representations to improve performance of the overall system.

Majority of the multi-modal fusion methods try to create a unified feature space. This is done either by mapping the current feature space to a higher dimension or by learning a latent representation after concatenating multi-modal features. Due to the recent advancements in new embedding methods such as [22] and complex multi dataset training procedures for deep learning, the features produced by such networks are highly optimal. Generating a unified representation from multiple, such features might require equally complex training procedures. Here we try to rethink the premise of the necessity of having a unified representation. We theoretically and experimentally show that learning linear transformation that increase separability of these features in their respective feature spaces can be an alternative to existing methods. In order to validate our claims, we apply our method to the problem of recognition of facial action units.

Facial expressions are one of the most important nonverbal-cues in any interpersonal communications. Facial expressions can be measured in two dimensions popularly. The Judgmental coding system describes emotions in a latent emotion space. The frequently used parameters in this scheme are the seven universal emotions namely Anger,Fear, Disgust, Happiness, Sadness, Surprise, and Contempt. A more elaborate way of describing emotions is using the FACS coding scheme. In their paper [7], define FACS as a measure of different facial muscle movements that contribute to the facial expressions either independently or in pairs. Each of the Action Units describes a movement or contraction of the facial muscle. Thus, they encode the anatomically visible changes rather than relying on the observers inference of emotions. The granularity of FACS is particularly beneficial in detecting even the subtle controlled or uncontrolled facial behavior. For several decades, they have been extensively used in forensics, neuro-marketing, healthcare, etc. Although Facial Action Units Recognition is well explored in the visible light domain (VLD), the RGB images suffer from illumination changes and can only capture the visual changes that occur as an effect of the AUs. There are, however, some physiological changes that the face undergoes during the occurrence of the Action Units, such as skin temperature changes, variation in the heart rate , blood pressure, and respiration rate that can't be captured using the visible images. In contrast, the Infrared images that allow detection of skin temperature variations and are invariant to illumination changes and skin tone variation from person to person, have been shown to be sensitive to AU movements [10]. Thus, the visible and thermal images encode complementary aspects of facial action units.

In this chapter, we demonstrate an alternative to the existing multi-modal fusion methods. We train a DenseNet Model on both the visible and thermal images and generate corresponding visible and thermal features. We present an idea of enhancing existing feature spaces by only applying scaling and translation perturbation. The perturbation that is to be applied to each feature, is learned by the network by jointly looking at all the feature representations. By doing so, we generate an enhanced feature representation of the original thermal and visible features. These enhanced features when combined, improve the overall performance of the system.

9.3.1 Multi-modal Conditional Feature Enhancement (MCFE)

The schematic of the proposed Multi-modal Conditional Feature enhancement (MCFE) method is shown in Fig. 9.5. MCFE consists of a feature extraction stage followed by a feature enhancement stage. Addressing the task of facial action unit recognition, we first train a deep CNN that learns to assign action unit labels for a given RGB frame focused on an individual's face. This network is optimized using a multi-task learning framework with class weighting incorporated, to solve the issue of class imbalance prevalent in such problems. While one network is trained on the visible spectrum, another network is trained similarly on the thermal spectrum for the same task. Each network learns a specific view of the task and we implement a novel multi-modal learning solution to *enhance* their corresponding representations by modeling their correlations in the shared subspace. Contrary to traditional fusion approaches, our approach doesn't attempt to create a unified fused representation of the modalities that is better equipped at solving the task. Rather, in our fusion approach, we emphasize transferring information that is uniquely learnt from the individual modalities to other view representations with the aim of improving the performance of each modality's feature representation oriented towards maximizing the combined system's performance. This method has an advantage over traditional fusion schemes, in that, instead of vastly increasing the search space for finding an

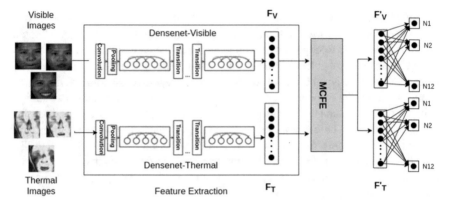

Fig. 9.5 Overview of the proposed MCFE framework. The system extracts Deep features (F_v and F_T) from paired visible and thermal images simultaneously using their corresponding DenseNet models. Further, using the proposed approach, the enhanced features (F'_v and F'_T) are produced

optimum representation that describes both modalities equally, it limits the problem by only determining the corrections/perturbations it needs to apply to each view's representation guided by the accompanying views. In doing so, we enhance not only the individual representations, but also the overall system performance which aggregates the performances of the individual modalities. In the following sections, we detail the approaches that we used for extracting action unit features and our novel fusion approach applied to multi-modal facial action unit recognition.

9.3.2 Feature Extraction

Research in facial expression recognition can be categorized mainly on the basis of feature extractors and classifiers. The ability of handcrafted features like SIFT [16], HOG [6], etc., to capture the complex non-linear transformations of the face caused by expressions are limited. CNNs on the other hand, have shown the ability to learn optimal features for vision based tasks like handwritten character recognition, face recognition, etc. A series of convolutional filters can extract features starting from abstract information like edges to complex patterns like faces in the subsequent layers. However, with deeper structures, the gradient vanishes as it reaches the beginning layers. Networks with short skip connections like ResNets [8] and Highway Networks [23] prevent this by providing an alternative and easier way for gradients to flow. Following this, DenseNet [9] was introduced wherein features from one layer are connected to features of all the preceding layers. As a result, the lower level abstract features are combined with the higher level granular features. Although DenseNets learn a representation similar to the deeper models, owing to its compact parameterization, it is less prone to over fitting, and enables feature reuse. To this end, we use the Densenet-121 architecture with modifications for extracting features from both visible and thermal images. The network consists of 4 dense blocks followed by an output layer consisting of 12 neurons for 12 classes. The sigmoid activation is used at the final classification layer. Typically the cross entropy loss is applied at the output layer. Consider N AU classes, then for each input, the multi-class cross entropy cost is calculated as

$$C = \sum_{i=1}^{N} (y_i' \log y_i + (1 - y_i')\log(1 - y_i)) \qquad (9.9)$$

In the above formulation, the individual components of loss corresponding to each AU is given equal weight. Most of the Facial Action Unit datasets are heavily imbalanced. Some of the action units have very low positive to negative sample ratio, otherwise called the occurrence rate. Therefore, in order to account for the under represented classes, the individual loss components needs to be weighted. However, calculating the weights for each class with respect to other classes for a multi-label classification problem can be quite complex as each data sample can contain more than one AU class. Therefore, to overcome this problem, we use a multi-task framework wherein a separate binary cross entropy loss is applied to each of the N output neurons and the weights applied to the loss components are weighted by the ratio of their respective positive and negative samples. The final output layer of the DenseNet is split into 12 output neurons. The cost function is calculated as

$$b_i = w_i * (y_i' \log y_i) + 1 * ((1 - y_i')\log(1 - y_i)) \qquad (9.10)$$

The negative samples are weighted by 1 and the positive samples are weighted by w_i given as

$$w_i = \frac{n_i}{p_i} \qquad (9.11)$$

where n_i is the total negative samples for AU_i and p_i is the total positive samples. The final loss is calculated as the sum individual binary cross entropy losses

$$C_i = \sum_{i=1}^{N} b_i \qquad (9.12)$$

Thus, the loss formulation takes into account the individual class distributions, while still learning the correlations between the different Action Units. The network is trained with random weight initialization instead of initializing with the typical ImageNet classification weights.

9.3.3 Deep Feature Enhancement

Deep multi-modal fusion has typically relied on learning the feature correlations among the modalities by stacking a number of fully connected layers applied on a merged representation (concatenation, sum, etc.) or by projecting each modality's feature space onto a common optimal subspace for the specific task. Such methods eventually arrive at a new shared representation for fusion, but is it necessary to construct a completely new representation? We address this question by proposing to employ the existing feature space and design a fusion scheme that, based upon a shared representation, learns to only modify or perturb the original features in such a way as to improve feature separability in their existing feature spaces.

Consider k input modalities x_i, $i = 1, .., k$ and their corresponding feature representations obtained as so

$$v_i = f(x_i; \theta_i) \tag{9.13}$$

where $v_i \in \mathbb{R}^{d_i}$, f may be an MLP, DNN or other feature extractors and θ_i are the parameters for the corresponding modalities which may be shared. We define a function g with parameters ∇ which transforms all the input modalities' features into a latent representation $l \in \mathbb{R}^n$ thus

$$l = g(v_1, v_2, ...v_k; \nabla) \tag{9.14}$$

Based on this latent representation, we compute M transformation factors (feature wise scaling and translation) $s_i = [s_i^1, .., s_i^M]$ and $t_i = [t_i^1, .., t_i^M]$ for each modality i as below (omitting the subscript i for brevity)

$$s^j = \sigma(W_s^{j^T} l + b_s^j) \tag{9.15}$$

$$t^j = \sigma(W_t^{j^T} l + b_t^j) \tag{9.16}$$

Since the above equations are for each modality, there are k weights and biases W_s^j and b_s^j corresponding to the scaling factors and k weights and biases W_t^j and b_t^j corresponding to the translation factors with $j = 1, .., M$ and σ denoting the sigmoid non-linearity. With these, we can construct M different variants of each feature vector v_i as:

$$e_i^j = (s_i^j \odot v_i) \oplus t_i^j \tag{9.17}$$

Finally, we choose 1 out of the M different enhanced features e_i^j by predicting *importance weights* ch_i^j for the M variants and running it through a softmax activation to pick the most relevant enhanced feature vector e_i^*

$$ch_i^j = softmax(W_c^{j^T} l + b_c^j) \tag{9.18}$$

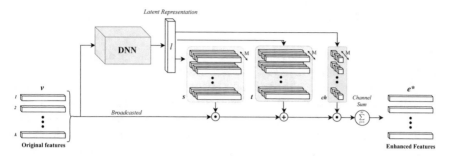

Fig. 9.6 MCFE Architecture. The system takes k modality features v and proposes M variants of element-wise scaling s and translation t parameters applied on the input features. It finally weights the aggregation of the proposals using ch via the channel sum operation to finally arrive at k enhanced features $e*$

$$e_i^* = \sum_{j=1}^{M} ch_i^j * e_i^j \qquad (9.19)$$

where W_c^j and b_c^j are the k weights and biases corresponding to the choice prediction function applied to the k modalities. This final enhanced feature representation is presented to the classification layer for improved performance on the task being solved. Figure 9.6 illustrates the general architecture of the proposed multi-modal conditional feature enhancement system.

The presence of M variants of transformation factors and consequently M different versions of the enhanced features enables the learning algorithm to explore a variety of improvements that can be applied to the original feature set. This is similar in concept to the multiple paths that are present in the Inception [25] deep network architecture, where each layer gets its data from several different *views* of the previous layer's output. The presence of such multiple paths of computation directly improves gradient flow to the prior layers since each path would begin to learn aspects unique to a particular view for better representing the solution space. Additionally, the ability to select the variant that is most suitable for the given sample of multi-modal features enables conditioning the fusion methodology applied on the features. This allows the gradients to pass through multiple parallel routes to the latent representation mapping function g, and thus allows faster convergence as observed in the experiments. It should be noted that the latent representation l being learnt is different from the traditional shared representations that prior works obtain during fusion. This is because in this formulation, l encapsulates information relevant only to making a decision as to how to *correct* every modality's feature representation independently rather than representing a common unified subspace which is used directly for the classification task. There are benefits for defining the latent representation in such a manner and this will be explored in Sect. 9.3.4.

9.3.4 Training MCFE for AU Recognition

MCFE is applied to the thermal and visible features arising out of the corresponding thermal and RGB frames extracted as described in Sect. 9.3.2. The latent mapping function g that operates on the two inputs, is implemented as a modified lightweight DenseNet CNN. The latent representation l is connected to 3 separate fully connected layers—one each for scaling, translation, and predicting choice. Upon computing the final enhanced feature set e_i* as described in Sect. 9.3.3 they are independently connected to separate AU classification layers (one each for thermal and visible spectrum) composed of 12 outputs each indicating the presence of a specific action unit for the given image. In order to orient the training of MCFE's parameters to the final goal of improving the overall system (thermal and visible combined) performance and not just the individual modality's performance, we perform a simple average of the predictions by the thermal and visible classification layers. By combining the predictions and optimizing against them, the network is encouraged to learn parameters that would enable a particular modality's enhanced feature to compensate for the shortcomings of the other feature.

In the naive approach, we can initialize the network with random weights. However, this would either fail to converge or provide unsatisfactory results. Intelligent initializations help the network perform significantly better than traditional approaches. We take two steps in this direction: (a) initialize the scaling layer to produce an output of 1s and the translation layer to produce an output of 0s; and (b) initialize the classification layers with the corresponding pre-trained weights learnt during the feature extraction step. Performing step (a) ensures that in the initial state of the network, the enhanced features are the same as the original features forming an identity map

$$e_i^j = (s_i^j \odot v_i) \oplus t_i^j$$
$$= (1 \odot v_i) \oplus 0 = v_i$$

which gives a good starting location to begin gradient descent. Since the overall error reduces in a specific direction, there can only be an improvement to the feature representation (or at worst can remain the same). Step (b) accomplishes the objective of reusing the information learnt from prior training to give a good target representation that the previous layers can attempt to produce from the original representation. One question that arises is how can we ensure that the enhanced features are indeed only *corrected* versions of the original features and not entirely a new set of features learnt by the network? We address this by freezing the classification layer weights from training (which are already pre-initialized). This enforces a constraint on the input to the classification layers (i.e., the enhanced features) such that it adheres to the general feature space representation that the classification layer has been trained to operate on. Imposing this constraint limits, the optimizer's task to finding δs to each individual feature so as to magnify or diminish their effect and in doing so make the feature space more separable with respect to the given classification layer parameters. This is a much simpler task to optimize for, than having to explore unconstrained combinations of feature spaces and, as we observe in the experiments, leads to much faster convergence with strong performance gains. Additionally, with the formulation of a single latent representation that is responsible for the computation of 3 separate quantities, it implements the information bottleneck principle which has been shown to produce better generalizations [21] and also mirrors the multi-task learning framework's principle of leveraging a single representation for accomplishing multiple disparate tasks.

9.3.5   Datasets for Experimental Analysis

We used the Multi-modal Spontaneous Emotion (MMSE) database to evaluate our performance. MMSE contains 2D and thermal videos of 140 participants from ten tasks, each eliciting different emotions. Among them, only four tasks were labeled for Facial Action Units. Expert AU coders annotated each frame using the Facial Action Coding System. The thermal sensor and the RGB camera were mounted on top of each other and their frame rates were set to 25 fps for synchronization. In our experiments, we used all 196, 793 visible frames and 195, 411 thermal frames. Out of the available images, only 133, 309 paired frames were available for our multi-modal learning experiments.

9.3.6   Experiment Settings

Pre-processing
All the input images in the dataset were aligned using the MTCNN framework [31] based on the 49 facial landmark points provided by the MMSE dataset. Further, the images were cropped to 170×170 and randomly rotated for data augmentation. The presence of each action unit was labeled as +1/0. The data samples with missing labels and faces were excluded from training.

Network Settings and Training for Feature Extraction
We adopted a three-fold cross validation protocol to train our networks. For each experiment, we split the dataset into three subject dependent partitions using two partitions for training and the remaining one partition for validation. Both CNNs in our implementation are trained using the weighted cross entropy loss defined in Eq. 9.10–9.12 The models are trained with SGD as the optimizer with learning rate initialized to $1e - 3$.

Network Settings and Training for MCFE
We use DenseNet-100×12 architecture, which has a depth of 100 and a growth rate of 12 for getting the latent representation from the features. Along with the DenseNet, the network consists of three fully connected layers corresponding to scaling,translation, and prediction, and two classification layers corresponding to

Table 9.3 Three-Fold cross validation results on the MMSE dataset over 12 action units. *Visible* and *Thermal* refer to the results obtained using the feature extraction method described in Sect. 9.3.2. *MFB* is multi-modal factorized binary pooling described in [30]. *MCFE* refers to the performance obtained by after fusion using method described in Sect. 9.3.3

| | | | | | | MMSE Three-fold validation results | | | | | | |
AU	1	2	4	6	7	10	12	14	15	17	23	24	Mean F1
Visible	0.306	0.264	0.267	0.845	0.874	0.902	0.882	0.809	0.358	0.413	0.484	0.282	0.557
Thermal	0.216	0.202	0.197	0.794	0.828	0.861	0.847	0.772	0.271	0.317	0.358	0.244	0.492
MFB	0.293	0.280	0.300	0.847	0.866	0.899	0.886	0.798	0.367	0.426	0.492	0.273	0.561
MCFE	0.304	0.299	0.312	0.851	0.864	0.906	0.884	0.815	0.354	0.421	0.473	0.314	0.566

the two modalities, resulting in a total of 24M parameters. The network is trained using Adam optimizer, with a learning rate of 0.1 and a batch size of 128 for 32 epochs.

9.3.7 Results

We show the results of our experiment in Table 9.3. We report our performance using the F1 metrics widely used in the literature of facial action unit recognition. For each experiment, we report the individual F1 scores of each action unit and also the average F1 score across all action units. We report the average F1 scores across three splits. We compare MCFE performance with a relevant state-of-the-art multi model fusion strategy—MFB [30]. We also compare our enhanced features with the original features from a simple sum score level fusion perspective.

From Table 9.3, we note that the fusion results of the proposed MCFE methods outperforms MFB overall. However, we also note that even though MFB performs slightly better on some action units, MCFE significantly outperforms MFB in the action units which are severely under represented in the data set. It is also interesting to note that results obtained by MCFE is without destroying the original feature space. The improvement can be attributed to the fact that the network learned good translation and scaling features to align the visible and thermal features so as to improve the overall performance of the system.

9.4 Conclusion

In this chapter, we discussed about template/set-based face verification and how it is deeply influenced by the aggregation or pooling strategy employed in generating representative template features. We presented a unimodal feature fusion approach of using metadata to judge the relative quality of every feature vector in a template for aggregation and investigate its ability to outperform related approaches. The M-FAN feature fusion approach produced significant gains over traditional pooling approaches on the IJB-A and Janus CS4 datasets proving its effectiveness. Moreover, this can be easily plugged into at the end of a face recognition pipeline to optimize the template feature generation process in order to produce improvements in the overall performance.

Additionally, we also discussed a method of improving facial action unit recognition using a novel approach to multi-modal feature fusion named MCFE. The fusion method is an alternative to the concept of learning a joint unified representation. Through theoretical and experimental validation, we find that the deep feature fusion method can learn the factors that help to better align the features in their respective feature spaces to maximize separability. The method also shows that such aligned features can be easily combined so as to improve the overall performance of the system.

References

1. Cai X, Wang C, Xiao B, Chen X, Zhou J (2012) Deep nonlinear metric learning with independent subspace analysis for face verification. In: Proceedings of the 20th ACM international conference on multimedia. ACM, pp 749–752
2. Chen J-C, Patel VM, Chellappa R (2016) Unconstrained face verification using deep CNN features. In: 2016 IEEE winter conference on applications of computer Vision (WACV), IEEE, pp. 1–9
3. Chen J-C, Ranjan R, Kumar A, Chen C-H, Patel VM, Chellappa R (2015) An end-to-end system for unconstrained face verification with deep convolutional neural networks. In: Proceedings of the IEEE international conference on computer vision workshops, pp 118–126
4. Chowdhury AR, Lin T-Y, Maji S, Learned-Miller E (2016) One-to-many face recognition with bilinear CNNS. In: 2016 IEEE winter conference on applications of computer vision (WACV). IEEE, pp 1–9
5. Crosswhite N, Byrne J, Stauffer C, Parkhi O, Cao Q, Zisserman A (2017) Template adaptation for face verification and identification. In: 2017 12th IEEE international conference on automatic face & gesture recognition (FG 2017). IEEE, pp 1–8
6. Dalal N, Triggs B (2005) Histograms of oriented gradients for human detection. In: 2005 IEEE computer society conference on computer vision and pattern recognition (CVPR'05), vol 1, pp 886–893
7. Ekman P, Friesen WV (1976) Measuring facial movement. Environ Psychol Nonverbal Behav 1(1):56–75
8. He K, Zhang X, Ren S, Sun J (2016) Deep residual learning for image recognition. In: Proceedings of the IEEE conference on computer vision and pattern recognition, pp 770–778
9. Huang G, Liu Z, Weinberger KQ, van der Maaten L (2017) Densely connected convolutional networks. In: Proceedings of the IEEE conference on computer vision and pattern recognition, vol 1, p 3
10. Jarlier S, Grandjean D, Delplanque S, N'diaye K, Cayeux I, Velazco MI, Sander D, Vuilleumier P, Scherer KR (2011) Thermal analysis of facial muscles contractions. IEEE Trans Affect Comput 2(1):2–9
11. Jiang H, Learned-Miller E (2017) Face detection with the faster R-CNN. In: 2017 12th IEEE international conference on automatic face & gesture recognition (FG 2017), IEEE, pp. 650–657

12. Klare BF, Klein B, Taborsky E, Blanton A, Cheney J, Allen K, Grother P, Mah A, Jain AK (2015) Pushing the frontiers of unconstrained face detection and recognition: Iarpa janus benchmark a. In: Proceedings of the IEEE conference on computer vision and pattern recognition, pp 1931–1939

13. Lahat D, Adalı T, Jutten C (n.d.) Multimodal data fusion: an overview of methods, challenges and prospects

14. Li M, Zhang T, Chen Y, Smola AJ (2014) Efficient mini-batch training for stochastic optimization. In: Proceedings of the 20th ACM SIGKDD international conference on Knowledge discovery and data mining. ACM, pp 661–670

15. Liu Y, Yan J, Ouyang W (2017) Quality aware network for set to set recognition. arXiv:1704.03373

16. Lowe DG (2004) Distinctive image features from scale-invariant keypoints. Int J Comput Vis 60(2):91–110

17. Lu C, Tang X (2015) Surpassing human-level face verification performance on LFW with gaussianface

18. Parkhi OM, Vedaldi A, Zisserman A et al (2015) Deep face recognition. In: BMVC, vol 1, p 6

19. Ranjan R, Sankaranarayanan S, Castillo CD, Chellappa R (2017) An all-in-one convolutional neural network for face analysis. In: 2017 12th IEEE international conference on automatic face & gesture recognition (FG 2017). IEEE, pp 17–24

20. Sankaranarayanan S, Alavi A, Castillo CD, Chellappa R (2016) Triplet probabilistic embedding for face verification and clustering. In: 2016 IEEE 8th international conference on biometrics theory, applications and systems (BTAS). IEEE, pp 1–8

21. Saxe AM, Bansal Y, Dapello J, Advani M, Kolchinsky A, Tracey BD, Cox DD (2018) On the information bottleneck theory of deep learning. In: International conference on learning representations

22. Schroff F, Kalenichenko D, Philbin J (2015) Facenet: a unified embedding for face recognition and clustering. In: Proceedings of the IEEE conference on computer vision and pattern recognition, pp 815–823

23. Srivastava RK, Greff K, Schmidhuber J (2015) Training very deep networks. In: Advances in neural information processing systems, pp 2377–2385

24. Sun Y, Liang D, Wang X, Tang X (2015) Deepid3: face recognition with very deep neural networks. arXiv:1502.00873

25. Szegedy C, Ioffe S, Vanhoucke V, Alemi AA (2017) Inception-v4, inception-resnet and the impact of residual connections on learning

26. Taigman Y, Yang M, Ranzato M, Wolf L (2014) Deepface: closing the gap to human-level performance in face verification. In: Proceedings of the IEEE conference on computer vision and pattern recognition, pp 1701–1708

27. Wolf L, Hassner T, Maoz I (2011) Face recognition in unconstrained videos with matched background similarity. In: 2011 IEEE Conference on computer vision and pattern recognition (CVPR), IEEE, pp 529–534

28. Xu C, Tao D, Xu C (2013) A survey on multi-view learning. arXiv:1304.5634

29. Yang J, Ren P, Chen D, Wen F, Li H, Hua G (2016) Neural aggregation network for video face recognition. arXiv:1603.05474

30. Yu Z, Yu J, Fan J, Tao D (n.d.) Multi-modal factorized bilinear pooling with co-attention learning for visual question answering. CoRR

31. Zhang K, Zhang Z, Li Z, Qiao Y (2016a) Joint face detection and alignment using multitask cascaded convolutional networks. IEEE Signal Process Lett 23(10):1499–1503

Chapter 10
Deep Learning for Video Face Recognition

Jiaolong Yang and Gang Hua

Abstract This chapter is concerned with face recognition based on videos or, more generally, sets of images, using deep learning techniques. We first briefly review some naive yet commonly used strategies pertaining to using frame-level features extracted by deep convolutional neural networks (CNNs) for video-level face recognition. Representative strategies include naive feature pooling and pairwise feature distance computation. Then, we present a method named neural aggregation network (NAN), which is a deep learning framework tailored for video-based representation and recognition. NAN can automatically learn the quality of faces in a video/image set and aggregate the frame-level deep features accordingly, yielding more discriminative video-level features. We conduct experimental evaluation on three video face recognition datasets. The results indicate that while previous deep learning-based methods with naive pooling or pairwise distances have obtained substantial improvements over traditional methods, the NAN method further outperforms them by an appreciable margin.

10.1 Introduction

The goal of video face recognition is to recognize human identities from face video clips or, more generally, face image sets. It has broad applications in video surveillance, biometric authentication, and video retrieval, to name a few. Compared to image-based face recognition, more information of the subjects can be exploited from the input videos, which naturally incorporate faces of the same subject in different poses and illumination conditions. In recent years, video face recognition,

J. Yang
Microsoft Research, No. 5 Danling Street, Beijing 100080, China

G. Hua (✉)
Microsoft Research, Redmond, Washington 98052, USA

especially in unconstrained settings (aka, "in the wild"), has caught more and more attention from the research community [1–13].

Similar to image-based face recognition, video face recognition can be categorized into two tasks: video face identification and video face verification. The former aims at recognizing the identity of a probe face video/image set from a gallery video face database where the identities are known, and the latter is to judge if two given face videos/image sets belong to the same person or not. For both tasks, the key is to build an appropriate representation for video face, such that it can effectively integrate the information across different frames together, maintaining beneficial while discarding noisy information.

Recently, tremendous success in face recognition has been achieved by the deep learning technique [10, 11, 14–19], with performance outclassing the traditional methods with handcrafted features, even surpassing human-level performance on some large-scale datasets [11, 14–16]. By training on large face data, a deep convolutional neural network (CNN) can embed a face image into a low-dimensional feature space where identity similarity can be effectively measured by Euclidean distance or angular distance.

CNNs operate on still images. To apply them for video-based face recognition, some simple strategies can be used. For example, one straightforward solution would be directly representing the video face as a set of frame-level face features extracted by a CNN. Such a representation comprehensively maintains the information across all frames. To compare two video faces, one can fuse the feature matching results across all pairs of frames between the two face videos (e.g., by taking the mean or minimum distance). Another naive approach is conducting a certain type of pooling (e.g., average and max pooling) to aggregate the frame-level features together to form a video-level representation. Due to the simplicity of the aforementioned solutions, they have been widely adopted in recent deep face recognition works for video-based recognition and good performance has been achieved. In later sections of this chapter, we will introduce them in details and evaluate them in the experiments.

Despite the simplicity and wide adoption, these solutions also have some drawbacks in terms of efficiency and effectiveness. For example, let n be the number of video frames, then the time complexity of video face similarity computation is $O(n^2)$ for the feature set-based representation, which is not desirable especially for large-scale recognition. Besides, such a representation would incur $O(n)$ space complexity per video face example, which demands a lot of memory storage and confronts efficient indexing. To make the recognition scalable to large-scale problems, it is more desirable to come with a compact, *fixed-size* feature representation at the video level, irrespective of the varied lengths of the videos. A fixed-size video-level representation would allow direct, *constant-time* computation of the similarity or distance without the need for frame-to-frame matching.

Such a video face representation can be achieved by frame-feature pooling. The naive pooling strategies such as average pooling assume that the frame features are of equal importance. However, a video (especially a long video sequence) or an image set may contain face images captured at various conditions of lighting, resolution, head pose, etc. A good pooling or aggregation strategy should *adaptively* weigh and combine the frame-level features across all frames. It should favor face images that are more discriminative (or more "memorizable") and prevent poor face images from jeopardizing the recognition accuracy.

To this end, we present an approach with adaptive weighting scheme to linearly combine all frame-level features from a video together to form a compact and discriminative face representation. Different from the previous methods, the frame feature weights are neither fixed nor set by any particular heuristics. Instead, a neural network is designed to automatically calculate the weights. The network is named Neural Aggregation Network (NAN), whose coefficients can be trained through supervised learning in a normal face recognition training task without the need for extra supervision signals.

The NAN is composed of two major modules that could be trained end to end or one by one separately. The first one is a feature embedding module which serves as a frame-level feature extractor using a deep CNN model. The other is the aggregation module that adaptively fuses the feature vectors of all the video frames together. NAN inherits the main advantages of pooling techniques, including the ability to handle arbitrary input size and producing order-invariant representations. Its key component is inspired by the Neural Turing Machine [20] and the work of [21], both of which applied an attention mechanism to organize the input through accessing an external memory. This mechanism can take an input of arbitrary size and work as a tailor emphasizing or suppressing each input element just via a weighted averaging, and very importantly it is order independent and has trainable parameters. In NAN, a simple network structure is designed which consists of two cascaded attention blocks associated with this attention mechanism for face feature aggregation.

Apart from building a video-level representation, the neural aggregation network can also serve as a subject-level feature extractor to fuse multiple data sources. For example, one can feed it with all available images and videos, or the aggregated video-level features of multiple videos from the same subject, to obtain a single feature representation with fixed size. In this way, the face recognition system not only enjoys the time and memory efficiency due to the compact representation, but also exhibits superior performance, as we will show in our experiments. Last but not least, NAN can serve as a general framework for learning content-adaptive pooling, so it is applicable to other computer vision tasks as well. The NAN method first appeared on a conference paper [13].

The reminder of this chapter is organized as follows. In Sect. 10.2, we briefly review the traditional, non-deep-learning methods proposed for video face recognition. We will focus on some recently proposed ones. In Sect. 10.3, we review the existing approaches using frame-level deep features extracted by CNNs for video face recognition. In Sect. 10.4, we present the neural aggregation network, including the network architecture and the training method. In Sect. 10.5, we experimentally evaluate the existing approaches as well as the NAN method with both video face verification and identification tasks, on three challenging datasets: the YouTube Face dataset [1], the IJB-A dataset [22], and the Celebrity-1000 dataset [7]. We show that while all deep learning-based methods perform quite well compared with transitional method, NAN consistently outperforms all the pairwise frame-feature distance computation or naive feature pooling strategies.

10.2 Traditional Methods

Face recognition based on videos or image sets has been actively studied in the past [1, 2, 8, 23–27]. In this chapter, we are concerned with the input being an orderless set of face images. Existing methods exploiting temporal dynamics will not be considered here. For set-based face recognition, many previous methods have attempted to represent the set of face images with appearance subspaces or manifolds and perform recognition via computing manifold similarity or distance [23, 26, 28–30]. These traditional methods may work well under constrained settings but usually cannot handle the challenging unconstrained scenarios where large appearance variations are present.

Along a different axis, some methods try to build video-level feature representation based on local features [2, 6, 8]. For example, the PEP methods [2, 6] take a part-based representation by extracting and clustering local features. Both appearance and spatial information are taken into account in the local features. The Video Fisher Vector Faces (VF^2) descriptor [8] uses Fisher Vector coding to aggregate local features across different video frames together to form a video-level representation. Despite large improvements have been achieved using these methods especially for the unconstrained scenarios, the features they use are still handcrafted and their performances have been lagged farther behind the deep learning-based approaches.

10.3 Existing Deep Learning-based Approaches

Recently, state-of-the-art face recognition methods have been dominated by deep convolution neural networks [10, 11, 16, 31, 32]. Let $\mathbf{x}$ be an input face image, a CNN can map it to a low-dimensional feature space by interleaved linear and non-linear transformations. This process can be denoted by

$$\mathbf{f} = CNN(\mathbf{x}). \tag{10.1}$$

Given two face images $\mathbf{x}$ and $\mathbf{x}'$, their identity (dis-)similarity can be measured by their feature angular distance $d(\mathbf{f}, \mathbf{f}') = \arccos \left(\dfrac{<\mathbf{f}, \mathbf{f}'>}{\|\mathbf{f}\| \cdot \|\mathbf{f}'\|} \right)$, or the Euclidean distance $d(\mathbf{f}, \mathbf{f}) = \|\mathbf{f} - \mathbf{f}\|$ with both $\mathbf{f}$ and $\mathbf{f}'$ normalized. The feature distance can then be used for face verification and identification tasks.

Now consider a face video sequence or a image set $\mathcal{X} = \{\mathbf{x}_1, \mathbf{x}_2, ..., \mathbf{x}_n\}$ with n total images. We can apply a CNN trained on single images to extract a frame-feature set $\mathcal{F} = \{\mathbf{f}_1, \mathbf{f}_2, ..., \mathbf{f}_n\}$. To achieve verification and identification, the key is

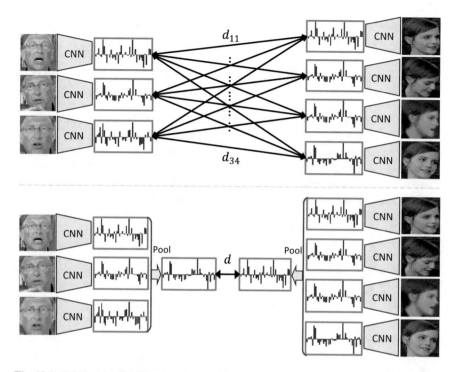

Fig. 10.1 Existing deep learning-based approaches using frame-level deep features for video face recognition. Top: matching based on pairwise frame-feature distances. Bottom: video-level feature representation by frame feature pooling

to build a proper distance measurement function $D(\{\mathbf{f}_i\}, \{\mathbf{f}'_j\})$ for two frame-feature sets extracted from two face videos, respectively. There are some simple distance functions commonly used by the existing deep learning approaches. They can be classified into two categories: those based on pairwise frame-feature distances and based on frame-feature pooling. Figure 10.1 illustrates these two categories, and we introduce them as follows.

10.3.1 Pairwise Distance-Based Methods

The pairwise distance-based methods measure identity similarity by comprehensively analyzing the distances between the frame pairs. It necessitates storing all image features of a video, i.e., with $O(n)$ space complexity, and usually computes the distance with $O(n^2)$ time complexity. We list some concrete examples as follows.

- *Mean pairwise distance*: The distance is computed as the mean of the pairwise frame-feature distances:

$$D(\{\mathbf{f}_i\}, \{\mathbf{f}'_i\}) = \frac{1}{n \cdot n'} \sum_{i,j} d(\mathbf{f}_i, \mathbf{f}_j). \tag{10.2}$$

Many CNN-based methods use mean pairwise distance for video face recognition. Representative works include DeepFace [10] and FaceNet [11].

- *Minimum or maximum pairwise distance*: The distance can be computed by enumerating all the pairwise frame-feature distances and selecting the minimal or maximum one. In other words, the video face similarity is measured based on the most similar or dissimilar frames between the videos/image sets:

$$D(\{\mathbf{f}_i\}, \{\mathbf{f}'_i\}) = \min_{i,j} d(\mathbf{f}_i, \mathbf{f}_j) \tag{10.3}$$

for minimum pairwise distance and

$$D(\{\mathbf{f}_i\}, \{\mathbf{f}'_i\}) = \max_{i,j} d(\mathbf{f}_i, \mathbf{f}_j) \tag{10.4}$$

for the maximum case. However, simply taking the minimum or maximum pairwise distance can be vulnerable to noise or outliers, making the recognition algorithm non-robust (as we will show later in the experiments, the results are clearly worse than other distance metrics). So the minimum and maximum pairwise distances are rarely used by existing methods. Instead, the following distance function was proposed.

- *Softmin pairwise distance*: This distance function is a robust version of minimal pairwise distance. It is computed by

$$D(\{\mathbf{f}_i\}, \{\mathbf{f}'_i\}) = \min_{i,j} \frac{e^{-\beta \cdot d(\mathbf{f}_i, \mathbf{f}_j)}}{\sum_{i,j} e^{-\beta \cdot d(\mathbf{f}_i, \mathbf{f}_j)}}, \tag{10.5}$$

where β is a positive scalar factor. Alternatively, one can use multiple scalar factors and compute the distance as

$$D(\{\mathbf{f}_i\}, \{\mathbf{f}'_i\}) = \min_{i,j} \sum_k \frac{e^{-\beta_k \cdot d(\mathbf{f}_i, \mathbf{f}_j)}}{\sum_{i,j} e^{-\beta_k \cdot D(\mathbf{f}_i, \mathbf{f}_j)}}. \tag{10.6}$$

The softmin pairwise distance function corresponds to the softmax similarity score advocated in the works of [18, 33, 34]. A single β is used in [34] while multiple are used in [18, 33]. The time complexity for computing this distance function is $O(m \cdot n^2)$, where m is the number of used scalar factors.

Of course, given two feature sets, many other set distance functions that can be potentially used, such as the Hausdorff distance or the distance between two affine/convex hulls spanned by the feature sets. These distance metrics can be found in some traditional face recognition methods such as [27]. Thus far, we are not aware of existing works using these sophisticated set distance functions for deep features.

10.3.2   Pooling-Based Methods

Instead of representing the face video as a set of deep frame features, some methods conduct feature pooling over the feature set. This way, the video face is represented by a single, fixed-sized vector and the distance can be computed in constant time: $D(\{\mathbf{f}_i\}, \{\mathbf{f}'_i\}) = d(\text{Pool}(\{\mathbf{f}_i\}), \text{Pool}\{\mathbf{f}_i\}) = d(\tilde{\mathbf{f}}, \tilde{\mathbf{f}}')$. The commonly used pooling methods are average pooling and max pooling.

- *Average pooling*: Average pooling averages the feature vectors in the set at each dimension, i.e.,

$$\text{Pool}(\{\mathbf{f}_i\}) = \tilde{\mathbf{f}}, \ \text{with } \tilde{\mathbf{f}}(d) = \frac{1}{n} \sum_i \mathbf{f}_i(d), \tag{10.7}$$

where $\mathbf{f}(d)$ is the element of vector $\mathbf{f}$ at the d-th dimension. Many existing deep face recognition methods use average pooling technique to obtain video-level features. Representative works include VGG-Face [16] and [31].
- *Max pooling*: Similarly, the feature max pooling can be expressed as

$$\text{Pool}(\{\mathbf{f}_i\}) = \tilde{\mathbf{f}}, \ \text{with } \tilde{\mathbf{f}}(d) = \max_i \mathbf{f}_i(d). \tag{10.8}$$

Max pooling for frame features is less common than average pooling. It is only adopted in a few works such as [32].

10.4   Neural Aggregation Network

As mentioned previously, it is more desirable to have a compact, fixed-size feature representation at the video level, similar to the pooling techniques. However, a good pooling or aggregation strategy should adaptively weigh and combine the frame-level features, as opposed to a naive handling such as average or max pooling. Here we present the neural aggregation network which has a learning-based feature weighting scheme to linearly combine the frame-level features.

Figure 10.2 shows the framework of the NAN. The network takes a set of face images of a person as input and outputs a single feature vector as its representation for the recognition task. It is built upon a modern deep CNN model for frame feature embedding, and becomes more powerful for video face recognition by adaptively aggregating all frames in the video into a compact vector representation.

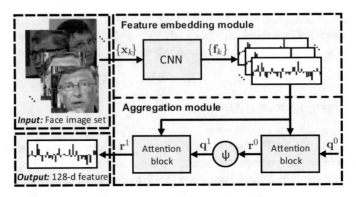

Fig. 10.2 The framework of Neural Aggregation Network. All input face images $\{x_k\}$ are processed by a feature embedding module with a deep CNN, yielding a set of feature vectors $\{f_k\}$. These features are passed to the aggregation module, producing a single, fixed-size vector r^1 to represent the input face images. This compact representation is used for recognition

10.4.1 Feature Embedding Module

The image embedding module of the NAN is a deep CNN, which embeds each frame of a video to a frame-level feature representation. In theory, any deep CNN with high-end performance can be applied here. In our implementation, we use the GoogLeNet [35] structure equipped with Batch Normalization (BN) [36] to test the performance of NAN. The GoogLeNet produces 128-dimension image features. We *normalized* these features to be unit vectors then fed into the aggregation module, which will be described in the following section.

10.4.2 Aggregation Module

Given a face video/image set $\mathcal{X} = \{x_1, x_2, ..., x_n\}$ and its corresponding normalized feature set $\mathcal{F} = \{f_1, f_2, ..., f_n\}$ extracted by the feature embedding module, the goal is to generate a set of weights $\{a_k\}_{k=1}^n$ for all the feature vectors, so that the aggregated feature representation for the video is a linear combination of the frame features:

$$\mathbf{r} = \sum_k a_k \mathbf{f}_k. \tag{10.9}$$

In this way, the aggregated feature vector has the same size as a single face image feature extracted by the CNN. Note that if $a_k \equiv \frac{1}{n}$, Eq. 10.9 will degrade to naive average pooling.

Three main principles have been considered in designing the aggregation module. First, the module should be able to process different numbers of images (i.e., different

K's), as the video data source varies from person to person. Second, the aggregation should be invariant to the image order, i.e., the result is unchanged when the image sequences are reversed or reshuffled. This way, the aggregation module can handle an arbitrary set of image or video faces without temporal information (e.g., those collected from different Internet locations). Third, the module should be adaptive to the input faces and has parameters trainable through supervised learning in a standard face recognition training task.

The solution is inspired by the *memory attention mechanism* described in [20, 21, 37]. The idea therein is to use a neural model to read external memories through a differentiable addressing/attention scheme. Such models are often coupled with Recurrent Neural Networks (RNN) to handle sequential inputs/outputs [20, 21, 37]. Although an RNN structure is not needed here, its memory attention mechanism is applicable to the aggregation task we consider here. In NAN, the frame-level face features are treated as the memory, and feature weighting is casted as a memory addressing procedure. Some "attention blocks" are employed in the aggregation module to achieve this.

10.4.2.1 Attention Blocks

An attention block reads all feature vectors from the feature embedding module, and generate linear weights for them. Specifically, let $\{\mathbf{f}_k\}$ be the face feature vectors, then an attention block filters them with a kernel $\mathbf{q}$ via dot product, yielding a set of corresponding significances $\{e_k\}$. They are then passed to a softmax operator to generate positive weights $\{a_k\}$ with $\sum_k a_k = 1$. These two operations can be described by the following equations, respectively:

$$e_k = \mathbf{q}^{\mathsf{T}} \mathbf{f}_k \tag{10.10}$$

$$a_k = \frac{\exp(e_k)}{\sum_j \exp(e_j)}. \tag{10.11}$$

It can be seen that the above algorithm essentially selects one point inside of the convex hull spanned by all the feature vectors.

In this way, the number of inputs $\{\mathbf{f}_k\}$ does not affect the size of aggregation $\mathbf{r}$, which is of the same dimension as a single feature $\mathbf{f}_k$. Besides, the aggregation result is invariant to the input order of $\mathbf{f}_k$: according to Eqs. 10.9, 10.10, and 10.11, permuting $\mathbf{f}_k$ and $\mathbf{f}_{k'}$ has no effects on the aggregated representation $\mathbf{r}$. Furthermore, an attention block is modulated by the filter kernel $\mathbf{q}$, which is trainable through standard backpropagation and gradient descent.

Single attention block—Universal face feature quality measurement. The adaptive aggregation can be by achieved by using one attention block. In this case, vector $\mathbf{q}$ is the parameter to learn. It has the same size as a single feature $\mathbf{f}$ and serves as a universal prior measuring the face feature quality.

Fig. 10.3 Face images in the IJB-A dataset, sorted by their scores (values of e in Eq. 10.10) from a *single attention block* trained in the face recognition task. The faces in the top, middle, and bottom rows are sampled from the faces with scores in the highest 5%, a 10% window centered at the median, and the lowest 5%, respectively

Table 10.1 Ablation study on the IJB-A dataset. TAR/FAR: true/false accept rate for verification. TPIR/FPIR: true/false positive identification rate for identification

	1:1 Verification TAR@FAR of:		1:N identification TPIR@FPIR of:	
Method	0.001	0.01	0.01	0.1
CNN+AvgPool	0.771	0.913	0.634	0.879
NAN single attention	0.847	0.927	0.778	0.902
NAN cascaded attention	0.860	0.933	0.804	0.909

To test the performance, we train the network for video face verification (see Sects. 10.4.3 and 10.5 for details) on the IJB-A dataset [22] with the extracted face features, and Fig. 10.3 shows the sorted scores of all the faces images in the dataset. It can be seen that after training, the network favors high-quality face images, such as those of high resolutions and with relatively simple backgrounds. It down-weights face images with blur, occlusion, improper exposure, and extreme poses. Table 10.1 shows that the network achieves higher accuracy than the average pooling strategy in the verification and identification tasks.

Cascaded two attention blocks—Content-aware aggregation. Although good performance can be achieved by the universal quality assessment, a content-aware aggregation can potentially perform even better. The intuition behind is that face image variation may be expressed differently at different geographic locations in the feature space (i.e., for different persons), and content-aware aggregation can learn to select features that are more discriminative for the identity of the input image set. To this end, we can employ two attention blocks in a cascaded and end-to-end fashion described as follows.

Let $\mathbf{q}^0$ be the kernel of the first attention block, and $\mathbf{r}^0$ be the aggregated feature with $\mathbf{q}^0$. We adaptively compute $\mathbf{q}^1$, the kernel of the second attention block, through

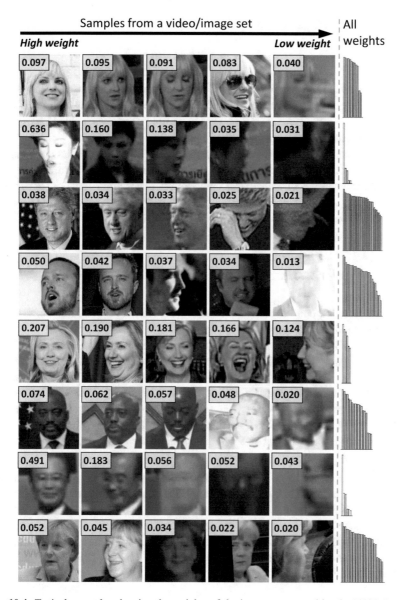

Fig. 10.4 Typical examples showing the weights of the images computed by the NAN. In each row, five face images are sampled from an image set and sorted based on their weights (numbers in the rectangles); the rightmost bar chart shows the sorted weights of all the images in the set (heights scaled)

a transfer layer taking $\mathbf{r}^0$ as the input:

$$\mathbf{q}^1 = \tanh(\mathbf{W}\mathbf{r}^0 + \mathbf{b}), \tag{10.12}$$

where $\mathbf{W}$ and $\mathbf{b}$ are the weight matrix and bias vector of the neurons, respectively, and $\tanh(x) = \frac{e^x - e^{-x}}{e^x + e^{-x}}$ imposes the hyperbolic tangent nonlinearity. The feature vector $\mathbf{r}^1$ generated by $\mathbf{q}^1$ will be the final aggregation results. Therefore, $(\mathbf{q}^0, \mathbf{W}, \mathbf{b})$ are now the trainable parameters of the aggregation module.

We train the network on the IJB-A dataset again, and Table 10.1 shows that the network obtained better results than using single attention block. Figure 10.4 shows some typical examples of the weights computed by the trained network for different videos or image sets. In all the remaining experimental results, we use the cascaded two attention block design in the NAN (as per Fig. 10.2).

10.4.3 Network Training

The NAN network can be trained for either face verification or face identification tasks with standard configurations.

For verification, a *Siamese* neural aggregation network structure [38] with two NANs sharing weights can be employed to minimize the average contrastive loss [39]: $l_{i,j} = y_{i,j}||\mathbf{r}_i^1 - \mathbf{r}_j^1||_2^2 + (1-y_{i,j})\max(0, m - ||\mathbf{r}_i^1 - \mathbf{r}_j^1||_2^2)$, where $y_{i,j} = 1$ if the pair (i, j) is from the same identity and $y_{i,j} = 0$ otherwise. The constant m is set to 2 in all the experiments in the next section. For identification, a fully connected layer is added on top of NAN followed by a softmax to minimize the average classification loss: $l_i = -\log p_{i,y_i}$ where y_i is the target label of the i-th video instance, $p_{i,y_i} = \frac{\exp(p_{i,y_i})}{\sum_z \exp(p_{i,z})}$, and $p_{i,z}$ is the z-th output of the fully connected layer.

The two modules of NAN can be trained either simultaneously in an end-to-end fashion, or separately one by one. While millions of still images can be obtained for training nowadays [10, 11, 16], it appears not practical to collect such amount of distinctive face videos or sets. So, we opt for first training the CNN in the feature embedding module on single images with the identification task, then training the aggregation module on top of the features extracted by CNN. More details can be found in Sect. 10.5.1.

10.5   Experiments

This section evaluates the performance of different methods as well as the NAN network. We will begin with introducing our training details, followed by reporting the results on three video face recognition datasets: the IARPA Janus Benchmark A (IJB-A) [22], the YouTube Face dataset [1], and the Celebrity-1000 dataset [7].

10.5.1   Training Details

To train the CNN, we use about 3M face images of 50K identities crawled from the Internet to perform image-based identification. The faces are detected using the JDA method [40], and aligned with the LBF method [41]. The input image size is 224 × 224. After training, the CNN is fixed and we focus on analyzing the effectiveness of different frame-feature pooling and set distance computation strategies as well as the neural aggregation module of NAN.

The aggregation module of NAN is trained on each video face dataset we tested on with standard backpropagation and an RMSProp solver [42]. An all-zero parameter initialization is used, i.e., we start from average pooling. The batch size, learning rate, and iteration are tuned for each dataset. As the network is quite simple and image features are compact (128-d), the training process is quite efficient: training on 5K video pairs with ∼1M images in total only takes less than 2 min on a CPU of a desktop PC.

10.5.2   Methods for Evaluation

We compare the results of simple aggregation strategies such as average pooling, the set-to-set similarity measurements leveraging pairwise frame-feature comparison, and the NAN. Unless specified, we simply use L_2 feature distance for face recognition (all features are normalized; thus it's equivalent to angular distance). It is possible to combine with an extra metric learning or template adaption technique [12] to further boost the performance on each dataset.

In these competing methods, *CNN+Min L_2*, *CNN+Max L_2*, *CNN+Mean L_2*, and *CNN+SoftMin L_2* measure the similarity of two video faces based on the L_2 feature distances of all frame pairs. They necessitate storing all image features of a video, i.e., with $O(n)$ space complexity. The first three use the minimum, maximum, and mean pairwise distance, respectively, thus having $O(n^2)$ complexity for similarity computation. *CNN+SoftMin L_2* corresponds to a SoftMax similarity score and is described in Sect. 10.3. It has $O(m \cdot n^2)$ complexity for distance computation where m is the number of scaling factor β used. In our experiments, we tested 20 combinations of β's including single or multiple values and report the best results obtained.

CNN+MaxPool and *CNN+AvePool* are, respectively, max pooling and average pooling along each feature dimension for aggregation. These two methods as well as the NAN produce a 128-d feature representation for each video and compute the similarity in $O(1)$ time.

Apart from these methods for which the training and testing results are obtained by us, we also compare with the prior art on each dataset, including both traditional- and deep learning-based methods.

10.5.3 Results on IJB-A Dataset

The IJB-A dataset [22] contains face images and videos captured from unconstrained environments. It features full pose variation and wide variations in imaging conditions thus is very challenging. There are 500 subjects with 5,397 images and 2,042 videos in total and 11.4 images and 4.2 videos per subject on average. We detect the faces with landmarks using STN [43] face detector, and then align the face image with similarity transformation.

In this dataset, each training and testing instance is called a "template," which comprises 1 to 190 mixed still images and video frames. Since one template may contain multiple medias and the dataset provides the media id for each image, another possible aggregation strategy is first aggregating the frame features in each media then the media features in the template [12, 44]. This strategy is also tested in this work with *CNN+AvePool* and our NAN. Note that media id may not be always available in practice.

We test the methods on both the "compare" protocol for *1:1 face verification* and the "search" protocol for *1:N face identification*. For verification, the true accept rates (TAR) versus false positive rates (FAR) are reported. For identification, the true positive identification rate (TPIR) versus false positive identification rate (FPIR) and the Rank-N accuracies are reported. Table 10.2 presents the numerical results of different methods, and Fig. 10.5 shows the receiver operating characteristics (ROC) curves for verification as well as the cumulative match characteristic (CMC) and decision error trade-off (DET) curves for identification. The metrics are calculated according to [22, 47] on the 10 splits.

In general, *CNN+Max L_2*, *CNN+Min L_2*, and *CNN+MaxPool* perform worst among the naive pooling and pairwise distance-based methods. *CNN+SoftMin L_2* performs slightly better than *CNN+MaxPool*. The use of media id significantly improves the performance of *CNN+AvePool*, but gives a relatively small boost to NAN. We believe NAN already has the robustness to templates dominated by poor images from a few media. Without the media aggregation, NAN outperforms all the naive pooling or pairwise distance-based methods by appreciable margins, especially on the low FAR cases. For example, in the verification task, the TARs of NAN at FARs of 0.001 and 0.01 are, respectively, 0.860 and 0.933, reducing the errors of the best results from these methods by about 39% and 23%, respectively.

Table 10.2 Identification and verification performance evaluation on the IJB-A dataset. For verification, the true accept rates (TAR) *versus* false positive rates (FAR) are reported. For identification, the true positive identification rate (TPIR) *versus* false positive identification rate (FPIR) and the Rank-N accuracies are presented

Method	1:1 Verification TAR					1:N Identification TPIR		
	FAR=0.001	FAR=0.01	FAR=0.1	FPIR=0.01	FPIR=0.1	Rank-1	Rank-5	Rank-10
B-CNN [32]	–	–	–	0.143 ± 0.027	0.341 ± 0.032	0.588 ± 0.020	0.796 ± 0.017	–
LSFS [45]	0.514 ± 0.060	0.733 ± 0.034	0.895 ± 0.013	0.383 ± 0.063	0.613 ± 0.032	0.820 ± 0.024	0.929 ± 0.013	–
DCNN$_{manual}$+metric[31]	–	0.787 ± 0.043	0.947 ± 0.011	–	–	0.852 ± 0.018	0.937 ± 0.010	0.954 ± 0.007
Triplet similarity [44]	0.590 ± 0.050	0.790 ± 0.030	0.945 ± 0.002	$0.556 \pm 0.065^{*}$	$0.754 \pm 0.014^{*}$	$0.880 \pm 0.015^{*}$	0.95 ± 0.007	$0.974 \pm 0.005^{*}$
Pose-aware models [18]	0.652 ± 0.037	0.826 ± 0.018	–	–	–	0.840 ± 0.012	0.925 ± 0.008	0.946 ± 0.007
Deep multi-pose [34]	–	0.876	0.954	0.52^{*}	0.75^{*}	0.846	0.927	0.947
DCNN$_{fusion}$ [46]	–	0.838 ± 0.042	0.967 ± 0.009	$0.577 \pm 0.094^{*}$	$0.790 \pm 0.033^{*}$	0.903 ± 0.012	0.965 ± 0.008	0.977 ± 0.007
Masi et al. [33]	0.725	0.886	–	–	–	0.906	0.962	0.977
Triplet embedding [44]	0.813 ± 0.02	0.90 ± 0.01	0.964 ± 0.005	0.753 ± 0.03	0.863 ± 0.014	0.932 ± 0.01	–	0.977 ± 0.005
VGG-face [16]	–	$0.805 \pm 0.030^{*}$	–	$0.461 \pm 0.077^{*}$	$0.670 \pm 0.031^{*}$	$0.913 \pm 0.011^{*}$	–	$0.981 \pm 0.005^{*}$
Template adaptation[12]	0.836 ± 0.027	0.939 ± 0.013	**0.979 ± 0.004**	0.774 ± 0.049	0.882 ± 0.016	0.928 ± 0.010	0.977 ± 0.004	**0.986 ± 0.003**
CNN+Max L_2	0.202 ± 0.029	0.345 ± 0.025	0.601 ± 0.024	0.149 ± 0.033	0.258 ± 0.026	0.429 ± 0.026	0.632 ± 0.033	0.722 ± 0.030
CNN+Min L_2	0.038 ± 0.008	0.144 ± 0.073	0.972 ± 0.006	0.026 ± 0.009	0.293 ± 0.175	0.853 ± 0.012	0.903 ± 0.010	0.924 ± 0.009
CNN+Mean L_2	0.688 ± 0.080	0.895 ± 0.016	0.978 ± 0.004	0.514 ± 0.116	0.821 ± 0.040	0.916 ± 0.012	0.973 ± 0.005	0.980 ± 0.004

(continued)

Table 10.2 (continued)

Method	1:1 Verification TAR					1:N Identification TPIR		
	FAR=0.001	FAR=0.01	FAR=0.1	FPIR=0.01	FPIR=0.1	Rank-1	Rank-5	Rank-10
CNN+SoftMin L_2	0.697 ± 0.085	0.904 ± 0.015	0.978 ± 0.004	0.500 ± 0.134	0.831 ± 0.039	0.919 ± 0.010	0.973 ± 0.005	0.981 ± 0.004
CNN+MaxPool	0.202 ± 0.029	0.345 ± 0.025	0.601 ± 0.024	0.079 ± 0.005	0.179 ± 0.020	0.757 ± 0.025	0.911 ± 0.013	0.945 ± 0.009
CNN+AvePool	0.771 ± 0.064	0.913 ± 0.014	0.977 ± 0.004	0.634 ± 0.109	0.879 ± 0.023	0.931 ± 0.011	0.972 ± 0.005	0.979 ± 0.004
CNN+AvePool[†]	0.856 ± 0.021	0.935 ± 0.010	0.978 ± 0.004	0.793 ± 0.044	0.909 ± 0.011	0.951 ± 0.005	0.976 ± 0.004	0.984 ± 0.004
NAN	0.860 ± 0.012	0.933 ± 0.009	$\mathbf{0.979 \pm 0.004}$	0.804 ± 0.036	0.909 ± 0.013	0.954 ± 0.007	0.978 ± 0.004	0.984 ± 0.003
NAN[†]	$\mathbf{0.881 \pm 0.011}$	$\mathbf{0.941 \pm 0.008}$	0.978 ± 0.003	$\mathbf{0.817 \pm 0.041}$	$\mathbf{0.917 \pm 0.009}$	$\mathbf{0.958 \pm 0.005}$	$\mathbf{0.980 \pm 0.005}$	$\mathbf{0.986 \pm 0.003}$

*Results cited from [12]
[†]First aggregating the images in each media then aggregate the media features in a template

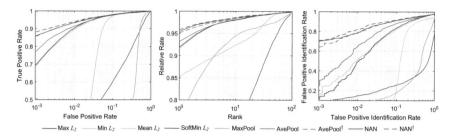

Fig. 10.5 Average ROC (Left), CMC (Middle), and DET (Right) curves of different methods on the IJB-A dataset over 10 test splits

With the media aggregation, the NAN achieves top performances compared to previous methods. It has a same verification TAR at FAR=0.1 and identification Rank-10 CMC as the method of [12], but outperforms it on all other metrics (e.g., 0.881 *versus* 0.836 TARs at FAR=0.01, 0.817 *versus* 0.774 TPIRs at FPIR=0.01, and 0.958 *versus* 0.928 Rank-1 accuracy).

Figure 10.4 has shown some typical examples of the weighting results. NAN exhibits the ability to choose high-quality and more discriminative face images while repelling poor face images.

10.5.4 Results on YouTube Face dataset

We then test the different methods on the YouTube Face (YTF) dataset [1] which is designed for unconstrained *face verification* in videos. It contains 3,425 videos of 1,595 different people, and the video lengths vary from 48 to 6,070 frames with an average length of 181.3 frames. Ten folds of 500 video pairs are available, and we follow the standard verification protocol to report the average accuracy with cross-validation. We again use the STN and similarity transformation to align the face images.

The face verification results of different methods are presented in Table 10.3, with their ROC curves shown in Fig. 10.6. Again, *CNN+Max L_2*, *CNN+Min L_2*, and *CNN+MaxPool* perform worse than *CNN+Mean L_2*, *CNN+SoftMin L_2*, and *CNN+AvePool*. *CNN+SoftMin L_2* performs best among these methods, with accuracy slighter higher than *CNN+Mean L_2* and *CNN+AvePool*. The NAN outperforms all these naive pooling or pairwise distance-based methods. The gaps between NAN and their best-performing ones are smaller compared to the results on IJB-A. This is because the face variations in this dataset are relatively small (compare the examples in Figs. 10.7 and 10.4), thus no much beneficial information can be extracted compared to naive average pooling or computing mean L_2 distances.

Compared to the previous methods, NAN achieves a mean accuracy of 95.72%, reducing the error of FaceNet by 12.3%. Note that FaceNet is also based on a

Table 10.3 Verification accuracy comparison of different methods on the YTF dataset

Method	Accuracy (%)	AUC
LM3L [48]	81.3 ± 1.2	89.3
DDML(combined)[9]	82.3 ± 1.5	90.1
EigenPEP [6]	84.8 ± 1.4	92.6
DeepFace-single [10]	91.4 ± 1.1	96.3
DeepID2+ [17]	93.2 ± 0.2	–
Wen et al. [19]	94.9	–
FaceNet [11]	95.12 ± 0.39	–
VGG-Face [16]	97.3	–
CNN+Max. L_2	91.96 ± 1.1	97.4
CNN+Min. L_2	94.96 ± 0.79	98.5
CNN+Mean L_2	95.30 ± 0.74	98.7
CNN+SoftMin L_2	95.36 ± 0.77	98.7
CNN+MaxPool	88.36 ± 1.4	95.0
CNN+AvePool	95.20 ± 0.76	98.7
NAN	95.72 ± 0.64	98.8

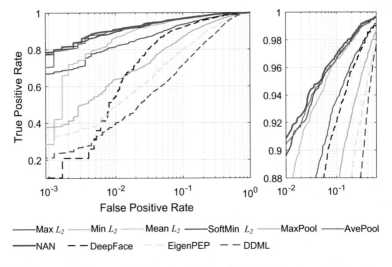

Fig. 10.6 Average ROC curves of different methods on the YTF dataset over the 10 splits

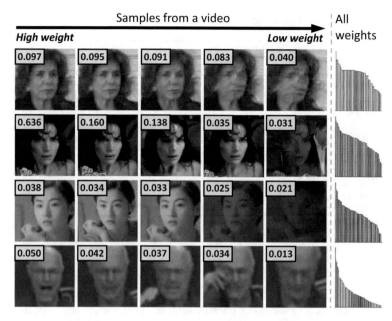

Fig. 10.7 Typical examples on the YTF dataset showing the weights of the video frames computed by NAN. In each row, five frames are sampled from a video and sorted based on their weights (numbers in the rectangles); the rightmost bar chart shows the sorted weights of all the frames (heights scaled)

GoogLeNet-style network, and it uses the average similarity of all pairs of 100 frames in each video (*i.e.*, 10K pairs) [11]. The VGG-Face [16] achieves an accuracy (97.3%) higher than NAN. However, that result is based on a further discriminative metric learning on YTF, without which the accuracy is only 91.5% [16].

10.5.5   Results on Celebrity-1000 Dataset

The Celebrity-1000 dataset [7] is designed to study the unconstrained video-based *face identification* problem. It contains 159,726 video sequences of 1,000 human subjects, with 2.4M frames in total (∼15 frames per sequence). We use the provided five facial landmarks to align the face images. Two types of protocols—open set and close set—exist on this dataset. In the open-set protocol, 200 subjects are used for training, while video sequences of the rest 800 subjects are used as the gallery set and probe set at the testing stage. There are four different settings with different numbers of probe and gallery subjects: 100, 200, 400, and 800. In the close-set protocol, the video sequences from all 1,000 subjects are divided into a training (gallery) subset and a testing (probe) subset. There are also four settings for close set: 100, 200, 500,

Table 10.4 Identification performance (rank-1 accuracies, %) on the Celebrity-1000 dataset for the *close-set* tests

Method	Number of subjects			
	100	200	500	1000
MTJSR [7]	50.60	40.80	35.46	30.04
Eigen-PEP [6]	50.60	45.02	39.97	31.94
CNN+Mean L_2	85.26	77.59	74.57	67.91
CNN+AvePool - VideoAggr	86.06	82.38	80.48	74.26
CNN+AvePool - SubjectAggr	84.46	78.93	77.68	73.41
NAN-VideoAggr	88.04	82.95	**82.27**	76.24
NAN- SubjectAggr	**90.44**	**83.33**	**82.27**	**77.17**

Table 10.5 Identification performance (rank-1 accuracies, %) on the Celebrity-1000 dataset for the *open-set* tests

Method	Number of subjects			
	100	200	400	800
MTJSR [7]	46.12	39.84	37.51	33.50
Eigen-PEP [6]	51.55	46.15	42.33	35.90
CNN+Mean L_2	84.88	79.88	76.76	70.67
CNN+AvePool - SubjectAggr	84.11	79.09	78.40	75.12
NAN - SubjectAggr	88.76	85.21	82.74	79.87

and 1000 subjects. More details about the protocols and the dataset can be found in [7].

Close-set tests:

For the close-set protocol, we first train the network on the video sequences with the identification loss. We take the FC layer output values as the scores and the subject with the maximum score as the result. We also train a linear classifier for *CNN+AvePool* to classify each video feature. As the features are built on video sequences, we call this approach "VideoAggr" to distinguish it from another approach to be described next. Each subject in the dataset has multiple video sequences, thus we can build a single representation for a subject by aggregating all available images

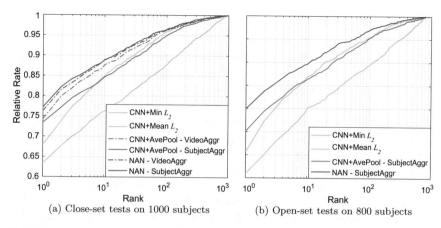

Fig. 10.8 The CMC curves of different methods on Celebrity-1000

in all the training (gallery) video sequences. We call this approach "SubjectAggr." This way, the linear classifier can be bypassed, and identification can be achieved simply by comparing the feature L_2 distances.

The results are presented in Table 10.4. Note that [6, 7] are not using deep learning and no deep network-based method reported result on this dataset. So we mainly compare NAN and the naive pooling or pairwise distance-based methods. It can be seen from Table 10.4 and Fig. 10.8a that NAN consistently outperforms these methods for both "VideoAggr" and "SubjectAggr." Significant improvements are achieved for the "SubjectAggr" approach. It is interesting to see that, "SubjectAggr" leads to a clear performance drop for *CNN+AvePool* compared to its "VideoAggr." This indicates that the naive aggregation gets even worse when applied on the subject level with multiple videos. However, our NAN can benefit from "SubjectAggr," yielding results consistently better than or on par with the "VideoAggr" approach and delivers a considerable accuracy boost compared to the naive approaches. This suggests our NAN works quite well on handling large data variations.

Open-set tests:

We then test the NAN with the *close-set* protocol. We first train the network on the provided training video sequences. In the testing stage, we take the "Subjec-tAggr" approach described before to build a highly compact face representation for each gallery subject. Identification is perform simply by comparing the L_2 distances between aggregated face representations.

The results in both Table 10.5 and Fig. 10.8b show that the NAN significantly reduces the error of *CNN+AvePool*. This again suggests that in the presence of large face variances, the widely used strategies such as average-pooling aggregation and the pairwise distance computation are far from optimal. In such cases, the learned NAN model is clearly more powerful, and the aggregated feature representation by it is more favorable for the video face recognition task.

10.6 Conclusions

In this chapter, we summarized the existing deep learning techniques for video-based face recognition. All these methods apply a deep CNN to extract image-level deep features for the video/image sets, and the key for recognition is how to measure the distance between two feature sets or how to aggregate the image-level features to obtain a video-level feature. For this, we reviewed some naive yet commonly used strategies, including feature pooling and pairwise feature distance computation. As shown in the experiments, most of these simple strategies are effective and the results are significantly better than traditional methods with handcrafted features. We further presented the neural aggregation network for video face representation and recognition. NAN fuses all input frames with a set of content-adaptive weights, resulting in a compact representation that is invariant to the input frame order. The aggregation scheme has small computation and memory footprints, but can generate quality face representations after training. The experimental results demonstrate that it outperforms the pairwise image feature distance computation or pooling-based methods by a wide margin.

References

1. Wolf L, Hassner T, Maoz I (2011) Face recognition in unconstrained videos with matched background similarity. In: IEEE conference on computer vision and pattern recognition (CVPR), pp 529–534
2. Li H, Hua G, Lin Z, Brandt J, Yang J (2013) Probabilistic elastic matching for pose variant face verification. In: IEEE conference on computer vision and pattern recognition (CVPR), pp 3499–3506
3. Wolf L, Levy N (2013) The SVM-minus similarity score for video face recognition. In: IEEE conference on computer vision and pattern recognition (CVPR), pp 3523–3530
4. Cui Z, Li W, Xu D, Shan S, Chen X (2013) Fusing robust face region descriptors via multiple metric learning for face recognition in the wild. In: IEEE conference on computer vision and pattern recognition (CVPR), pp 3554–3561
5. Mendez-Vazquez H, Martinez-Diaz Y, Chai Z (2013) Volume structured ordinal features with background similarity measure for video face recognition. In: International conference on biometrics (ICB)
6. Li H, Hua G, Shen X, Lin Z, Brandt J (2014) Eigen-PEP for video face recognition. In: Asian conference on computer vision (ACCV), pp 17–33
7. Liu L, Zhang L, Liu H, Yan S (2014) Toward large-population face identification in unconstrained videos. IEEE Trans Circuits Syst Video Technol 24(11):1874–1884
8. Parkhi OM, Simonyan K, Vedaldi A, Zisserman A (2014) A compact and discriminative face track descriptor. In: IEEE conference on computer vision and pattern recognition (CVPR), pp 1693–1700
9. Hu J, Lu J, Tan Y-P (2014) Discriminative deep metric learning for face verification in the wild. In: IEEE conference on computer vision and pattern recognition (CVPR), pp 1875–1882
10. Taigman Y, Yang M, Ranzato M, Wolf L (2014) DeepFace: closing the gap to human-level performance in face verification. In: IEEE conference on computer vision and pattern recognition (CVPR), pp 1701–1708

11. Schroff F, Kalenichenko D, Philbin J (2015) FaceNet: a unified embedding for face recognition and clustering. In: IEEE conference on computer vision and pattern recognition (CVPR), pp 815–823
12. Crosswhite N, Byrne J, Parkhi OM, Stauffer C, Cao Q, Zisserman A (2016) Template adaptation for face verification and identification. arXiv:1603.03958
13. Yang J, Ren P, Zhang D, Chen D, Wen F, Li H, Hua G (2017) Neural aggregation network for video face recognition. In: IEEE conference on computer vision and pattern recognition (CVPR)
14. Sun Y, Wang X, Tang X (2014) Deep learning face representation from predicting 10,000 classes. In: IEEE conference on computer vision and pattern recognition (CVPR), pp 1891–1898
15. Lu C, Tang X (2015) Surpassing human-level face verification performance on LFW with gaussianface. In: AAAI conference on artificial intelligence (AAAI), pp 3811–3819
16. Parkhi OM, Vedaldi A, Zisserman A (2015) Deep face recognition. In: British machine vision conference (BMVC), vol 1, no 3, p 6
17. Sun Y, Wang X, Tang X (2015) Deeply learned face representations are sparse, selective, and robust. In: IEEE conference on computer vision and pattern recognition (CVPR), pp 2892–2900
18. Masi I, Rawls S, Medioni G, Natarajan P (2016) Pose-aware face recognition in the wild. In: IEEE conference on computer vision and pattern recognition (CVPR), pp 4838–4846
19. Wen Y, Zhang K, Li Z, Qiao Y (2016) A discriminative feature learning approach for deep face recognition. In: European conference on computer vision (ECCV), pp 499–515
20. Graves A, Wayne G, Danihelka I (2014) Neural turing machines. CoRR. arXiv:1410.5401
21. Vinyals O, Bengio S, Kudlur M (2016) Order matters: sequence to sequence for sets. In: International conference on learning representation
22. Klare BF, Klein B, Taborsky E, Blanton A, Cheney J, Allen K, Grother P, Mah A, Burge M, Jain AK (2015) Pushing the frontiers of unconstrained face detection and recognition: Iarpa janus benchmark a. In: IEEE conference on computer vision and pattern recognition (CVPR), pp 1931–1939
23. Lee K-C, Ho J, Yang, M-H, Kriegman D (2003) Video-based face recognition using probabilistic appearance manifolds. In: IEEE conference on computer vision and pattern recognition (CVPR)
24. Shakhnarovich G, Fisher JW, Darrell T (2002) Face recognition from long-term observations. In: European conference on computer vision (ECCV), pp 851–865
25. Zhou SK, Rama C (2005) Beyond one still image: face recognition from multiple still images or a video sequence, pp 547–575
26. Arandjelovic O, Shakhnarovich G, Fisher J, Cipolla R, Darrell T (2005) Face recognition with image sets using manifold density divergence. In: IEEE conference on computer vision and pattern recognition (CVPR)
27. Cevikalp H, Triggs B (2010) Face recognition based on image sets. In: IEEE conference on computer vision and pattern recognition (CVPR), pp 2567–2573
28. Kim T-K, Arandjelović O, Cipolla R (2007) Boosted manifold principal angles for image set-based recognition. Pattern Recognit 40(9):2475–2484
29. Wang R, Shan S, Chen X, Gao W (2008) Manifold-manifold distance with application to face recognition based on image set. In: IEEE Conference on computer vision and pattern recognition (CVPR), pp 1–8
30. Turaga P, Veeraraghavan A, Srivastava A, Chellappa R (2011) Statistical computations on grassmann and stiefel manifolds for image and video-based recognition. IEEE Trans Pattern Anal Mach Intell 33(11):2273–2286
31. Chen J-C, Ranjan R, Kumar A, Chen C-H, Patel V, Chellappa R (2015) An end-to-end system for unconstrained face verification with deep convolutional neural networks. In: IEEE international conference on computer vision workshops, pp 118–126
32. Chowdhury AR, Lin T-Y, Maji S, Learned-Miller E (2016) One-to-many face recognition with bilinear CNNS. In: IEEE winter conference on applications of computer vision (WACV)

33. Masi I, Trãn ATA, Leksut JT, Hassner T, Medioni G (2016) Do we really need to collect millions of faces for effective face recognition? In: European conference on computer vision (ECCV)
34. AbdAlmageed W, Wu Y, Rawls S, Harel S, Hassner T, Masi I, Choi J, Lekust J, Kim J, Natarajan P, et al (2016) Face recognition using deep multi-pose representations. In: IEEE winter conference on applications of computer vision (WACV)
35. Szegedy C, Liu W, Jia Y, Sermanet P, Reed S, Anguelov D, Erhan D, Vanhoucke V, Rabinovich A (2015) Going deeper with convolutions. In: IEEE conference on computer vision and pattern recognition (CVPR), pp 1–9
36. Ioffe S, Szegedy C (2015) Batch normalization: accelerating deep network training by reducing internal covariate shift. arXiv:1502.03167
37. Sukhbaatar S, Weston J, Fergus R, et al (2015) End-to-end memory networks. In: Advances in neural information processing systems (NIPS), pp 2440–2448
38. Chopra S, Hadsell R, LeCun Y (2005) 'Learning a similarity metric discriminatively, with application to face verification. In: IEEE conference on computer vision and pattern recognition (CVPR), vol 1, pp 539–546
39. Hadsell R, Chopra S, LeCun Y (2006) Dimensionality reduction by learning an invariant mapping. In: IEEE conference on computer vision and pattern recognition (CVPR), vol 2, pp 1735–1742
40. Chen D, Ren S, Wei Y, Cao X, Sun J (2014) Joint cascade face detection and alignment. In: European conference on computer vision (ECCV), pp 109–122
41. Ren S, Cao X, Wei Y, Sun J (2014) Face alignment at 3000 fps via regressing local binary features. In: IEEE conference on computer vision and pattern recognition (CVPR), pp 1685–1692
42. Tieleman T, Hinton G (2012) RMSProp: divide the gradient by a running average of its recent magnitude. Tech Rep
43. Chen D, Hua G, Wen F, Sun J (2016) Supervised transformer network for efficient face detection. In: European conference on computer vision (ECCV), pp 122–138
44. Sankaranarayanan S, Alavi A, Castillo C, Chellappa R (2016) Triplet probabilistic embedding for face verification and clustering. arXiv:1604.05417
45. Wang D, Otto C, Jain AK (2015) Face search at scale: 80 million gallery. arXiv:1507.07242
46. Chen J-C, Patel VM, Chellappa R (2016) Unconstrained face verification using deep cnn features. In: IEEE winter conference on applications of computer vision (WACV)
47. Grother P, Ngan M (2014) Face recognition vendor test (FRVT) performance of face identification algorithms. In: Technical report NIST IR 8009. National Institute of Standards and Technology
48. Hu J, Lu J, Yuan J, Tan Y-P (2014) Large margin multi-metric learning for face and kinship verification in the wild. In: Asian conference on computer vision (ACCV), pp 252–267

Chapter 11
Thermal-to-Visible Face Synthesis and Recognition

Xing Di, He Zhang, and Vishal M. Patel

Face is one of the most widely used biometrics. One key advantage of using faces as a biometric is that they do not require the cooperation of the test subject. Various face recognition (FR) systems have been developed over the last two decades. Recent advances in machine learning and computer vision methods have provided robust systems that achieve significant gains in performance of face recognition systems [5, 19]. Deep learning methods, enabled by the vast improvements in processing hardware coupled with the ubiquity of face data and algorithmic development, have led to significant improvements in face recognition accuracy, particularly in unconstrained scenarios [4, 5, 19, 20, 27]. Also, largely driven by social network companies, progress in face recognition research, development, and deployment have focused on faces collected in visible regimes of the electromagnetic spectrum.

Thermal imaging has been proposed for night-time and low-light face recognition when external illumination is not feasible due to various collection considerations. The infrared spectrum can be divided into a reflection-dominated region consisting of the near-infrared (NIR) and shortwave-infrared (SWIR) bands, and an emission-dominated thermal region consisting of the midwave infrared (MWIR) and longwave infrared (LWIR) bands [23]. Thermal face images, while having a strong signature at night time, are not carefully maintained in biometric-enabled watch lists and so must be compared with visible-light face images to enable face recognition in low lighting conditions. This introduces the additional complexity of performing cross-domain matching (see Fig. 11.1).

X. Di · V. M. Patel (✉)
Department of Electrical and Computer Engineering, Johns Hopkins University, Barton 211 3400 N. Charles St, Baltimore, MD 21218, USA
e-mail: vpatel36@jhu.edu

H. Zhang
Adobe, CA, USA

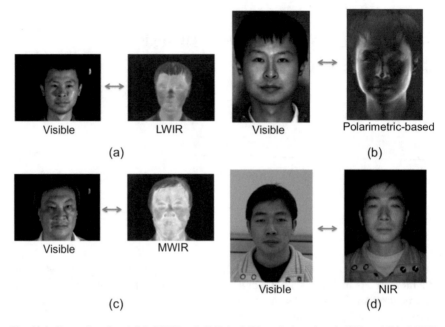

Visible   LWIR   Visible   Polarimetric-based

(a)   (b)

Visible   MWIR

Visible   NIR

(c)   (d)

Fig. 11.1 Examples of **a** visible-LWIR pair [22], **b** visible-polarimetric pair [28], **c** visible-MWIR pair [22], and **d** visible-NIR pair [22]

It is well known that deep learning methods are tuned to perform well on data of a particular type on which they are trained and often lose effectiveness when presented with data from a different domain. Domain adaptation and transfer learning methods have been proposed to bridge this gap between data from different source and target domains [18]. Transfer learning usually involves fine-tuning a pre-trained neural network model (that is trained on source domain data) on target domain data and this requires availability of sufficient samples from target domain. For applications such as thermal-to-visible face matching, there are limited databases available with corresponding visible and thermal face imagery and, as discussed earlier, a large modality and performance gap exists between the two domains. Furthermore, with limited training data, training a Siamese-like network for cross-domain face verification may result in over-fitting to the limited training samples. Hence, this kind of cross-domain verification network may not be able to generalize well to testing sets. In this chapter, we focus on an alternative solution for cross-domain verification. Rather than directly learning a domain adaptive network, one can leverage a network to synthesize visible faces first from thermal faces and then use a pre-trained face recognition network for face verification. We provide a review of such methods [7, 8, 26, 30, 31] in this chapter.

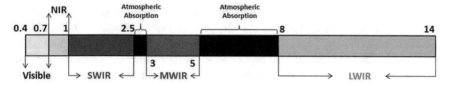

Fig. 11.2 Electromagnetic spectrum map: face recognition research challenges across the band. The illustrated wavelengths are in μm

11.1 The Infrared Spectrum

As discussed earlier, most face recognition systems depend on the usage of face images captured in the visible range of the electromagnetic spectrum, i.e., 380–750 *nm*. However, in real-world scenarios (military and law enforcement), we deal with harsh environmental conditions characterized by unfavorable lighting and pronounced shadows. Such an example is a night time environment [2], where human recognition based solely on various detectors (see Fig. 11.2).

Specifically, the infrared (IR) spectrum is comprised of the active IR and the thermal (passive) IR band. The active IR band (0.7−2.5 μm) is divided into the NIR (near-infrared) and the SWIR (shortwave-infrared) spectrum. The SWIR has a longer wavelength range than NIR and is more tolerant to low levels of obscurants like fog and smoke. Differences in appearance between images sensed in the visible and the active IR bands are due to the properties of the object being imaged. The passive IR band is further divided into the Midwave (MWIR) and the Longwave Infrared (LWIR) bands. MWIR ranges from 3 to 5 μm, while LWIR ranges from 7 to 14 μm. Both MWIR and LWIR cameras can sense temperature variations across the face at a distance, and produce thermograms in the form of 2D images. However, while both pertain to the thermal spectrum, they reveal different image characteristics of the facial skin. The difference between MWIR and LWIR is that MWIR has both reflective and emissive properties, whereas LWIR consists primarily of emitted radiation. The importance of FR outside the visible spectrum has been recently discussed in our previous work in [1, 3].

11.2 GAN-Based Synthesis of Visible Faces From Thermal Faces

Thermal-to-visible synthesis is an important step of a cross-modal face matching system. The large modality gap caused by differences in the physics of image acquisition and formation processes, the measured visible face signatures are very different from that of the thermal face signatures. This in turn makes the cross-domain face recognition challenging. Various methods have been developed in the literature for bridging this gap.

One possible solution to this problem is to improve the cross-domain face recognition performance by designing novel neural network architectures that minimize the distribution change between the domains. However, it is still of great importance to guarantee that the human examiners can identify whether the given thermal and the visible image share the same identity or not. Consider the thermal face images shown in the even columns of Fig. 11.1. The corresponding visible images are also shown in the odd columns of Fig. 11.1. As can be seen from these images, it is extremely difficult for either human examiners or existing face recognition systems to determine whether these images share the same identity. Hence, methods that can automatically generate high-quality visible images from their corresponding thermal images are needed.

11.2.1 Generative Adversarial Networks (GANs)

Generative Adversarial Networks were proposed by Goodfellow et al. in [11] to synthesize realistic images by effectively learning the distribution of training images. The authors adopted a game theoretic min-max optimization framework to train two models: a generative model, G, and a discriminative model, D. The success of GANs in synthesizing realistic images has led to researchers exploring the adversarial loss for numerous computer vision and image processing applications such as style transfer, image in-painting, image-to-image translation, image super-resolution, and image restoration. Inspired by the success of these methods, many approaches have been proposed that make use of the adversarial loss to learn the distribution of visible face images for their accurate estimation.

In order to learn a good generator G so as to fool the learned discriminator D and to make the discriminator D good enough to distinguish synthesized visible image from real ground truth, the method alternatively updates G and D following the structure proposed in [13]. Given an input polarimetric image X, conditional GAN aims to learn a mapping function to generate output image Y by solving the following optimization problem:

$$\min_{G} \max_{D} \; \mathbb{E}_{X \sim p_{data(X)}}[\log(1 - D(X, G(X)))] + \mathbb{E}_{X \sim p_{data(X,Y)}}[\log D(X, Y)].$$

$$(11.1)$$

11.2.2 GAN-Based Synthesis Network

In a recent work [30], a unified GAN-based synthesis network was proposed that can directly learn an end-to-end mapping between the thermal image and its corresponding visible image. The proposed network contains an encoder-decoder structure, where the learned visible features can be regarded as the output of the encoder part

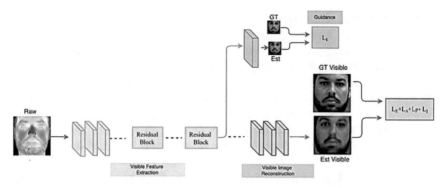

Fig. 11.3 An overview of the GAN-based image synthesis method proposed in [30]. It consists of three modules. **a** Visible feature extraction module, **b** guidance sub-network, and **c** visible image reconstruction module. Firstly, the visible feature is extracted from the raw thermal image. Then, to make sure that the learned feature can better reconstruct the visible image, a guidance sub-network is involved into the optimization. Finally, the guided feature is used to reconstruct the photorealistic visible image using the combination of different losses

and input for the decoder part. To guarantee the reconstructability of the encoded features and to make sure that the leaned features contain geometrically meaningful information, a guidance sub-network was introduced at the end of the visible feature extraction part. The overall network architecture is shown in Fig. 11.3.

To overcome the side effect of blurry results brought by the traditional Euclidean loss (L_E loss) and to discriminate the generated visible face images from their corresponding ground truth, a GAN structure is deployed. Even though GAN's structure can generate more reasonable results compared to the tradition L_E loss, it has been shown that the results generated by the traditional GAN contain undesirable facial artifacts, resulting in a less photorealistic image [30]. To address this issue and meanwhile generate visually pleasing results, the perceptual loss is included in the network, where the perceptual loss is evaluated on the pre-trained VGG-16 model.

As the ultimate goal of the synthesis method is to guarantee that human examiners can identify the person given his/her synthesized face images, it is also important to involve the discriminative information into consideration. Similar to the perceptual loss, an identity preserving loss is also proposed that is evaluated on a certain layer of the fine-tuned VGG-thermal model.

An encoder-decoder structure is adapted as the basis in the generator part. Basically, the generator can be divided into two parts. Firstly, a set of convolutional layers with stride 2 combined with a set of residual blocks are regarded as visible feature estimation part. Specifically, the residual blocks are composed of two convolutional layers with 3×3 kernels and 64 feature maps followed by batch-normalization layers and Parametric Rectified Linear Unit (PReLU) as the activation function. Then, a set of transpose convolutional layers with stride 2 are denoted as the visible image reconstruction procedure. To make sure that the transformed features contain enough semantic information, a guided sub-part is enforced in the network. Meanwhile, to make the generated visible face images indistinguishable from the ground truth visible face images, a CNN-based differentiable discriminator is used as a guidance to guide the generator in generating better visual results. For the discriminator, Patch-GANs [13] are used to discriminate whether the given images are real or fake.

The network is trained using the following loss functions: the Euclidean $L_{E(G)}$ loss enforced on the recovered visible image, the L_E loss enforced on the guidance part, the adversarial loss to guarantee more sharp results, the perceptual loss to preserve more photorealistic details, and the identity loss to preserve more distinguishable information for the outputs. The overall loss function is defined as follows:

$$L_{\text{all}} = L_E + L_{E(G)} + \lambda_A L_A + \lambda_P L_P + \lambda_I L_I, \tag{11.2}$$

where L_E denotes the Euclidean loss, $L_{E(G)}$ denotes the Euclidean loss on the guidance sub-network, L_A represents the adversarial loss, L_P indicates the perceptual loss, and L_I is the identity loss. Here, λ_A, λ_P, and λ_I are the corresponding weights. The L_E and the adversarial losses are defined as follows:

$$L_E, L_{E(G)} = \frac{1}{WH} \sum_{w=1}^{W} \sum_{h=1}^{H} \|\phi_G(I)^{w,h} - Y_t^{w,h}\|_2, \tag{11.3}$$

$$L_A = -\log(\phi_D(\phi_G(I)^{c,w,h}), \tag{11.4}$$

where I is the input thermal image, Y_t is the ground truth visible image, $W \times H$ is the dimension of the input image, ϕ_G is the generator sub-network G, and ϕ_D is the discriminator sub-network D. As the perceptual loss and the identity losses are evaluated on a certain layer of the given CNN model, both can be defined as follows:

$$L_{P,I} = \frac{1}{C_i W_i H_i} \sum_{c=1}^{C_i} \sum_{w=1}^{W_i} \sum_{h=1}^{H_i} \|V(\phi_E(I))^{c,w,h} - V, (Y_t)^{c,w,h}\|_2, \tag{11.5}$$

where Y_t is the ground truth visible image, ϕ_E is the proposed generator, V represents a non-linear CNN transformation, and C_i, W_i, H_i are the dimensions of a certain high-level layer V, which differs for perceptual and identity losses.

Multiple experiments were evaluated on different experimental protocols using the ARL polarimetric thermal dataset [30] and it was demonstrated that this method is able to produce good quality visible images from input thermal images. A polarimetric thermal image consists of three Stokes images: S_0, S_1, S_2, and degree-of-linear-polarization (DoLP) image, where S_0 indicates the conventional total intensity thermal image, S_1 captures the horizontal and vertical polarization-state information, S_2 captures the diagonal polarization-state information, and DoLP describes the portion of an electromagnetic wave that is linearly polarized [12]. These Stokes images along with the visible and the polarimetric images corresponding to a subject in the ARL dataset [12] are shown in Fig. 11.4. A polarimetric signature/image in this work

Fig. 11.4 Sample Stokes as well as polarimetric and visible images corresponding to a subject in the ARL dataset [12]

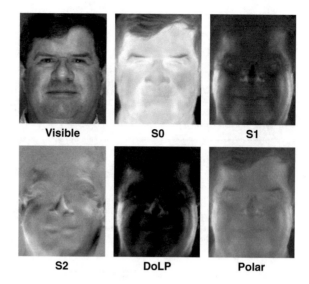

was defined as consisting of three Stokes images (S_0, S_1, S_2) as its three channels, analogous to the RGB channels in visible color imagery.

11.3 Synthesis of High-Quality Visible Faces From Polarimetric Thermal Faces Using GANs

Though the GAN-based method described in previous section is able to effectively synthesize photorealistic visible face images from polarimetric thermal images, the results are still far from optimal [30]. One possible reason is that [30] concatenate the Stokes images into a single-input sample without any additional attempts to capture multi-channel information inherently present in the different Stokes (modalities) images from the thermal infrared band. In order to efficiently leverage the multimodal information provided by the polarimetric thermal images, a novel GAN-based multistream feature-level fusion method for synthesizing visible images from thermal images was recently proposed in [31].

In particular, the GAN-based method proposed in [31] consists of a generator, a discriminator sub-network, and a deep guided sub-network (see Fig. 11.5). The generator is composed of a multi-stream encoder-decoder network based on dense-residual blocks, the discriminator is designed to capture features at multiple scales for discrimination and the deep guided subnet aims to guarantee that the encoded features contain geometric and texture information to recover the visible face. To further enhance the network's performance, it is guided by the perceptual loss and an identity preserving loss in addition to the adversarial loss. It was shown that once the face images are synthesized, any off-the-shelf face recognition and verification

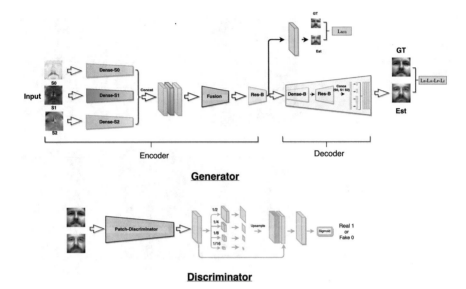

Fig. 11.5 An overview of the GAN-based multi-stream encoder-decoder network proposed in [31]. The generator contains a multi-stream feature-level fusion encoder-decoder network. In addition, a deep-guided subnet is stacked at the end of the encoding part. The discriminator is composed of a multi-scale patch-discriminator structure

networks trained on the visible-only face data can be used for matching. In particular, features from the second last fully connected layer of the VGG-face network [17] are extracted and the cosine distance is used to calculate the scores.

11.4 Thermal-to-Visible Face Verification Via Attribute-Preserved Synthesis

In a recent work [8], a different approach was proposed to the problem of thermal to visible matching. Figure 11.6 compares the traditional cross-modal verification problem with that of the new attribute-preserved cross-modal verification approach proposed in [8]. Given a visible and thermal pair, the traditional approach first extracts some features from these images and then verifies the identity based on the extracted features [15] (see Fig. 11.6b). In contrast, a novel framework was proposed in which the authors make use of the attributes extracted from the visible image to synthesize the attribute-preserved visible image from the input thermal image for matching (see Fig. 11.6b). In particular, a pre-trained VGG-Face model [17] is used to extract the attributes from the visible image. Then, a novel Attribute-Preserved Generative Adversarial Network (AP-GAN) is proposed to synthesize the visible image from

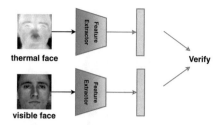

(a) Traditional Heterogeneous Face Verification

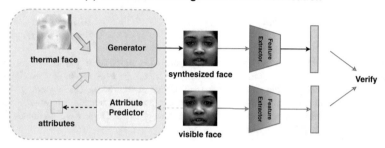

(b) Proposed Heterogeneous Face Verification

Fig. 11.6 a Traditional heterogeneous face verification approaches use the features directly extracted from different modalities for verification [30]. **b** The heterogeneous face verification approach proposed in [8] uses a thermal face and semantic attributes to synthesize a visible face. Finally, deep features extracted from the synthesized and visible faces are used for verification

the thermal image guided by the extracted attributes. Finally, a deep network is used to extract features from the synthesized and the input visible images for verification.

The AP-GAN model proposed in [8] was inspired by the recent image generation from attributes/text works [6, 21, 30]. The AP-GAN consists of two parts: (i) a multimodal compact bilinear (MCB) pooling-based generator [9, 10] and (ii) a triplet-pair discriminator. The generator fuses the extracted attribute vector with the image feature vector in the latent space. On the other hand, the discriminator uses triplet pairs (real image/true attributes, fake image/true attributes, fake image/wrong attributes) to not only discriminate between real and fake images but also to discriminate between the image and the attributes. In order to generate high-quality and attribute-preserved images, the generator is optimized by a multi-purpose objective function consisting of adversarial loss [11], L_1 loss, perceptual loss [14], identity loss [30], and attribute-preserving loss. The entire AP-GAN framework is shown in Fig. 11.7. Extensive experiments were conducted using the ARL Polarimetric face dataset and it was shown that this method achieves significant synthesis and matching improvements over the state-of-the-art methods.

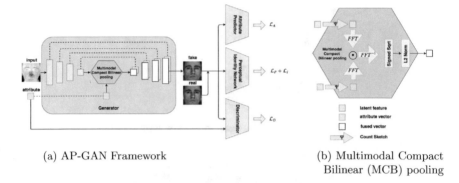

(a) AP-GAN Framework   (b) Multimodal Compact
                                            Bilinear (MCB) pooling

Fig. 11.7 a A U-net-based generator with MCB pooling is proposed in [8] to fuse the semantic attribute information with the image feature in the latent space. A triplet pair is adopted for
the discriminator in order to discriminate fake/real images as well as the corresponding semantic
attributes. In order to generate high-quality and attribute-preserving images, a multi-purpose loss
is optimized for training the network. **b** The architecture of MCB. Here, FFT indicates the fast
Fourier transform and FFT^{-1} indicates the inverse FFT

11.5 Self-attention Guided Synthesis

In [7], a new approach to the problem of thermal-to-visible matching was proposed
by exploring the complementary information of different modalities. Figure 11.8
gives an overview of this approach [7]. Given a thermal-visible pair $(\mathbf{x}_t, \mathbf{x}_v)$, these
images are first transformed into their spectrum counterparts using two trained gener-
ators as $\hat{\mathbf{x}}_v = G_{t \to v}(\mathbf{x}_t)$, $\hat{\mathbf{x}}_t = G_{v \to t}(\mathbf{x}_v)$. Then a feature extractor network $Feat$, in
particular, the VGG-Face model [17], is used to extract features $f_{\mathbf{x}_t} = Feat(\mathbf{x}_t)$,
$f_{\hat{\mathbf{x}}_v} = Feat(\hat{\mathbf{x}}_v)$, $f_{\mathbf{x}_v} = Feat(\mathbf{x}_v)$, and $f_{\hat{\mathbf{x}}_t} = Feat(\hat{\mathbf{x}}_t)$. These features are then
fused to generate the gallery template $g_{\mathbf{x}_v} = (f_{\mathbf{x}_v} + f_{\hat{\mathbf{x}}_t})/2$ and the probe template
$g_{\mathbf{x}_t} = (f_{\mathbf{x}_t} + f_{\hat{\mathbf{x}}_v})/2$. Finally, the cosine similarity score between these feature templates is calculated for verification.

Note that CycleGAN-based networks [32] can be used to train these generators. However, experiments in [7] have shown that CycleGAN often fails to capture
the geometric or structural patterns around the eye and mouth regions of the face.
One possible reason could be that the network relies heavily on convolutions to
model the dependencies across different image regions. The long-range dependencies are not well captured by the local receptive field of convolutional layers [29].
For improvement, the self-attention techniques were adopted from SAGAN [29].
The self-attention module was applied right before the last convolutional layer of the
generator and the discriminator. Given the feature maps, this module learns the attention maps by itself with a softmax function and then the learned attention maps are
multiplied with the feature maps to output the self-attention guided feature maps. In
addition, the generator is optimized by an objective function consisting of the adversarial loss [11], L_1 loss, perceptual loss [14], identity loss [30], and cycle-consistency
loss [32]. The entire synthesis framework is shown in Fig. 11.9.

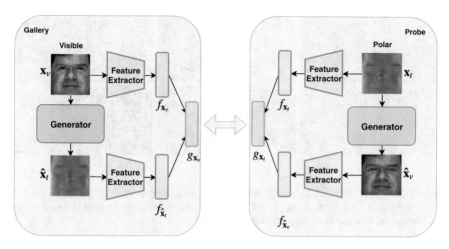

Fig. 11.8 An overview of the cross-modal face verification method proposed in [7]. Given a visible gallery image $\mathbf{x}_v$, a generator network is used to synthesize the corresponding thermal image $\hat{\mathbf{x}}_t$. Similarly, given a polarimetric thermal probe image $\mathbf{x}_t$, a different generator network is used to synthesize the corresponding visible image $\hat{\mathbf{x}}_v$. Pre-trained CNNs are used to extract features from the original and the synthesized images. These features are then fused to generate the gallery template $g_{\mathbf{x}_v}$ and the probe template $g_{\mathbf{x}_t}$. Finally, the cosine similarity score between these feature templates is calculated for verification

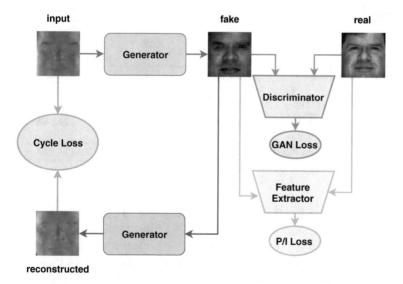

Fig. 11.9 Self-attention guided synthesis of visible images from polarimetric thermal input [7]. In order to minimize the domain gap between different modalities, the input thermal/visible images are directly mapped into the visible/thermal modality. In order to obtain the image-level style, the pixel GAN loss (blue) and cycle consistency loss (green) are introduced. The feature-level semantic information is captured by the identity and perceptual losses (yellow). Similar architecture was also used for synthesizing thermal images from visible images

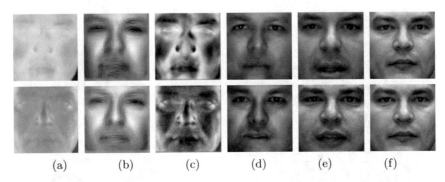

(a)   (b)   (c)   (d)   (e)   (f)

Fig. 11.10 The synthesized samples from different methods: Riggan et al. [25], Mahendran et al. [16], Zhang et al. [30], Di et al. [8], ground truth. The first row results correspond to the S0 image, and the second row results correspond to the Polar image

Table 11.1 Protocol I Verification performance comparison among the thermal-to-visible synthesis and matching methods for both polarimetric thermal (Polar) and conventional thermal (S0) cases on the ARL dataset

Method	AUC (Polar)	AUC (S0)	EER (Polar)	EER (S0)
Raw (%)	50.35	58.64	48.96	43.96
Mahendran et al. [16] (%)	58.38	59.25	44.56	43.56
Riggan et al. [25] (%)	75.83	68.52	33.20	34.36
GAN-VFS [30] (%)	79.90	79.30	25.17	27.34
Riggan et al. [24] (%)	85.42	82.49	21.46	26.25
AP-GAN [8]	88.93% ± 1.54%	84.16% ± 1.54%	19.02% ± 1.69%	23.90% ± 1.52%
Multi-stream GAN [31]	**96.03%**	85.74%	**11.78%**	23.18%
Di et al. [7]	93.68% ± 0.97%	**89.20% ± 1.56%**	13.46% ± 1.92%	**18.77% ± 1.36%**

Figure 11.10 compares the synthesis performance of different methods on the ARL polarimetric thermal dataset [12]. As can be seen from this figure, in general, CNN-based methods are able to provide good quality visible faces from thermal faces. In particular, the attribute-based synthesis method [8] provides the best quality results. Table 11.1 compares the face verification performance corresponding to different synthesis methods on the ARL polarimetric thermal dataset. As can be seen from this method, multi-stream GAN-based synthesis and the recently proposed self-attention guided synthesis methods produce the best cross-modal face verification results on the ARL polarimetric thermal dataset on Protocol I [7].

11.6 Conclusion

In this chapter, we provided a review of recent deep CNN-based thermal-to-visible face synthesis methods. In particular, GAN-based methods were reviewed and it was shown that they can provide high-quality synthesis and matching performance on different thermal modalities.

References

1. Bourlai T, Hornak L (2016) Face recognition outside the visible spectrum. Image Vis Comput 55(1):14–17
2. Bourlai T, Kalka N, Cao D, Decann B, Jafri Z, Nicolo F, Whitelam C, Zuo J, Adjeroh D, Cukic B, Dawson J, Hornak L, Ross A, Schmid NA (2011) Ascertaining human identity in night environments. Distrib Video Sens Netw 451–467
3. Bourlai, T., Ross, A., Chen, C. & Hornak, L. (2012), A study on using middle-wave infrared images for face recognition, *in* 'SPIE Biometric Technology for Human Identification IX'
4. Chen JC, Patel VM, Chellappa R (2016) Unconstrained face verification using deep CNN features. In: 2016 IEEE winter conference on applications of computer vision (WACV), pp 1–9
5. Chen J-C, Ranjan R, Sankaranarayanan S, Kumar A, Chen C-H, Patel VM, Castillo CD, Chellappa R (2017) Unconstrained still/video-based face verification with deep convolutional neural networks. Int J Comput Vis
6. Di X, Patel VM (2017) Face synthesis from visual attributes via sketch using conditional VAES and GANS. arXiv:1801.00077
7. Di X, Riggan BS, Hu S, Short NJ, Patel VM (2019) Polarimetric thermal to visible face verification via self-attention guided synthesis. In: IAPR international conference on biometrics (ICB)
8. Di X, Zhang H, Patel VM (2018) Polarimetric thermal to visible face verification via attribute preserved synthesis. In: IEEE international conference on biometrics: theory, applications, and systems (BTAS)
9. Fukui A, Park DH, Yang D, Rohrbach A, Darrell T, Rohrbach M (2016) Multimodal compact bilinear pooling for visual question answering and visual grounding. arXiv:1606.01847
10. Gao Y, Beijbom O, Zhang N, Darrell T (2016) Compact bilinear pooling, In: Proceedings of the IEEE conference on computer vision and pattern recognition, pp 317–326
11. Goodfellow I, Pouget-Abadie J, Mirza M, Xu B, Warde-Farley D, Ozair S, Courville A, Bengio Y (2014) Generative adversarial nets. In: NIPS, pp 2672–2680
12. Hu S, Short NJ, Riggan BS, Gordon C, Gurton KP, Thielke M, Gurram P, Chan AL (2016) A polarimetric thermal database for face recognition research. In: Proceedings of the IEEE conference on computer vision and pattern recognition workshops, pp 119–126
13. Isola P, Zhu J, Zhou T, Efros AA (2017) Image-to-image translation with conditional adversarial networks. In: 2017 IEEE conference on computer vision and pattern recognition (CVPR), pp 5967–5976
14. Johnson J, Alahi A, Fei-Fei L (2016) Perceptual losses for real-time style transfer and super-resolution. In: European conference on computer vision. Springer, pp 694–711
15. Klare B, Jain AK (2010) Heterogeneous face recognition: matching NIR to visible light images. In: 2010 20th international conference on pattern recognition (ICPR). IEEE, pp 1513–1516
16. Mahendran A, Vedaldi A (2015) Understanding deep image representations by inverting them. In: IEEE conference on computer vision and pattern recognition
17. Parkhi OM, Vedaldi A, Zisserman A (2015) Deep face recognition. In: Proceedings of the British machine vision conference (BMVC)

18. Patel VM, Gopalan R, Li R, Chellappa R (2015) Visual domain adaptation: a survey of recent advances. IEEE Signal Process Mag 32(3):53–69
19. Ranjan R, Sankaranarayanan S, Bansal A, Bodla N, Chen JC, Patel VM, Castillo CD, Chellappa R (2018) Deep learning for understanding faces: machines may be just as good, or better, than humans. IEEE Signal Process Mag 35(1):66–83
20. Ranjan R, Sankaranarayanan S, Castillo CD, Chellappa R (2017) An all-in-one convolutional neural network for face analysis. In: 2017 12th IEEE international conference on automatic face gesture recognition (FG 2017), pp 17–24
21. Reed S, Akata Z, Yan X, Logeswaran L, Schiele B, Lee H (2016) Generative adversarial text to image synthesis. arXiv:1605.05396
22. Riggan BS, Reale C, Nasrabadi NM (2015) Coupled auto-associative neural networks for heterogeneous face recognition. IEEE Access 3:1620–1632
23. Riggan BS, Short NJ, Hu S (2016) Optimal feature learning and discriminative framework for polarimetric thermal to visible face recognition. In: 2016 IEEE winter conference on applications of computer vision (WACV), pp 1–7
24. Riggan BS, Short NJ, Hu S (2018) Thermal to visible synthesis of face images using multiple regions. In: IEEE winter conference on applications of computer vision (WACV)
25. Riggan BS, Short NJ, Hu S, Kwon H (2016) Estimation of visible spectrum faces from polarimetric thermal faces. In: 2016 IEEE 8th international conference on biometrics theory, applications and systems (BTAS). IEEE, pp 1–7
26. Riggan BS, Short NJ, Sarfraz MS, Hu S, Zhang H, Patel VM, Rasnayaka S, Li J, Sim T, Iranmanesh SM, Nasrabadi NM (2018) ICME grand challenge results on heterogeneous face recognition: polarimetric thermal-to-visible matching. In: 2018 IEEE international conference on multimedia expo workshops (ICMEW), pp 1–4
27. Sankaranarayanan S, Alavi A, Castillo CD, Chellappa R (2016) Triplet probabilistic embedding for face verification and clustering. In: 2016 IEEE 8th international conference on biometrics theory, applications and systems (BTAS), pp 1–8
28. Short N, Hu S, Gurram P, Gurton K, Chan A (2015) Improving cross-modal face recognition using polarimetric imaging. Opt Lett 40(6):882–885
29. Zhang H, Goodfellow I, Metaxas D, Odena A (2018) Self-attention generative adversarial networks. arXiv:1805.08318
30. Zhang H, Patel VM, Riggan B, Hu S (2018) Generative adversarial network-based synthesis of visible faces from polarimetric thermal faces. In: International joint conference on biometrics (IJCB 2017)
31. Zhang H, Riggan BS, Hu S, Short NJ, Patel VM (2019) Synthesis of high-quality visible faces from polarimetric thermal faces using generative adversarial networks. Int J Comput Vis 127(6):845–862
32. Zhu J, Park T, Isola P, Efros AA (2017) Unpaired image-to-image translation using cycle-consistent adversarial networks. In: 2017 IEEE international conference on computer vision (ICCV), pp 2242–2251

Chapter 12
Obstructing DeepFakes by Disrupting Face Detection and Facial Landmarks Extraction

Yuezun Li and Siwei Lyu

Abstract Recent years have seen fast development in synthesizing realistic human faces using AI technologies. AI-synthesized fake faces can be weaponized to cause negative personal and social impact. In this work, we develop technologies to defend individuals from becoming victims of recent AI-synthesized fake videos by sabotaging would-be training data. This is achieved by disrupting deep neural network (DNN)-based face detection and facial landmark extraction method with specially designed imperceptible adversarial perturbations to reduce the quality of the detected faces. We empirically show the effectiveness of our methods in disrupting state-of-the-art DNN-based face detectors and facial landmark extractors on several datasets.

12.1 Introduction

The recent advances in machine learning and the availability of vast volume of online personal images and videos have drastically improved the synthesis of highly realistic human faces in images [26, 27] and videos [8, 30, 67, 72]. While there are interesting and creative applications of the AI face synthesis systems, they can also be weaponized. Due to the strong association of faces to the identity of an individual, well-crafted AI-synthesized fake videos can create illusions of a person's presence and activities that do not occur in reality (see Fig. 12.1), which can lead to serious political, social, financial, and legal consequences [11]. The potential threats range from revenge pornographic videos of a victim whose face is synthesized and spliced

Y. Li (✉) · S. Lyu
University at Buffalo, State University of New York, Buffalo, USA
e-mail: yuezunli@buffalo.edu

S. Lyu
e-mail: siweilyu@buffalo.edu

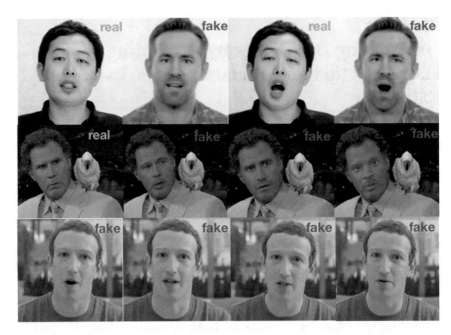

Fig. 12.1 Examples of AI-synthesized impersonation videos (DeepFakes). (top) Head puppetry entails synthesizing a video of a target person's head using a video of a source person's head, so the synthesized target appears to behave the same way as the source. (middle) Face swapping (DeepFake) involves generating a video of the target with the faces replaced by synthesized faces of the source while keeping the same facial expressions. (bottom) Lip syncing is to create a falsified video by only manipulating the lip region

in to realistic videos of state leaders seeming to make inflammatory comments they never actually made or a high-level executive commenting about her company's performance to influence the global stock market.

Foreseeing this threat, many forensic techniques aiming to detect AI-synthesized faces in images or videos have been proposed recently [1, 20, 36, 37, 47, 85]. However, given the speed and reach of the propagation of online media, even the currently best forensic techniques will largely operate in a postmortem fashion, applicable only after AI-synthesized fake face images or videos emerge. In this work, we aim to develop *proactive* approaches to protect individuals from becoming the victims of such attacks. The protection methods are complementary to the forensic tools, and our solution is to add specially designed patterns known as the *adversarial perturbations* that are imperceptible to human eyes but can result in detection failures.

The rationale of our method is the following. High-quality AI face synthesis models need large number of, typically in the range of thousands, sometimes even millions, training face images collected using automatic face detection methods, i.e., the *face sets*. Adversarial perturbations "pollute" a face set to have few actual faces and many non-faces with low or no utility as training data for AI face synthesis models. Since the face detectors can be used to crop out the faces and facial landmarks can be used to align the cropped faces to a standard coordinate for training, we focus our study on adversarial perturbations to deep neural network (DNN)-based face detectors, e.g., [25, 50, 66, 71, 74, 83, 87], and facial landmark extractors, e.g., [54, 65, 88]. Concretely, we first describe a new white-box adversarial perturbation generation method, assuming the knowledge of the structure and parameters of the underlying DNN models. We then discuss the feasibility of extending the white-box method to attack black-box models of face detectors and facial landmark extractors.

12.2   Background and Related Works

AI Face Synthesis Methods. Synthesizing realistic faces using algorithms has always been an important task in computer vision and computer graphics. Even with sophisticated computer graphics systems (e.g., 3D Studio Max and Maya) and high-resolution 3D surface models to render high-quality realistic human faces. However, the process is lengthy, costly, and technically demanding for ordinary users.

This has been significantly changed with the recent development of Generative Adversarial Networks (GANs) [4, 13, 18, 21, 24, 26, 28, 41, 63, 70], which lead to more realistic synthesized faces with considerable reduction in time and cost. For instance, fake videos generated with face swapping, commonly known as the DeepFakes, can be used to create realistic impersonation videos. Specifically, to make a DeepFake video, the faces of a *target* individual are replaced by the faces of a *donor* individual synthesized by a DNN model trained using the target and the donor's faces, retaining the target's facial expressions and head poses. Similarly, whole face or upper-body reenactment can be synthesized with AI algorithms such as Face2Face [72], ReenactGAN [77], and DeepPortrait [30], and even whole body can be reenacted in [9] more recently.

Face Detection. The first efficient and effective face detector [73] uses Haar-type features in a cascaded classifier based on AdaBoost. Subsequently, more robust, effective, and efficient face detectors are proposed in the literature based on various feature types such as LBP [52], SURF [34, 35], and DPMs [55]. Using the HOG feature [12], software package DLib represents the state of the art for the pre-DNN face detection methods.

Recently, DNN-based face detectors have become mainstream with their high performance and robustness with regard to variations in pose, expression, and occlusion. There has been a plethora of DNN-based face detectors, e.g., [17, 25, 33, 50, 56–58, 66, 71, 74, 80, 81, 83, 87]. Regardless the idiosyncrasies of different detectors, they all follow a similar work flow. First *face proposals*, which are potential candidate regions corresponding to faces, are identified. Using face proposals avoids the more expensive sliding window search. Each face proposal is then further classified by a trained classifier, known as the *backbone* network to determine if the proposal is a face. The prohibitive cost of searching optimal network structures and architectures makes the choice of the backbone network limited to two well-tested DNN models, namely, the VGG network [64] or the ResNet [22]. As reported on the leader board of face detection challenge on the WIDER dataset [82], the top performance is achieved from 23 DNN-based methods, all using VGG or ResNet as backbone networks, differing in the numerical values of the parameters and model variants (VGG16, VGG19 [64], ResNet50, or ResNet101 [22]).

Facial landmark extractors. The facial landmark extractors detect and locate key points of important facial parts such as the tips of the nose, eyes, eyebrows, mouth, and jaw outline. Earlier facial landmark extractors are based on simple machine learning methods such as the ensemble of regression trees (ERT) [29] as in the Dlib package [31]. The more recent ones are based on CNN models, which have achieved significant improved performance over the traditional methods, e.g., [6, 23, 54, 65, 76, 88]. The current CNN-based facial landmark extractors typically contain two stages of operations. In the first stage, a set of heat-maps (feature maps) are obtained to represent the spatial probability of each landmark. In the second stage, the final locations of facial landmarks are extracted based on the peaks of the heat-maps. In this work, we mainly focus on attacking the CNN-based facial landmark extractors because of their better performance.

Adversarial Perturbations. Adversarial perturbations are intentionally designed noises that are imperceptible to human observers, yet can seriously reduce the deep neural network performance if added to the input image. Many methods [2, 7, 14, 19, 32, 45, 48, 49, 53, 68, 86] have been proposed to impair image classifiers by adding adversarial perturbations on the entire image. Recently, there have been several works on adversarial perturbation generation for general object detectors [10, 16, 38, 44, 78]. Most of these works are for the white-box setting and generate adversarial perturbations for increasing mis-detection. Several recent works [40, 43, 51] have been proposed to use adversarial perturbations to disrupt object detectors and classifiers. Other works [60, 84] target GAN-based models under the white-box setting. Comparing to the existing methods, our method attacks the face detection step, which is the first step in most AI-based face synthesis system. To date, there are not many studies of the vulnerability of face detectors. Compared to the only existing method [5], which uses a GAN model [19] to increase mis-detection of VGG16-based Faster-RCNN face detector, our method optimizes specifically crafted loss function with respect to image, to increase both the mis-detection and false detection of different face detectors under different attack settings. Moreover, to date, there is no existing work to attack CNN-based facial landmark extractors using adversarial perturbations. Compared to the attack to image CNN-based classifiers, which aims to change the prediction of a single label, disturbing facial landmark extractors is more challenging as we need to simultaneously perturb the spatial probabilities of multiple facial landmarks to make the attack effective.

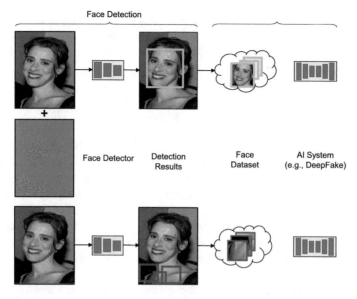

Fig. 12.2 Overview of the proposed method of disrupting AI face synthesis. Our aim is to use the adversarial perturbations (amplified by 30 for better visualization) to distract DNN-based face detectors, such that the quality of the obtained face set as training data to the AI face synthesis is reduced

12.3   Attacking Face Detectors

12.3.1   White-Box Adversarial Perturbation Generation

In this section, for the purpose of disrupting training of AI face synthesis models, we develop adversarial perturbation generation for face detectors. We start with a white-box adversarial perturbation generation (Sect. 12.3.1), where we assume access to the DNN model of the face detector. We then extend the white-box method to the more general case where the DNN model is not completely accessible, i.e., the gray-box (Sect. 12.3.2) and (Sect. 12.3.3). In Sect. ??, we introduce a new performance evaluation metric, data utility quality (DUQ), that is specifically designed to evaluate the effectiveness of adversarial perturbation generation in disrupting the training process of AI face synthesis models. Section ?? summarizes comparison of our method with existing works. We perform extensive experimental evaluations on three widely used benchmark datasets, i.e., WIDER [82], 300-W [62] and UMDFaces [3] to demonstrate the effectiveness of our method in disrupting state-of-the-art DNN-based face detectors (measured by DUQ) with low visual distortions (measured by SSIM), as well as its robustness with regard to JPEG compression, additive noise, blurring, and median filtering. Figure 12.2 illustrates the overview of attacking face detectors.

We use $[n]$ as a shorthand for $\{1, \ldots, n\}$, $|A|$ is the cardinality of set A, and A/B as the set of elements in A but not in B. We use $\mathbf{b}$ to represent a rectangular region (bounding box) in an image. The *intersection over union* (IoU) of two bounding boxes $\mathbf{b}$ and $\mathbf{b}'$ is computed as

$$\text{IoU}(\mathbf{b}, \mathbf{b}') = \frac{\text{area}(\mathbf{b} \cap \mathbf{b}')}{\text{area}(\mathbf{b} \cup \mathbf{b}')}.$$

IoU takes values in the range of $[0, 1]$, and it is one when $\mathbf{b} = \mathbf{b}'$ and zero when $\mathbf{b}$ and $\mathbf{b}'$ do not overlap. For image $\mathbf{I}$, we use $\mathbf{I}(\mathbf{b})$ to denote sub-image of $\mathbf{I}$ restricted to region $\mathbf{b}$. For a particular DNN-based face detector with backbone network $\mathbf{F}$, we denote $\mathbf{P}(\mathbf{I}) = \{\mathbf{b}_1^p, \mathbf{b}_2^p, \ldots, \mathbf{b}_n^p\}$ as the set of bounding boxes of n face proposals obtained on an input image $\mathbf{I}$. Each proposal $\mathbf{b}$ comes with a corresponding *prediction score* $\mathbf{F}(\mathbf{I}(\mathbf{b})) \in [0, 1]$ given by the backbone network. Proposals with prediction scores $\mathbf{F}(\mathbf{I}(\mathbf{b})) \geq \theta_d$, with θ_d as a threshold set internally in the face detector, are added to the set of detected faces on $\mathbf{I}$, $\mathbf{D}(\mathbf{I}) = \{\mathbf{b}_1^d, \mathbf{b}_2^d, \ldots, \mathbf{b}_m^d\}$. Usually, $\mathbf{D}(\mathbf{I}) \subseteq \mathbf{P}(\mathbf{I})$ and $m \ll n$.

The white-box adversarial perturbation generation for image $\mathbf{I}$ seeks a perturbed image $\mathbf{I}' = \mathbf{I} + \mathbf{z}$, such that when processed by the same DNN-based face detector with backbone network $\mathbf{F}$, $\mathbf{I}'$ has more mis-detections and false detections than $\mathbf{I}$. Specifically, this is achieved by finding $\mathbf{z}$ that reduces the prediction score of the original detections in $\mathbf{D}(\mathbf{I})$, while increasing the prediction score of proposals in $\mathbf{P}(\mathbf{I})/\mathbf{D}(\mathbf{I})$. To be specific, we consider set $\tilde{\mathbf{P}}(\mathbf{I}) = \{\mathbf{b} | \mathbf{b} \in \mathbf{P}(\mathbf{I})/\mathbf{D}(\mathbf{I}) \wedge \mathbf{F}(\mathbf{I}(\mathbf{b})) \in [\theta_p, \theta_d)\}$ for candidates of non-face regions. The threshold θ_p is to avoid adding adversarial perturbations to structure-less regions and reduce the number of non-face proposals to be considered. The overall problem is a constrained optimization problem with regards to $\mathbf{z}$, as

$$\max_{\|\mathbf{z}\|_2 \leq \epsilon} \sum_{\mathbf{b} \in \mathbf{D}(\mathbf{I})} \log(1 - \mathbf{F}((\mathbf{I} + \mathbf{z})(\mathbf{b}))) + \sum_{\mathbf{b} \in \tilde{\mathbf{P}}(\mathbf{I})} \log \mathbf{F}((\mathbf{I} + \mathbf{z})(\mathbf{b})). \quad (12.1)$$

In Eq. (12.1), we use a slightly abused notation $(\mathbf{I} + \mathbf{z})(\mathbf{b})$ to denote the image region $\mathbf{b}$ in the perturbed image. We use $L(\mathbf{z})$ to denote the objective function of Eq. (12.1). The first term in $L(\mathbf{z})$ corresponds to increasing mis-detection, as maximizing it leads to decreasing the prediction score of original detection. For our purpose of sabotaging training face set of AI face synthesis models, we also need to increase the false detection rate, which is given by the second term in $L(\mathbf{z})$. The constraint $\|\mathbf{z}\|_2 \leq \epsilon$ ensures that the adversarial perturbation does not introduce large visual distortion to the image, with ϵ being a parameter of the problem.

Solution to Eq. (12.1) is obtained with a projected gradient ascent algorithm. Starting with initial value of $\mathbf{z}$, which is chosen as samples from uniform distributions in $[-\epsilon/2, \epsilon/2]$, at the t-th iteration, we update the current estimation of $\mathbf{I}_t$ by first moving it along the direction of the gradient (or sub-gradient when the network involves non-differentiable activation functions such as ReLU or leaky ReLU) of $L(\mathbf{z})$, with a small step size $\gamma_t > 0$, as

$$\mathbf{z}_{t+1} = \mathbf{z}_t + \gamma_t \nabla L(\mathbf{z}_t). \tag{12.2}$$

The gradient can be computed using the chain rule as

$$\nabla L(\mathbf{z}) = -\sum_{\mathbf{b} \in \mathbf{D}(\mathbf{I})} \frac{1}{1 - \mathbf{F}((\mathbf{I} + \mathbf{z})(\mathbf{b}))} \frac{\partial F((\mathbf{I} + \mathbf{z})(\mathbf{b}))}{\partial \mathbf{z}}$$

$$+ \sum_{\mathbf{b} \in \tilde{\mathbf{P}}(\mathbf{I})} \frac{1}{\mathbf{F}((\mathbf{I} + \mathbf{z})(\mathbf{b}))} \frac{\partial F((\mathbf{I} + \mathbf{z})(\mathbf{b}))}{\partial \mathbf{z}}. \tag{12.3}$$

Derivative $\frac{\partial F((\mathbf{I} + \mathbf{z})(\mathbf{b}))}{\partial \mathbf{z}}$ is the gradient of the backbone network $\mathbf{F}$ with regard to its input, which can be computed with the backpropagation algorithm [61]. Step size γ_t is determined to ensure that the update satisfies the constraint as

$$\gamma_t = \underset{\gamma}{\text{argmax}} \left\{ \gamma \left| \begin{array}{l} \|\mathbf{z}_t + \gamma \nabla L(\mathbf{z}_t)\|_2 \le \epsilon \wedge \\ L(\mathbf{z}_t + \gamma \nabla L(\mathbf{z}_t)) > L(\mathbf{z}_t). \end{array} \right. \right\}. \tag{12.4}$$

The solution is obtained by a 1D line search procedure.

We can obtain adversarial perturbation by repeating Eqs.(12.2)-(12.4) until the algorithm converges to a local minimum of $L(\mathbf{z})$. However, face proposals extracted from the perturbed image, $\mathbf{I}'$, may be different with those extracted from the original image, $\mathbf{I}$, due to the adversarial perturbation. Therefore, simply running the iterative algorithm given by Eqs. (12.2)–(12.4) until convergence may not achieve the desired perturbation. To solve this problem, we use a technique known as *warm start* [32]. Specifically, instead of running the iterations given by Eqs. (12.2)–(12.4) until convergence, we run only one round of the update and obtain an intermediate image $\tilde{\mathbf{I}}$. Then we set $\mathbf{I} = \tilde{\mathbf{I}}$, and run the DNN-based face detector $\mathbf{F}$ on the updated $\mathbf{I}$ to initiate a new round of optimization using Eqs. (12.2)–(12.4). This procedure is repeated until either (1) the perturbation introduced in the image violates the constraint $\|\mathbf{z}\| \le \epsilon$, or (2) no detections in $\mathbf{D}$ can be identified in the perturbed image, i.e., all original detections in $\mathbf{I}$ are missing in $\mathbf{I}'$.

12.3.2 Gray-Box Adversarial Perturbation Generation

Using model gradient makes the search for adversarial perturbation efficient but it also entails a dependence on the details of the underlying DNN model. This limits the applicability of white-box attack. On the other hand, as mentioned previously in Sect. 12.2, current DNN-based face detectors only use variants of a few standard architectures, adversarial perturbation developed for one DNN-based face detector can be transferred to the other DNN-based face detectors using the same backbone network but with different parameters, when the generation method is less reliant on the exact numerical values of the model parameters. This can be achieved by randomizing the projected gradient ascent scheme in Eq. (12.2), as

$$\mathbf{z}_{t+1} = \mathbf{z}_t + \gamma_t (\nabla L(\mathbf{z}_t) + \mathbf{n}_t), \tag{12.5}$$

where $\mathbf{n}_t$ is a sample from i.i.d. zero-mean white Gaussian noise with a small standard deviation. This is equivalent to a randomization to the gradient, so the update in Eq. (12.5) does not follow the exact gradient obtained from a fixed set of parameters, which increases the robustness of the solution to different model parameters.

12.3.3 Black-Box Adversarial Perturbation Generation

The more challenging setting is when we have no knowledge about the backbone network structures or parameters, other than that the face detector is based on a DNN model. In this work, we define black-box attack to DNN-based face detectors as where we have no knowledge of the network structure or the parameters other than that it is based on a DNN model.

As mentioned previously, existing DNN-based face detectors use two basic types of backbone networks, and future generation of DNN-based face detectors is expected to follow the same trend. So, combining gray-box adversarial perturbations obtained from different DNN-based face detectors with known backbone networks is the basis for the black-box generation method. Specifically, denote $\mathbf{F}_k, k = 1, \ldots, K$, as different backbone networks for DNN-based face detectors, and $L_k(\mathbf{z})$ as the objective function for the artifacts of F_k as defined in Eq. (12.1), we find adversarial perturbation for image $\mathbf{I}$ by solving

$$\max_{\|\mathbf{z}\|_2 \leq \epsilon} \sum_{k=1}^{K} L_k(\mathbf{z}). \tag{12.6}$$

Equation (12.6) is then optimized using the algorithms described in Sects. 12.3.1 and 12.3.2, with warm start and randomly perturbed gradient update.

12.4 Attacking Facial Landmark Extractors

In this section, we describe a white-box method to obstruct the creation of DeepFakes based on disrupting the facial landmark extraction. The facial landmarks are key locations of important facial parts including tips and middle points of eyes, nose, mouth, eye brows as well as contours, see Fig. 12.3. Our method attacks the facial landmark extractors by adding adversarial perturbations [19, 69], which are image noises purposely designed to mislead DNN-based facial landmark extractors [6, 54, 65]. The overall procedure of our method is illustrated in Fig. 12.3.

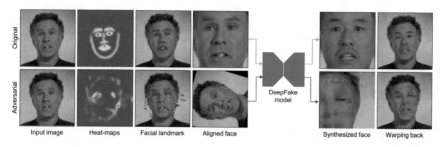

Fig. 12.3 The overview of our method on obstructing DeepFake generation by disrupting the facial landmark extraction. The top row shows the original DeepFake generation, and the bottom row corresponds to the disruption after facial landmarks are disrupted. The landmark extractor we use is FAN [6] and the "Heat-maps" is visualized by summing all heat-maps. Note that training of the DeepFake generation model is also affected by disrupted facial landmarks, but is not shown here

Let $\mathbf{F}$ denote the mapping function of a CNN-based landmark extractor of which the parameters we have access to, and $\{h_1, \ldots, h_k\} = \mathbf{F}(\mathbf{I})$ be the set of heat-maps of running $\mathbf{F}$ on input image $\mathbf{I}$. Our goal is to find an image $\mathbf{I}^{\mathrm{adv}}$, which can lead the prediction of landmark locations to a large error, while visually similar to as original image $\mathbf{I}$. The difference $\mathbf{I}^{\mathrm{adv}} - \mathbf{I}$ is the adversarial perturbation. We denote the heat-maps from the perturbed image $\{\hat{h}_1, \ldots, \hat{h}_k\} = \mathbf{F}(\mathbf{I}^{\mathrm{adv}})$.

We introduce a loss function that aims to enlarge the error between predicted heat-maps and original heat-maps, while constraining the pixel distortion in a certain budget as

$$\operatorname{argmin}_{\mathbf{I}^{\mathrm{adv}}} L(\mathbf{I}^{\mathrm{adv}}, \mathbf{I}) = \sum_{i=1}^{k} \frac{h_i^{\top} \hat{h}_i}{\|h_i\| \|\hat{h}_i\|}, \tag{12.7}$$
$$s.t. \ \|\mathbf{I}^{\mathrm{adv}} - \mathbf{I}\|_{\infty} \leq \epsilon,$$

where ϵ is a constant. We use cosine distance to measure the error as it can naturally normalize the loss range in $[-1, 1]$. Minimizing this loss function increases the error between predicted and original heat-maps, which will disrupt the facial landmark locations.

We use the MI-FGSM method [14] to optimize problem Eq. (12.7). Specifically, let t denote the iteration number and $\mathbf{I}_t^{\mathrm{adv}}$ denote the adversarial image obtained at iteration t. The start image is initialized as $\mathbf{I}_0^{\mathrm{adv}} = \mathbf{I}$. $\mathbf{I}_{t+1}^{\mathrm{adv}}$ is obtained by considering the momentum and gradient as

$$m_{t+1} = \lambda \cdot m_t + \frac{\nabla_{\mathbf{I}^{\mathrm{adv}}}(L(\mathbf{I}_t^{\mathrm{adv}}, \mathbf{I}))}{\|\nabla_{\mathbf{I}^{\mathrm{adv}}}(L(\mathbf{I}_t^{\mathrm{adv}}, \mathbf{I}))\|_1},$$
$$\mathbf{I}_{t+1}^{\mathrm{adv}} = \mathtt{clip}\{\mathbf{I}_t^{\mathrm{adv}} - \alpha \cdot \mathtt{sign}(m_{t+1})\}, \tag{12.8}$$

where $\nabla_{\mathbf{I}^{\mathrm{adv}}}(L(\mathbf{I}_t^{\mathrm{adv}}, \mathbf{I}))$ is the gradient of L with respect to the input image $\mathbf{I}_t^{\mathrm{adv}}$ at iteration t; m_t is the accumulated gradient and λ is the decay factor of momentum; α is the step size and $\mathtt{sign}$ returns the signs of each component of the input vector; and $\mathtt{clip}$ is the truncation function to ensure the pixel value of the resulting image is in

[0, 255]. The algorithm stops when the maximum number of iteration T is reached or the distortion threshold ϵ is reached.

12.5 Experiments

12.5.1 Attacking Face Detection

Datasets. We validate our method using three datasets.

- *sub-WIDER*: We construct a dataset of 909 images from the validation set of the WIDER dataset [82], which is one of the largest benchmarks for face detection. We exclude faces with small sizes, heavy occluded, or unusual orientations as they are not relevant for training AI face synthesis methods.
- *300-W*: This dataset has 600 images each containing a single face from the test set of the *300 Faces In-the-Wild Challenge* 300-W [62].[1]
- *sub-UMDFaces*: This dataset is constructed from 500 images randomly sampled from the UMDFaces dataset [3].

DNN-based Face Detectors. We consider several state-of-the-art DNN-based face detectors as the target of our experiments. For Faster-RCNN [59]-based face detectors, we consider two different backbone networks: VGG16 [64] and ResNet101 [22], which are denoted by *Fv16* and *Fr101*, respectively. For SSD [42]-based face detectors, we consider two state-of-the-art face detectors: PyramidBox [71] and SFD [87]. We use ResNet50-based PyramidBox and VGG16-based SFD, which are denoted by *Pr50* and *Sv16*, respectively. All of these face detectors are trained on complete WIDER training set.

Evaluation Metrics. Performance of face detection methods is usually evaluated using *average precision* (AP) [82]. However, for evaluating the effectiveness of adversarial perturbation, AP does not directly reflect the quality reduction of face detection over a dataset in terms of the number of mis-detections and false detections caused by the adversarial perturbation. To this end, we introduce a new metric, *data utility quality* (DUQ), to evaluate the utility of adversarial perturbation in contaminating a face set.

For an image $\mathbf{I}$ and its perturbed version $\mathbf{I}'$, we define set $\mathbf{D}_T(\mathbf{I})$ that contains all detections in the perturbed image $\mathbf{I}'$ that have significant overlaps with detections in the unperturbed image $\mathbf{D}(\mathbf{I})$, as $\mathbf{D}_T(\mathbf{I}) = \{\mathbf{b}|\mathbf{b} : \mathbf{b} \in \mathbf{D}(\mathbf{I}') \wedge \max_{\mathbf{b}' \in \mathbf{D}(\mathbf{I})} \mathrm{IoU}(\mathbf{b}, \mathbf{b}') > \rho\}$. In other words, these are detections in the original image that survive the adversarial perturbation. Correspondingly, set $\mathbf{D}(\mathbf{I}')/\mathbf{D}_T(\mathbf{I})$ includes detections that are not present in the unperturbed image, i.e., they are potentially false detections. Over a set of images, DUQ is defined as

---

[1] Since ground truth faces are not labeled in 300-W, we use the detection results of Dlib as the ground truth detection, which is also the protocol used in a compared work [5].

$$DUQ = \frac{\sum_{\mathbf{I}} |\mathbf{D}_T(\mathbf{I})| - \sum_{\mathbf{I}} |\mathbf{D}(\mathbf{I}')/\mathbf{D}_T(\mathbf{I})|}{\sum_{\mathbf{I}} |\mathbf{D}(\mathbf{I})|}. \tag{12.9}$$

DUQ takes value in the range $(-\infty, 1]$, where a value 1 indicates no effect of adversarial perturbation, i.e., no mis-detections and no false detections are generated. A lower DUQ corresponds to a lower purity of the detection face set when used as training data for AI face synthesis algorithms, in particular, a negative DUQ suggests a significant number of false detections that are not in the original image have been included in the face set.

We use DUQ as the major performance metric to evaluate the effectiveness of various adversarial perturbation generation schemes in lowering utility of resulting face sets as training data for AI face synthesis models. We also use SSIM [75] to assess the visual quality of images after adversarial perturbation are added. SSIM takes value in [0, 1], and higher value corresponds to better visual quality.

Implementation Details and Running Time. Some of the key constants are set empirically as follows. The upper bound of distortion (Mean Square Error) is set to $\epsilon = 5 \times 10^{-5}$. The step size in each iteration is set to $\gamma_t = \frac{30}{||\nabla L(\mathbf{z}_t)||_2}$. The confidence score threshold θ_d is set to 0.5. Since the amount of non-face proposals in set $\tilde{\mathbf{P}}(\mathbf{I})$ is large, we only pick the top 1000 ones in terms of confidence score into optimization for better efficiency. The IoU threshold ρ is set to 0.5 in DUQ.

Our adversarial perturbation generation method is implemented with unoptimized code `Python` and `pyTorch`. All experiments are performed on a machine which is equipped an Intel(R) Xeon(R) CPU E5-2620 v3 @ 2.40 GHz with 24 cores and 96 GB RAM. The GPU we use is a NVIDIA TITAN × (Pascal) with 12 GB memory. The average time for generating successful adversarial perturbation for an image is 4.69 s. The running time will be further improved if we optimize the code for practical deployment.

Baselines and Compared Methods. We evaluate the performance of our method and compared with three algorithms.

- **Random**: The is a simple baseline algorithm that adds random Gaussian noise to image.
- **SSOD**: This algorithm is based on a recent adversarial perturbation generator for general object detectors [78]. SSOD was originally trained and evaluated on Pascal VOC dataset [15], which has no label corresponding to human faces. To facilitate comparison, we use the original code of SSOD and refine it on the face datasets using the same parameter settings as in our experiments.
- **NNCO**: This algorithm is from [5]. NNCO uses a GAN model with a generator of adversarial perturbations targeting VGG16-based Faster-RCNN face detectors trained and tested both on dataset 300-W. We compare our method with NNCO by both applying them on VGG16 face detector.

12.5.1.1   White-Box Adversarial Perturbation Generation

Table 12.1 shows the performance of compared methods for Faster-RCNN- and SSD-based face detectors with different backbone networks before and after adversarial perturbations. We show both the effectiveness of the adversarial perturbation (DUQ) and image quality (SSIM). Note the ground truth of datasets is used as the correct detection set to compute DUQ of original and perturbed image for efficacy demonstration. Figure 12.5 provides four visual examples of the results of adversarial perturbation of face images.

As these results show, face detection is significantly affected after the adversarial perturbations generated with our method are added to these images. For example, DUQ of Fv16 is reduced to -4.89 from 0.81 on sub-WIDER dataset with a minor reduction of SSIM (0.02), and this shows that our method conceals true detections in the original unperturbed image and introduces a large number of false detections (also, see Fig. 12.5). In contrast, SSOD can only reduce DUQ of all face detectors on all datasets to around 0, as it only considers reducing true detections. On the other hand, adding random Gaussian noise has almost no effect on face detectors, suggesting that the dependency structure in the adversarial perturbation is essential. The performance for the SSD-based DNN face detectors is similar.

Furthermore, when compared with method NNCO, our method can reduce the DUQ of Fv16 from $0.81, 0.64, 0.55$ to $-4.89, -9.07, -7.52$ on three datasets, respectively, while NNCO only reduces DUQ to $0.17, 0.06, 0.12$. Moreover, compared with NNCO, our method achieves better image quality. Figure ?? shows a visual comparison of the perturbed image generated by our method and NNCO with enlarged area. Note the visible artifacts generated by NNCO, which are not present in the perturbed image generated with our method. This is corroborated by the quantitative results when comparing the SSIM scores of our method ($0.98, 0.98, 0.98$) and those of NNCO ($0.92, 0.91, 0.92$) on all datasets, respectively.

12.5.1.2   Black-Box Adversarial Perturbation Generation

Following the method described in Sect. 12.3.3, we generate adversarial perturbations using Eq. (12.6) for the combination of four face detectors for black-box attack of unknown face detectors. Specifically, we use the following three DNN-based face detectors as the unknown face detectors:

- Faster-RCNN-based face detector with backbone network ResNet50 (denoted as *Fr50*);
- SSD-based face detector SSH [50] with backbone networks VGG16 (denoted as *SSHv16*);
- SSD-based face detector SSH [50] with backbone networks ResNet50 (*SSHr50*).

All the experiments are conducted on sub-WIDER dataset. We denote our adversarial perturbation generation method targeting the ensemble of known face detectors as $Ours_{(Fv16+Fr101+Pr50+Sv16)}$ as described in Sect. 12.3.3. For comparison, we adapt

Table 12.1 Performance of adversarial perturbation generation against Faster-RCNN- and SSD-based face detectors on three datasets. Fv16 and Fr101 denote faster-RCNN-based face detector with backbone network VGG16 [64] and ResNet101 [22]. Pr50 denotes PyramidBox [71] with backbone network ResNet50 and Sv16 denotes SFD [87] with backbone network VGG16

Adversarial perturbation generation method		Sub-WIDER				300-W				Sub-UMDFaces			
		Fv16	Fr101	Pr50	Sv16	Fv16	Fr101	Pr50	Sv16	Fv16	Fr101	Pr50	Sv16
DUQ	Original	0.81	0.82	0.92	0.88	0.64	0.64	0.76	0.81	0.55	0.54	0.64	0.64
	Random	0.81	0.82	0.92	0.81	0.64	0.64	0.76	0.81	0.54	0.54	0.64	0.64
	NNCO [5]	0.17	–	–	–	0.06	–	–	–	0.12	–	–	–
	SSOD [78]	−0.06	−0.21	−0.74	−0.53	−0.02	−0.03	−0.81	−1.01	−0.18	−0.65	−0.88	-0.99
	Ours	**−4.89**	**−7.98**	**−19.18**	**−9.40**	**−9.07**	**−7.95**	**−17.91**	**−8.80**	**−7.52**	**−9.91**	**−26.88**	**−8.80**
SSIM	NNCO [5]	0.92	–	–	–	0.91	–	–	–	0.92	–	–	–
	SSOD [78]	1.0	0.99	0.97	0.99	0.94	0.98	0.95	0.98	1.0	1.0	0.95	0.98
	Ours	0.98	0.96	0.96	0.98	0.98	0.97	0.94	0.96	0.98	0.97	0.93	0.96

Table 12.2 Performance evaluated in DUQ for black-box adversarial perturbation generation. Rows denote perturbed images generated from different cases. Columns denote different face detectors. Fv16 and Fr101 denote Faster-RCNN-based face detector with backbone network VGG16 [64] and ResNet101 [22]. Pr50 denotes PyramidBox face detector [71] with backbone network ResNet50 and Sv16 denotes SFD face detector [87] with backbone network VGG16. The last three columns, Fr50, SSHv16, SSHr50, are black-box face detectors, which are Faster-RCNN-based face detector with backbone network ResNet50 and SSH [50] with backbone network VGG16 and ResNet50

	SSIM	Known face detectors				Unknown face detectors		
		Fv16	Fr101	Pr50	Sv16	Fr50	SSHv16	SSHr50
Original	1.0	0.81	0.82	0.92	0.88	0.87	0.93	0.94
SSOD(Fv16+Fr101+Pr50+Sv16)	0.95	0.14	0.34	0.26	0.04	0.57	0.65	0.66
Ours(Fv16+Fr101+Pr50+Sv16)	0.91	−4.94	−5.69	−14.14	−6.37	**−1.72**	**−0.26**	**−0.47**

the adversarial perturbation generation scheme for general object detectors [78] to face detectors, and denote the corresponding method as $SSOD_{(Fv16+Fr101+Pr50)}$. SSOD can be extended to the black-box setting by generating adversarial perturbations for different face detectors independently, and then uses their summation as the final perturbations to any unknown face detectors. On the other hand, our method is based on an optimization problem, Eq. (12.6), and also considers false detections.

Table 12.2 shows the performance evaluated in DUQ of SSOD and our method. In addition, Fig. 12.4 shows three examples of black-box attack on Fr50, SSHv16, and SSHr50. As the adversarial perturbations are obtained by considering the four known DNN-based face detectors, they are expected to work well in those cases for all methods, which is confirmed by their performance under the white-box setting. When applied to the unknown DNN-based face detectors, both methods tend to be effective in reducing DUQ of the resulting face set. However, our method shows more reduction and usually leads to negative DUQ scores, suggesting that it generates more

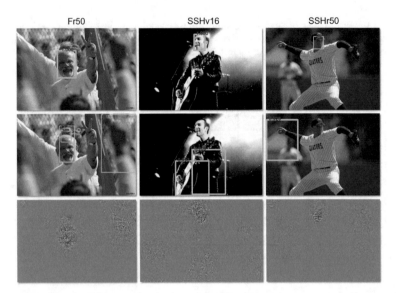

Fig. 12.4 Visual examples of black-box attack on Fr50, SSHv16, and SSHr50 face detectors, respectively. The top row corresponds to detection results on original images. The middle row corresponds to the detection results on images after adversarial perturbation is added to the original image. The bottom row shows the actual noise added, which are amplified by 30 for better visualization

false detections, and thus is more effective in reducing the quality of the face set as training data for AI face synthesis system. This is further corroborated by the visual results shown in Fig. 12.4.

12.5.2 Attacking Landmark Extractors

Our method is validated on three state-of-the-art CNN-based facial landmark extractors, namely, FAN [6], HRNet [65], and AVS-SAN [54]. FAN is constructed by multiple stacked hourglass structures, where we use one hourglass structure for simplicity. HRNet is composed by parallel high-to-low resolution sub-networks and repeats the information exchange across multi-resolution sub-networks. AVS-SAN first disentangles face images to style and structure space, which is then used as augmentation to train the network. We use implementations of all three methods trained on the WLFW dataset [76].

Datasets. To demonstrate the effectiveness of our method on obstructing DeepFake generation, we conduct experiments on Celeb-DF dataset [39], which contains high-quality DeepFake videos of 59 celebrities. Each video contains one subject with various head pose and facial expression. We choose this dataset as the pretrained

Table 12.3 The NME and SSIM scores of our method on different landmark extractors. The landmark extractors shown in leftmost column denote where the adversarial perturbation is from and the ones shown in the top row denotes which landmark extractors are attacked

NME↑				SSIM				
Attacks	FAN	HRNet	AVS-SAN	Attacks	SSIM$_I$ ↑	SSIM$_W$ ↓		
						FAN	HRNet	AVS-SAN
None	0.03	0.00	0.09					
FAN	0.87	0.05	0.09	FAN	0.81	0.68	0.89	0.89
HRNet	0.04	0.87	0.09	HRNet	0.78	0.89	0.67	0.88
AVS-SAN	0.06	0.04	0.92	AVS-SAN	0.78	0.87	0.87	0.69

DeepFake models are available to us, which can be used to test our method. In our experiment, we utilize the DeepFake method described in [39] to synthesize fake videos using original and adversarial images, respectively. We randomly select six identities, corresponding to 36 videos in total. Since the adjacent frames in a video show little variations, we apply Our method to the key frames of each video, i.e., 600 frames in total, for evaluation. Since the Celeb-DF dataset does not have the ground truth of facial landmarks, we use the results of HRNet as the ground truth due to its superior performance.

Evaluations. We use two metrics to evaluate our method, namely, Normalized Mean Error (NME) [65] and Structural Similarity (SSIM) [75]. The relation of these metrics is shown in Fig. ??.

- NME is the average Euclidean distance between landmarks on adversarial image and the ground truth, which is then normalized by the distance between the leftmost key point in left eye and the rightmost key point in right eye. Higher NME score indicates less accurate landmark detection, which is the objective of our method.
- The SSIM metric simulates perceptual image quality. We use this indicator to demonstrate our method can affect the visual quality of DeepFake. As shown in Fig. ??, we compute mask-SSIM [46] of original and adversarial input images (SSIM$_I$) and then compute the SSIM of the synthesized results (SSIM$_W$). The lower score indicates the image quality is degraded. Ideally, the attacking method should have large SSIM$_I$ that the adversarial perturbation does not affect the quality of input image, and small SSIM$_W$ that the synthesis quality is degraded.

Baselines. To better analyze our method, we adapt other two methods FGSM [69] and I-FGSM [19] from attacking image classifiers to our task. Specifically, the FGSM is a single-step optimization method as $\mathbf{I}_1^{adv} = \texttt{clip}\{\mathbf{I}_0^{adv} - \alpha \cdot \texttt{sign}(\nabla_{\mathbf{I}_0^{adv}}(L(\mathbf{I}_0^{adv}, \mathbf{I})))\}$, while I-FGSM is an iterative optimization method without considering momentum as $\mathbf{I}_{t+1}^{adv} = \texttt{clip}\{\mathbf{I}_t^{adv} - \alpha \cdot \texttt{sign}(\nabla_{\mathbf{I}^{adv}}(L(\mathbf{I}_t^{adv}, \mathbf{I})))\}$. The step size α and iteration number T of I-FGSM are set as same in our method. We use these two adapted methods as our baseline methods, which are denoted as *Base1* and *Base2*, respectively.

Fig. 12.5 Visual examples of our method attacking Fv16, Fr101, Pr50, and Sv16, respectively. The top row corresponds to detection results on original images. The middle row corresponds to the detection results on images after adversarial perturbation is added to the original image. The bottom row show the actual noise added, which are amplified by 30 for better visualization

Table 12.4 The NME and SSIM performance of different attacking methods

	NME↑		
Attacks	FAN	HRNet	AVS-SAN
None	0.03	0.00	0.09
Base1	0.05	0.04	0.10
Base2	0.85	0.88	0.92
LB	0.87	0.87	0.92

	SSIM$_I$ ↑ / SSIM$_W$ ↓		
Attacks	FAN	HRNet	AVS-SAN
Base1	0.52/0.73	0.46/0.71	0.49/0.69
Base2	0.88/0.71	0.88/0.70	0.86/0.73
LB	0.81/0.68	0.78/0.67	0.78/0.69

12.5.2.1 Results

Table 12.3 shows the NME and SSIM performance of our method. The landmark extractors shown in leftmost column denote where the adversarial perturbation is from and the ones shown in the top row denotes which landmark extractor is attacked. "None" denotes no perturbations are added to image. Our method can notably increase the NME score and decrease the SSIM$_W$ score in white-box attack (e.g., the value in the row of "FAN" and the column of "FAN"), which indicates our method can effectively disrupt facial landmarks extraction and subsequently affect the visual quality of the synthesized faces. We also compare our method with two

Table 12.5 The NME and SSIM performance of black-box attack. See text for details

NME↑

Attacks		FAN	HRNet	AVS-SAN
None		0.03	0.00	0.09
FAN	LB_{trans}	0.22	0.03	0.09
	LB_{mix}	0.24	0.04	0.09
HRNet	LB_{trans}	0.04	0.10	0.09
	LB_{mix}	0.04	0.14	0.09
AVS-SAN	LB_{trans}	0.04	0.03	0.55
	LB_{mix}	0.05	0.03	0.56

SSIM

Attacks		$SSIM_I$ ↑	$SSIM_W$ ↓		
			FAN	HRNet	AVS-SAN
FAN	LB_{trans}	0.91	0.88	0.94	0.94
	LB_{mix}	0.90	0.86	0.94	0.93
HRNet	LB_{trans}	0.92	0.95	0.94	0.95
	LB_{mix}	0.90	0.95	0.91	0.94
AVS-SAN	LB_{trans}	0.89	0.94	0.93	0.82
	LB_{mix}	0.88	0.93	0.93	0.81

baselines, Base1 and Base2, in Table 12.4. We can observe the Base1 method merely has effect on the NME performance but can largely degrade the quality of adversarial images compared to Base2 and our method (LB). The Base2 method can also achieve the competitive performance with our method in NME but is slightly degraded in SSIM.

Following existing works attacking image classifiers [14, 69], which achieves the black-box attack by transferring the adversarial perturbations from a known model to an unknown model (transferability), we also test the black-box attack using the adversarial perturbation generated from one landmark extractor to attack other extractors. However, the results show that the adversarial perturbations have merely effect on different extractors.

As shown in Table 12.3, the transferability of our method is weak. To improve the transferability, we employ the strategies commonly used in black-box attack on image classifiers: (1) Input transformation [79]: we randomly resize the input image and then padding around with zero at each iteration (denoted as LB_{trans}); (2) Attacking mixture [79]: we alternatively use Base2 and our method to increase the diversity in optimization (denoted as LB_{mix}). Table 12.5 shows the results of black-box attack, which reveals that the strategies effective in attacking image classifiers do not work on attacking landmark extractors. This is probably due to the mechanism of landmark extractors is more complex than image classifiers, as the landmark extractors need to output a series of points instead of labels and only a minority of points shifted does not affect the overall prediction.

12.6 Conclusion

AI-synthesized fake faces are becoming a problem encroaching our trust to online media. As most AI-based face synthesis algorithms require automatic face detection as an indispensable pre-processing step in preparing training data, an effective protection scheme can be obtained by disrupting the face detection methods. In this chapter, we describe a *proactive* protection method to deter bulk reuse of automatically detected face and face alignment for the production of AI-synthesized faces. Our method exploits the sensitivity of DNN-based face detectors and facial landmark extractors, and use adversarial perturbation to contaminate the face sets.

We expect this technology to spawn counter-measures from the forgery makers. In particular, operations that can destroy or reduce the adversarial perturbation are expected to be developed. It is thus our continuing effort to improve the robustness of the adversarial perturbation generation method. Another important direction to further explore is a more generic black-box attack scheme that does not limit to DNN-based face detectors and facial landmark extractors, and do not rely on the differentiability of the underlying model. Furthermore, we will also work on improving the running time efficiency of the current method so it can scale up to large number of images.

References

1. Afchar D, Nozick V, Yamagishi J, Echizen I (2018) MesoNet: a compact facial video forgery detection network. In: IEEE international workshop on information forensics and security (WIFS)
2. Baluja S, Fischer I (2018) Learning to attack: adversarial transformation networks. In: Association for the advancement of artificial intelligence (AAAI)
3. Bansal A, Nanduri A, Castillo CD, Ranjan R., Chellappa R (2016) Umdfaces: an annotated face dataset for training deep networks. arXiv preprint arXiv:1611.01484v2
4. Berthelot D, Schumm T, Metz L (2017) Began: boundary equilibrium generative adversarial networks. arXiv preprint arXiv:1703.10717
5. Bose AJ, Aarabi P (2018) Adversarial attacks on face detectors using neural net based constrained optimization. In: IEEE international workshop on multimedia signal processing (MMSP)
6. Bulat A, Tzimiropoulos G (2017) How far are we from solving the 2D & 3D face alignment problem? (and a dataset of 230,000 3D facial landmarks). In: ICCV
7. Carlini N, Wagner D (2017) Towards evaluating the robustness of neural networks. In: IEEE symposium on security and privacy (sp)
8. Chan C, Ginosar S, Zhou T, Efros AA (2018) Everybody dance now. arXiv preprint arXiv:1808.07371
9. Chan C, Ginosar S, Zhou T, Efros AA (2019) Everybody dance now. In: IEEE international conference on computer vision (ICCV)
10. Chen ST, Cornelius C, Martin J, Chau DH (2018) Robust physical adversarial attack on faster R-CNN object detector. arXiv preprint arXiv:1804.05810
11. Chesney R, Citron DK, Deep fakes: a looming challenge for privacy, democracy, and national security. 107 California Law Review (2019, Forthcoming); U of Texas Law, Public Law Research Paper No. 692; U of Maryland Legal Studies Research Paper No. 2018-21
12. Dalal N, Triggs B (2005) Histograms of oriented gradients for human detection. In: IEEE conference on computer vision and pattern recognition (CVPR)
13. Denton EL, Chintala S, Fergus R, et al (2015) Deep generative image models using a laplacian pyramid of adversarial networks. In: Conference on neural information processing systems (NeurIPS)
14. Dong Y, Liao F, Pang T, Su H, Zhu J, Hu X, Li J (2018) Boosting adversarial attacks with momentum. In: IEEE conference on computer vision and pattern recognition (CVPR)
15. Everingham M, Eslami SA, Van Gool L, Williams CK, Winn J, Zisserman A (2015) The pascal visual object classes challenge: a retrospective. Int J Comput Vis (IJCV)

16. Eykholt K, Evtimov I, Fernandes E, Li B, Rahmati A, Tramer F, Prakash A, Kohno T, Song D (2018) Physical adversarial examples for object detectors. arXiv preprint arXiv:1807.07769
17. Farfade SS, Saberian MJ, Li LJ (2015) Multi-view face detection using deep convolutional neural networks. In: ACM on international conference on multimedia retrieval
18. Goodfellow I, Pouget-Abadie J, Mirza M, Xu B, Warde-Farley D, Ozair S, Courville A, Bengio Y (2014) Generative adversarial nets. In: Conference on neural information processing systems (NeurIPS)
19. Goodfellow IJ, Shlens J, Szegedy C (2015) Explaining and harnessing adversarial examples. In: International conference on learning representations (ICLR)
20. Güera D, Delp EJ (2018) Deepfake video detection using recurrent neural networks. In: AVSS
21. Gulrajani I, Ahmed F, Arjovsky M, Dumoulin V, Courville AC (2017) Improved training of wasserstein gans. In: Conference on neural information processing systems (NeurIPS)
22. He K, Zhang X, Ren S, Sun J (2016) Deep residual learning for image recognition. In: IEEE conference on computer vision and pattern recognition (CVPR)
23. Hu T, Qi H, Xu J, Huang Q (2018) Facial landmarks detection by self-iterative regression based landmarks-attention network. In: AAAI
24. Isola P, Zhu JY, Zhou T, Efros AA (2017) Image-to-image translation with conditional adversarial networks. In: IEEE conference on computer vision and pattern recognition (CVPR)
25. Jiang H, Learned-Miller E (2017) Face detection with the faster R-CNN. In: IEEE international conference on automatic face & gesture recognition (FG)
26. Karras T, Aila T, Laine S, Lehtinen J (2018) Progressive growing of GANs for improved quality, stability, and variation. In: International conference on learning representations (ICLR)
27. Karras T, Laine S, Aila T (2018) A style-based generator architecture for generative adversarial networks. arXiv preprint arXiv:1812.04948
28. Karras T, Laine S, Aila T (2019) A style-based generator architecture for generative adversarial networks. In: IEEE conference on computer vision and pattern recognition (CVPR)
29. Kazemi V, Sullivan J (2014) One millisecond face alignment with an ensemble of regression trees. In: CVPR
30. Kim H, Carrido P, Tewari A, Xu W, Thies J, Niessner M, Pérez P, Richardt C, Zollhöfer M, Theobalt C (2018) Deep video portraits. ACM Trans Graph (TOG)
31. King DE (2009) Dlib-ml: a machine learning toolkit. J Mach Learn Res 10:1755–1758
32. Kurakin A, Goodfellow I, Bengio S (2017) Adversarial examples in the physical world. In: International conference on learning representations (ICLR)
33. Li H, Lin Z, Shen X, Brandt J, Hua G (2015) A convolutional neural network cascade for face detection. In: IEEE conference on computer vision and pattern recognition (CVPR)
34. Li J, Wang T, Zhang Y (2011) Face detection using surf cascade. In: IEEE international conference on computer vision workshops (ICCV Workshops)
35. Li J, Zhang Y (2013) Learning surf cascade for fast and accurate object detection. In: IEEE conference on computer vision and pattern recognition (CVPR)
36. Li Y, Chang MC, Lyu S (2018) In Ictu Oculi: exposing AI generated fake face videos by detecting eye blinking. In: IEEE international workshop on information forensics and security (WIFS)
37. Li Y, Lyu S (2019) Exposing deepfake videos by detecting face warping artifacts. In: IEEE conference on computer vision and pattern recognition workshops (CVPRW)
38. Li Y, Tian D, Chang M, Bian X, Lyu S (2018) Robust adversarial perturbation on deep proposal-based models. In: British machine vision conference (BMVC)
39. Li Y, Yang X, Sun P, Qi H, Lyu S (2020) Celeb-DF: a large-scale challenging dataset for deepfake forensics. In: CVPR
40. Liu B, Ding M, Zhu T, Xiang Y, Zhou W (2018) Using adversarial noises to protect privacy in deep learning era. In: IEEE global communications conference (GLOBECOM)
41. Liu MY, Breuel T, Kautz J (2017) Unsupervised image-to-image translation networks. In: Conference on neural information processing systems (NeurIPS)
42. Liu W, Anguelov D, Erhan D, Szegedy C, Reed S, Fu CY, Berg AC (2016) SSD: single shot multibox detector. In: European conference on computer vision (ECCV)

43. Liu Y, Zhang W, Yu N (2017) Protecting privacy in shared photos via adversarial examples based stealth. Secur Commun Netw
44. Lu J, Sibai H, Fabry E (2017) Adversarial examples that fool detectors. arXiv:1712:02494
45. Luo B, Liu Y, Wei L, Xu Q (2018) Towards imperceptible and robust adversarial example attacks against neural networks. In: Association for the advancement of artificial intelligence (AAAI)
46. Ma L, Jia X, Sun Q, Schiele B, Tuytelaars T, Van Gool L (2017) Pose guided person image generation. In: NeurIPS
47. Matern F, Riess C, Stamminger M (2019) Exploiting visual artifacts to expose deepfakes and face manipulations. In: IEEE winter applications of computer vision workshops (WACVW)
48. Moosavi-Dezfooli SM, Fawzi A, Fawzi O, Frossard P (2017) Universal adversarial perturbations. In: IEEE conference on computer vision and pattern recognition (CVPR)
49. Moosavi-Dezfooli SM, Fawzi A, Frossard P (2016) Deepfool: a simple and accurate method to fool deep neural networks. In: IEEE conference on computer vision and pattern recognition (CVPR)
50. Najibi M, Samangouei P, Chellappa R, Davis LS (2017) SSH: single stage headless face detector. In: IEEE international conference on computer vision (ICCV)
51. Oh SJ, Fritz M, Schiele B (2017) Adversarial image perturbation for privacy protection a game theory perspective. In: IEEE international conference on computer vision (ICCV)
52. Ojala T, Pietikäinen M, Mäenpää T (2002) Multiresolution gray-scale and rotation invariant texture classification with local binary patterns. IEEE Trans Pattern Anal Mach Intell (TPAMI)
53. Papernot N, McDaniel P, Jha S, Fredrikson M, Celik ZB, Swami A (2016) The limitations of deep learning in adversarial settings. In: EuroS&P
54. Qian S, Sun K, Wu W, Qian C, Jia J (2019) Aggregation via separation: boosting facial landmark detector with semi-supervised style translation. In: ICCV
55. Ramanan D, Zhu X (2012) Face detection, pose estimation, and landmark localization in the wild. In: IEEE conference on computer vision and pattern recognition (CVPR)
56. Ranjan R, Patel VM, Chellappa R (2015) A deep pyramid deformable part model for face detection. In: IEEE international conference on biometrics theory, applications and systems (BTAS)
57. Ranjan, R., Patel VM, Chellappa R (2019) Hyperface: a deep multi-task learning framework for face detection, landmark localization, pose estimation, and gender recognition. IEEE Trans Pattern Anal Mach Intell (TPAMI)
58. Ranjan, R., Sankaranarayanan, S., Castillo, C.D., Chellappa, R.: An all-in-one convolutional neural network for face analysis. In: IEEE international conference on automatic face & gesture recognition (FG)
59. Ren S, He K, Girshick R, Sun J (2017) Faster R-CNN: towards real-time object detection with region proposal networks. IEEE Trans Pattern Anal Mach Intell (TPAMI)
60. Ruiz N, Sclaroff S (2020) Disrupting deepfakes: adversarial attacks against conditional image translation networks and facial manipulation systems. arXiv preprint arXiv:2003.01279
61. Rumelhart DE, Hinton GE, Williams RJ, et al (1988) Learning representations by back-propagating errors. Cognit Model
62. Sagonas C, Tzimiropoulos G, Zafeiriou S, Pantic M (2013) 300 faces in-the-wild challenge: the first facial landmark localization challenge. In: IEEE international conference on computer vision workshops (ICCVW)
63. Salimans T, Goodfellow I, Zaremba W, Cheung V, Radford A, Chen X (2016) Improved techniques for training GANS. In: Conference on neural information processing systems (NeurIPS)
64. Simonyan K, Zisserman A (2014) Very deep convolutional networks for large-scale image recognition. arXiv preprint arXiv:1409.1556
65. Sun K, Xiao B, Liu D, Wang J (2019) Deep high-resolution representation learning for human pose estimation. In: CVPR
66. Sun X, Wu P, Hoi SC (2018) Face detection using deep learning: an improved faster RCNN approach. Neurocomputing

67. Suwajanakorn S, Seitz SM, Kemelmacher-Shlizerman I (2017) Synthesizing obama: learning lip sync from audio. ACM Trans Graph (TOG)
68. Szegedy C, Zaremba W, Sutskever I, Bruna J, Erhan D, Goodfellow I, Fergus R (2013) Intriguing properties of neural networks. arXiv 1312:6199
69. Szegedy C, Zaremba W, Sutskever I, Bruna J, Erhan D, Goodfellow I, Fergus R (2014) Intriguing properties of neural networks. In: ICLR
70. Taigman Y, Polyak A, Wolf L (2016) Unsupervised cross-domain image generation. arXiv preprint arXiv:1611.02200
71. Tang X, Du DK, He Z, Liu J (2018) Pyramidbox: a context-assisted single shot face detector. In: European conference on computer vision (ECCV)
72. Thies J, Zollhofer M, Stamminger M, Theobalt C, Niessner M (2016) Face2Face: real-time face capture and reenactment of RGB videos. In: IEEE conference on computer vision and pattern recognition (CVPR)
73. Viola P, Jones M (2001) Rapid object detection using a boosted cascade of simple features. In: IEEE conference on computer vision and pattern recognition (CVPR)
74. Wang H, Li Z, Ji X, Wang Y (2017) Face R-CNN. arXiv preprint arXiv:1706.01061
75. Wang Z, Bovik AC, Sheikh HR, Simoncelli EP, et al (2004) Image quality assessment: from error visibility to structural similarity. IEEE Trans Image Process (TIP)
76. Wu W, Qian C, Yang S, Wang Q, Cai Y, Zhou Q (2018) Look at boundary: a boundary-aware face alignment algorithm. In: CVPR
77. Wu W, Zhang Y, Li C, Qian C, Change Loy C (2018) Reenactgan: learning to reenact faces via boundary transfer. In: ECCV
78. Xie C, Wang J, Zhang Z, Zhou Y, Xie L, Yuille A (2017) Adversarial examples for semantic segmentation and object detection. In: IEEE international conference on computer vision (ICCV)
79. Xie C, Zhang Z, Zhou Y, Bai S, Wang J, Ren Z, Yuille AL (2019) Improving transferability of adversarial examples with input diversity. In: CVPR
80. Yang B, Yan J, Lei Z, Li SZ (2015) Convolutional channel features. In: IEEE international conference on computer vision (ICCV)
81. Yang S, Luo P, Loy CC, Tang X (2015) From facial parts responses to face detection: a deep learning approach. In: IEEE international conference on computer vision (ICCV)
82. Yang S, Luo P, Loy CC, Tang X (2016) Wider face: a face detection benchmark. In: IEEE Conference on computer vision and pattern recognition (CVPR)
83. Yang S, Xiong Y, Loy CC, Tang X (2017) Face detection through scale-friendly deep convolutional networks. arXiv preprint arXiv:1706.02863
84. Yang X, Dong Y, Pang T, Zhu J, Su H (2020) Towards privacy protection by generating adversarial identity masks. arXiv preprint arXiv:2003.06814
85. Yang, X., Li, Y., Lyu, S.: Exposing deep fakes using inconsistent head poses. In: ICASSP (2019)
86. Zeng X, Liu C, Qiu W, Xie L, Tai YW, Tang CK, Yuille AL (2017) Adversarial attacks beyond the image space. arXiv:1711.07183
87. Zhang S, Zhu X, Lei Z, Shi H, Wang X, Li SZ (2017) S3FD: single shot scale-invariant face detector. In: IEEE international conference on computer vision (ICCV)
88. Zou X, Zhong S, Yan L, Zhao X, Zhou J, Wu Y (2019) Learning robust facial landmark detection via hierarchical structured ensemble. In: ICCV

Chapter 13
Multi-channel Face Presentation Attack Detection Using Deep Learning

Anjith George and Sébastien Marcel

Abstract Face recognition has emerged as a widely used biometric modality. However, its vulnerability to presentation attacks remains a significant security threat. Although Presentation Attack Detection (PAD) methods attempt to remedy this problem, often they fail in generalizing to unseen attacks and environments. As the quality of presentation attack instruments improves over time, achieving reliable PA detection using only visual spectra remains a major challenge. We argue that multi-channel systems could help solve this problem. In this chapter, we first present an approach based on a multi-channel convolutional neural network for the detection of presentation attacks. We further extend this approach to a one-class classifier framework by introducing a novel loss function that forces the network to learn a compact embedding for the *bonafide* class while being far from the representation of attacks. The proposed framework introduces a novel way to learn a robust PAD system from *bonafide* and available (known) attack classes. The superior performance in unseen attack samples in publicly available multi-channel PAD database *WMCA* shows the effectiveness of the proposed approach. Software, data, and protocols for reproducing the results are made publicly available.

13.1 Introduction

Biometrics provides a secure and convenient means for access control. Facial biometrics is one of the most convenient modalities for biometric authentication due to its non-intrusive nature. Even though facial recognition systems achieve human perfor-

A. George (✉) · S. Marcel
Idiap Research Institute, Centre du Parc, Rue Marconi 19, 1920 Martigny, CH, Switzerland
e-mail: anjith.george@idiap.ch

S. Marcel
e-mail: sebastien.marcel@idiap.ch

mance in identifying people in many difficult [32] datasets, most facial recognition systems are still vulnerable to presentation attacks (PA), also known as spoofing[1] [29, 41, 42]. Simply showing a printed photo to an unprotected facial recognition system might be enough to fool the system [2]. Vulnerability to presentation attacks limits the reliable deployment of such systems for applications in unsupervised conditions.

According to the ISO [29] standard, a presentation attack is defined as:

> A presentation to the biometric data capture subsystem with the goal of interfering with the operation of the biometric system.

Presentation attacks include both "impersonation" and "obfuscation" of identity. Impersonation refers to attacks in which the attacker wants to be recognized as a different person, while, in "obfuscation" attacks, the goal is to hide the identity of the attacker. The biometric characteristic or object used in a presentation attack is known as a presentation attack instrument (PAI).

Often, features such as color, texture [8, 39], motion [2], and physiological cues [28, 55] and CNN-based methods [20] are used for detection of attacks like 2D prints and replays. However, detection of sophisticated attacks like 3D masks and partial attacks is challenging and poses a serious threat to the reliability of face recognition systems. Most of the presentation attack detection (PAD) methods available in prevailing literature try to solve the problem for a limited number of presentation attack instruments and on visible spectrum images [42]. Though some success has been achieved in addressing 2D presentation attacks, performance of the algorithms in realistic 3D masks and other kinds of attacks is poor. With the increase in quality of attack instruments, it becomes harder to discriminate between *bonafide* and PAs in the visible spectrum alone. Moreover, considering a real-world situation with a wide variety of 2D, 3D, and partial attacks, PAD in visual spectra alone is challenging and inadequate for security-critical applications. Partial attacks refer to attacks where the attack instrument covers only a part of the face. These attacks are much harder to detect as they appear similar to *bonafide* in most of the face regions, and they can fool holistic liveliness detection systems easily. Multi-channel methods have been proposed as an alternative [5, 21–23, 26, 44, 45, 54, 57], since they use complementary information from different channels to improve the discrimination between *bonafide* and attacks. In the multi-channel scenario, the additional channels used can be any modality which can provide complementary representation such as depth, infrared, and thermal channels. Multi-channel PAD approaches are more promising in the context of a wide variety of attacks since they make PAD systems harder to fool.

Even with the use of multiple channels, one of the main issues with PAD is its poor generalization to unseen attacks [23]. This is particularly important, since at the time of developing a PAD system, anticipating all possible attacks is impossible. Malicious attackers can always come up with new attacks to fool the PAD systems. In such situations, PAD systems which are robust against unseen attacks are of

---

[1] The term spoofing should be deprecated in favor of presentation attacks to comply with ISO standards.

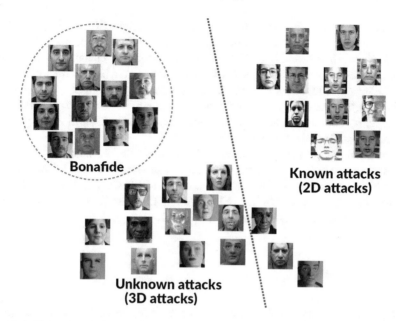

Fig. 13.1 Illustration of the embedding space with known and unknown attack classes. The red dotted line shows the learned decision boundary when only *bonafide* and known attacks are present in the training set, this results in misclassification of unknown attacks. If a decision boundary of the *bonafide* class (green dotted lines) is learned, known and unknown attacks can be classified correctly

paramount importance. Moreover, while it is comparatively easy to collect data for attacks like 2D prints and replays, making replicas of challenging presentation attack instruments (PAI) like silicone mask is often very costly [6] and resource-intensive. In this context, it will be ideal to have a framework which can be trained with *bonafide* alone, or with a combination of *bonafide* and easy-to-manufacture PAIs.

In real-world scenarios, it can be assumed that all presentation attacks are unseen, as it is not possible to foretell all the variations a PAD system could encounter a priori. A toy example of the decision boundary in an unseen attack scenario is illustrated in Fig. 13.1. Performances in typical PAD databases may not be representative of the performance of a PAD system in real-world conditions. This necessitates the PAD algorithms to be robust against unseen attacks. Since it is easy (in effort and cost) to collect data from more straightforward attacks compared to complex PAIs, we try to learn the representation leveraging the information from PA classes which are available at the training stage (while not over-fitting on the available attacks). To achieve this, we propose a one-class classifier-based framework, where the feature representation is learned with a CNN to have discriminative properties. The core of the framework is a multi-channel CNN trained to learn the embedding using a specific loss function. The Multi-Channel Convolutional Neural Network (MC-CNN) architecture efficiently combines multi-channel information for robust detection of presentation attacks. The network uses a pre-trained LightCNN model as the base network, which obviates the requirement to train the framework from scratch. In MC-CNN, only low-level LightCNN features across multiple channels are re-trained, while high-level layers of pre-trained LightCNN remain unchanged. In combination with the new loss function, the network aims at learning a compact representation for the *bonafide* class while leveraging the discriminative information for PAD task.

The source code and protocols to reproduce the results are made available publicly and are accessible at the following link.[2]

The rest of the chapter is organized as follows. Section 13.2 describes the related work with a particular focus on unseen attack detection. Section 13.3 outlines the proposed framework. Extensive evaluations, comparison with baseline methods, and ablation studies are shown in Sect. 13.4. Section 13.5 discusses the importance of the results, and Sect. 13.6 presents the conclusions.

13.2 Related Work

Most of the work related to face presentation attack detection addresses detection of 2D attacks, specifically print and 2D replay attacks. A brief review of recent PAD methods is given in this section.

13.2.1 Feature-Based Approaches for Face PAD

For PAD using visible spectrum images, several methods such as detecting motion patterns [2], color texture and histogram-based methods in different color spaces, and variants of Local Binary Patterns (LBP) in grayscale [8] and color images [9], [39] have shown good performance. Image quality-based feature [18] is one of the successful methods available in prevailing literature. Methods identifying moiré patterns [49], and image distortion analysis [59], use the alteration of the images due to the replay artifacts. Most of these methods treat PAD as a binary classification problem which may not generalize well for unseen attacks [46].

Chingovska et al. [10] studied the amount of client-specific information present in features used for PAD. They used this information to build client-specific PAD methods. Their method showed a 50% relative improvement and better performance in unseen attack scenarios.

Arashloo et al. [3] proposed a new evaluation scheme for unseen attacks. Authors have tested several combinations of binary classifiers and one-class classifiers. The performance of one-class classifiers was better than binary classifiers in the unseen attack scenario. BSIF-TOP was found successful in both one-class and two-class scenarios. However, in cross-dataset evaluations, image quality features were more

---

[2] Source code: https://gitlab.idiap.ch/bob/bob.paper.oneclass_mccnn_2019.

useful. Nikisins et al. [46] proposed a similar one-class classification framework using one-class Gaussian Mixture Models (GMM). In the feature extraction stage, they used a combination of Image Quality Measures (IQM). The experimental part involved an aggregated database consisting of replay attack [9], replay mobile [11], and MSU-MFSD [59] datasets.

Heusch and Marcel [27] recently proposed a method for using features derived from remote photoplethysmography (rPPG). They used the long-term spectral statistics (LTSS) of pulse signals obtained from available methods for rPPG extraction. The LTSS features were combined with SVM for PA detection. Their approach obtained better performance than state-of-the-art methods using rPPG in four publicly available databases.

13.2.2 CNN-Based Approaches for Face PAD

Recently, several authors have reported good performance in PAD using convolutional neural networks (CNN). Gan et al. [19] proposed a 3D CNN-based approach, which utilized the spatial and temporal features of the video. The proposed approach achieved good results in the case of 2D attacks, prints, and videos. Yang et al. [64] proposed a deep CNN architecture for PAD. A preprocessing stage including face detection and face landmark detection is used before feeding the images to the CNN. Once the CNN is trained, the feature representation obtained from CNN is used to train a SVM classifier and used for final PAD task. Boulkenafet et al. [7] summarized the performance of the competition on mobile face PAD. The objective was to evaluate the performance of the algorithms under real-world conditions such as unseen sensors, different illumination, and presentation attack instruments. In most of the cases, texture features extracted from color channels performed the best. Li et al. [34] proposed a 3D CNN architecture, which utilizes both spatial and temporal nature of videos. The network was first trained after data augmentation with a cross-entropy loss, and then with a specially designed generalization loss, which acts as a regularization factor. The Maximum Mean Discrepancy (MMD) distance among different domains is minimized to improve the generalization property.

There are several works involving various auxiliary information in the CNN training process, mostly focusing on the detection of 2D attacks. Authors use either 2D or 3D CNNs. The main problem of CNN-based approaches mentioned above is the lack of training data, which is usually required to train a network from scratch. One broadly used solution is fine-tuning, rather than a complete training of the networks trained for face recognition, or image classification tasks. Another issue is poor generalization in cross-database, and unseen attack tests. To circumvent these issues, some researchers have proposed methods to train a CNN using auxiliary tasks, which is shown to improve generalization properties. These approaches are discussed below.

Liu et al. [36] presented a novel method for PAD with auxiliary supervision. Instead of training a network end to end directly for PAD task, they used CNN-RNN model to estimate the depth with pixel-wise supervision and estimate remote photoplethysmography (rPPG) with sequence-wise supervision. The estimated rPPG and depth were used for PAD task. The addition of the auxiliary task improved the generalization capability.

Atoum et al. [4] proposed a two-stream CNN for 2D presentation attack detection by combining a patch-based model and holistic depth maps. For the patch-based model, an end-to-end CNN was trained. In the depth estimation, a fully convolutional network was trained using entire face image. The generated depth map was converted to feature vector by finding the mean values in the $N \times N$ grid. The final PAD score was obtained by fusing the scores from the patch and depth CNNs.

Shao et al. [56] proposed a deep convolutional network-based architecture for 3D mask PAD. They tried to capture the subtle differences in facial dynamics using the CNN. Feature maps obtained from the convolutional layer of a pre-trained VGG network was used to extract features in each channel. Optical flow was estimated using the motion constraint equation in each channel. Further, the dynamic texture was learned using the data from different channels. The proposed approach achieved an AUC (area under curve) score of 99.99% in 3DMAD dataset.

13.2.3   One-Class Models for Face PAD

Most of these methods handle the PAD problem as binary classification, which results in classifiers over-fitting to the known attacks resulting in poor generalization to unseen attacks. We focus the further discussion on the detection of unseen attacks. However, it is imperative that methods working for unseen attacks must perform accurately for known attacks as well. One naive solution for such a task is one-class classifiers (OCC). OCC provides a straightforward way of handling the unseen attack scenario by modeling the distribution of the *bonafide* class alone.

Arashloo et al. [3] and Nikisins et al. [46] have shown the effectiveness of one-class methods against unseen attacks. Even though these methods performed better than binary classifiers in an unseen attack scenario, the performance in known attack protocols was inferior to that of binary classifiers. Xiong et al. [62] proposed unseen PAD methods using auto-encoders and one-class classifiers with texture features extracted from images. However, the performance of the methods compared to recent CNN-based methods is very poor. CNN-based methods outperform most of the feature-based baselines for PAD task. Hence, there is a clear need of one-class classifiers or anomaly detectors in the CNN framework. One of the drawbacks of one-class model is that they do not use the information provided by the known attacks. An anomaly detector framework which utilizes the information from the known attacks could be more efficient.

Perera and Patel [50] presented an approach for one-class transfer learning in which labeled data from an unrelated task is used for feature learning. They used two loss functions, namely, descriptive loss, and compactness loss to learn the representations. The data from the class of interest is used to calculate the compactness loss whereas an external multi-class dataset is used to compute the descriptive loss.

Accuracy of the learned model in classification using another database is used as the descriptive loss. However, in the face PAD problem, this approach would be challenging since the *bonafide* and attack classes appear very similar.

Fatemifar et al. [16] proposed an approach to ensemble multiple one-class classifiers for improving the generalization of PAD. They introduced a class-specific normalization scheme for the one-class scores before fusion. Seven regions, three one-class classifiers, and representations from three CNNs were used in the pool of classifiers. Though their method achieved better performance as compared to client independent thresholds, the performance is inferior to CNN-based state-of-the-art methods. Specifically, many CNN-based approaches have achieved 0% HTER in Replay-Attack and Replay-Mobile datasets. Moreover, the challenging unseen attack scenario is not evaluated in this work.

Pérez-Cabo et al. [51] proposed a PAD formulation from an anomaly detection perspective. A deep metric learning model is proposed, where a triplet focal loss is used as a regularization for "metric-softmax," which forces the network to learn discriminative features. The features learned in such a way is used together with an SVM with RBF kernel for classification. They have performed several experiments on an aggregated RGB-only datasets showing the improvement made by their proposed approach. However, the analysis is mostly limited to RGB-only models and 2D attacks. Challenging 3D and partial attacks is not considered in this work. Specifically, the effectiveness in challenging unknown attacks (2D vs 3D) is not evaluated.

Recently, Liu et al. [37] proposed an approach for the detection of unknown spoof attacks as Zero-Shot Face Anti-spoofing (ZSFA). They proposed a Deep Tree Network (DTN) which partitions the attack samples into semantic sub-groups in an unsupervised manner. Each tree node in their network consists of a Convolutional Residual Unit (CRU) and a Tree Routing Unit (TRU). The objective is to route the unknown attacks to the most proper leaf node for correctly classifying it. They have considered a wide variety of attacks in their approach and their approach achieved superior performance compared to the considered baselines.

Jaiswal et al. [30] proposed an end-to-end deep learning model for PAD which used unsupervised adversarial invariance. In their method, the discriminative information and nuisance factors are disentangled in an adversarial setting. They showed that by retaining only discriminative information, the PAD performance improved for the same base architecture. Mehta et al. [43] trained an Alexnet model with a combination of cross-entropy and focal losses. They extracted the features from Alexnet and trained a two-class SVM for PAD task. However, results in challenging datasets such as OULU and SiW were not reported.

Recently, Joshua and Jain [14] utilized multiple GANs for spoof detection in fingerprints. Their method essentially consisted of training a DCGAN [53] using only the *bonafide* samples. At the end of the training, the generator is discarded, and the discriminator is used as the PAD classifier. They combined the results from different GANs operating on different features. However, this approach may not work well for face images as the recaptured images look very similar to the *bonafide* samples.

13.2.4 Multi-channel-Based Approaches for Face PAD

In general, most of the visible spectrum-based PAD methods try to detect the subtle differences in image quality when it is recaptured. However, this method could fail as the quality of capturing devices and printers improves. For 3D attacks, the problem is even more severe. As the technology to make detailed masks is available, it becomes very hard to distinguish between *bonafide* and presentation attacks by just using visible spectrum imaging. Many researchers have suggested using multi-spectral and extended range imaging to solve this issue [54, 57].

Raghavendra et al. [54] presented an approach using multiple spectral bands for face PAD. The main idea is to use complementary information from different bands. To combine multiple bands they observed a wavelet-based feature level fusion, and a score fusion methodology. They experimented with detecting print attacks prepared using different kinds of printers. They obtained better performance with score level fusion as compared to the feature fusion strategy.

Erdogmus and Marcel [15] evaluated the performance of a number of face PAD approaches against 3D masks using 3DMAD dataset. This work demonstrated that 3D masks could fool PAD systems easily. They achieved HTER of 0.95 and 1.27% using simple LBP features extracted from color and depth images captured with Kinect.

Steiner et al. [57] presented an approach using multi-spectral SWIR imaging for face PAD. They considered four wavelengths—935, 1060, 1300, and 1550 nm. In their approach, they trained a SVM for classifying each pixel as a skin pixel or not. They defined a Region Of Interest (ROI) where the skin is likely to be present, and skin classification results in the ROI is used for classifying PAs. The approach obtained 99.28% accuracy in per pixel skin classification.

Dhamecha et al. [13] proposed an approach for PAD by combining the visible and thermal image patches for spoofing detection. They classified each patch as either *bonafide* or attack and used the *bonafide* patches for subsequent face recognition pipeline.

In [6], Bhattacharjee et al. showed that it is possible to spoof commercial face recognition systems with custom silicone masks. They also proposed to use mean temperature of face region for PAD.

Bhattacharjee et al. [5] presented a preliminary study of using multi-channel information for PAD. In addition to visible spectrum images, they considered thermal, near-infrared, and depth channels. They showed that detecting rigid masks and 2D attacks is simple in thermal and depth channels, respectively. Most of the attacks can be detected with a similar approach with combinations of different channels, where the features and combinations of channels to use are found using a learning-based approach.

Wang et al. [58] proposed multimodal face presentation attack detection with a ResNet-based network using both spatial and channel attentions. Specifically, the approach was tailored for the *CASIA-SURF* [67] database which contained RGB, near-infrared, and depth channels. The proposed model is a multi-branch model where the individual channels and fused data are used as inputs. Each input channel has its own feature extraction module and the features extracted are concatenated in a late fusion strategy. Followed by more layers to learn a discriminative representation for PAD. The network training is supervised by both center loss and softmax loss. One key point is the use of spatial and channel attention to fully utilize complementary information from different channels. Though the proposed approach achieved good results in the *CASIA-SURF* database, the challenging problem of unseen attack detection is not addressed.

Parkin et al. [47] proposed a multi-channel face PAD network based on ResNet. Essentially, their method consists of different ResNet blocks for each channel followed by fusion. Squeeze and excitation modules (SE) are used before fusing the channels, followed by remaining residual blocks. Further, they add aggregation blocks at multiple levels to leverage inter-channel correlations. Their approach achieved state-of-the-art results in *CASIA-SURF* [67] database. However, the final model presented in is a combination of 24 neural networks trained with different attack-specific folds, pre-trained models, and random seeds, which would increase the computation greatly.

13.2.5   Challenges in PAD

In general, presentation attack detection in real-world scenario is challenging. Most of the PAD methods available in prevailing literature try to solve the problem for a limited number of presentation attack instruments. Though some success has been achieved in addressing 2D presentation attacks, the performance of the algorithms in realistic 3D masks and other kinds of attacks is poor.

As the quality of attack instruments evolves, it becomes increasingly difficult to discriminate between *bonafide* and PAs in the visible spectrum alone. In addition, more sophisticated attacks, like 3D silicone masks, make PAD in visual spectra challenging. These issues motivate the use of multiple channels, making PAD systems harder to by-pass.

We argue that the accuracy of the PAD methods can get better with a multi-channel acquisition system. Multi-channel acquisition from consumer-grade devices can improve the performance significantly. Hybrid methods, combining both extended hardware and software, could help in achieving good PAD performance in real-world scenarios. We extend the idea of a hybrid PAD framework and develop a multi-channel framework for presentation attack detection. Even with multi-channel methods, to achieve robustness against unseen attacks, the classifier part should move away from the typical binary classification formulation. One-class classifiers could be a good alternative for binary classification in the PAD task. However, the features used for one-class classifiers should be discriminative and compact to outperform binary classification.

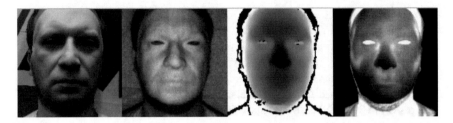

Fig. 13.2 Preprocessed images from a rigid mask attack; channels showed are grayscale, infrared, depth, and thermal, respectively. Channels were preprocessed with face detection, alignment, and normalization

13.3 Proposed Method

A Multi-Channel Convolutional Neural Network (MC-CNN)-based approach using a new loss function is proposed for PAD. Different stages of the framework are described below.

13.3.1 Preprocessing

Face detection is performed in the color channel using the MTCNN algorithm [66]. Once the face bounding box is obtained, face landmark detection is performed in the detected face bounding box using Supervised Descent Method (SDM) [63]. Alignment is accomplished by transforming image, such that the eye centers and mouth center are aligned to predefined coordinates. The aligned face images are converted to grayscale, and resized to the resolution of 128×128 pixels. An example of the result of this first stage in the preprocessing pipeline is shown in Fig. 13.2.

The preprocessing stage for non-RGB channels requires the images from different channels to be aligned both spatially and temporally with the color channel. For these channels, the facial landmarks detected in the color channel are reused, and a similar alignment procedure is performed. A normalization using Mean Absolute Deviation (MAD) [33] is performed to cast the type of non-RGB facial images to 8-bit format.

13.3.2 Network Architecture

Many of previous work in face presentation attack detection utilize transfer learning from pre-trained face recognition networks. This is required since the data available for PAD task is often of a very limited size, being insufficient to train a deep architecture from scratch.

The features learned in the low level of CNN networks are usually similar to Gabor filter masks, edges, and blobs [65]. Deep CNNs compute more discriminant features as the depth increases [40]. It has been observed in different studies [35, 65], that is, features, which are closer to the input are more general, while features in the higher levels contain task-specific information. Hence, most of the literature in the transfer learning attempts to adapt the higher level features for the new tasks.

Recently, Freitas Pereira et al. [17] showed that the high-level features in deep convolutional neural networks, trained in visual spectra, are domain independent, and they can be used to encode face images collected from different image-sensing domains. Their idea was to use the shared high-level features for heterogeneous face recognition task, re-training only the lower layers. In their method, they split the parameters of the CNN architecture into two, the higher level features are shared among the different channels, and the lower level features (known as Domain-Specific Units (DSU)) are adapted separately for different modalities. The objective was to learn the same face encoding for different channels, by adapting just the DSUs. The network was trained using contrastive loss (with Siamese architecture) or triplet loss. Re-training of only low-level features has the advantage of modifying a minimal set of parameters.

We extend the idea of domain-specific units (DSU) for multi-channel PAD task. Instead of forcing the representation from different channels to be same, we leverage the complementary information from a joint representation obtained from multiple channels. We hypothesize that the joint representation contains discriminatory information for PAD task. By concatenating the representation from different channels, and using fully connected layers, a decision boundary for the appearance of *bonafide* and attack presentations can be learned via backpropagation. The lower layer features, as well as the higher level fully connected layers, are adapted in the training.

In this work, we utilize a LightCNN model [61], which was pre-trained on a large number of face images for face recognition. The LightCNN network is especially interesting as the number of parameters is much smaller than in other networks used for face recognition. LightCNN achieves a reduced set of parameters using a Max-Feature Map (MFM) operation as an alternative to Rectified Linear Units (ReLU), which suppresses low activation neurons in each layer.

The block diagram of the proposed framework is shown in Fig. 13.3. The pre-trained LightCNN model produces a 256-dimensional embedding, which can be used as face representation. The LightCNN model is extended to accept four channels. The 256-dimensional representation from all channels is concatenated, and two fully connected layers are added at the end for PAD task. The first fully connected layer has ten nodes, and the second one has one node. A sigmoidal activation function is used in each fully connected layer. The higher level features are more related to the task to be solved. Hence, the fully connected layers added on top of the concatenated representations are tuned exclusively for PAD task. Reusing the weights from a network pre-trained for face recognition on a large set of data, we avoid plausible over-fitting, which can occur due to limited amount of training data.

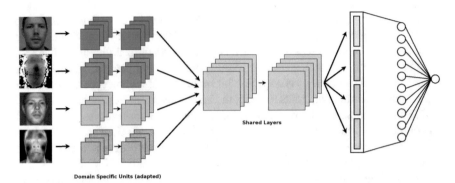

Fig. 13.3 Block diagram of the basic multichannel network. The gray color blocks in the CNN part represent layers which are not re-trained, and other colored blocks represent re-trained/adapted layers

Binary Cross Entropy (BCE) is used as the loss function to train the model using the ground truth information for PAD task.

Several experiments were done by adapting the different blocks of layers, starting from the low-level features. The final fully connected layers are adapted for PAD task in all the experiments.

While doing the adaptation, the weights are always initialized from the weights of the pre-trained layers. Apart from the layers adapted, the parameters for the rest of the network remain shared.

The layers corresponding to the color channel are not adapted since the representation from the color channel can be reused for face recognition, hence making the framework suitable for simultaneous face recognition and presentation attack detection.

13.3.3 One-Class Contrastive Loss (OCCL)

From a practical viewpoint, it is not possible to anticipate all the possible types of attacks and to have them in the training set. This, in turn, make the PAD task an unseen classification problem in a broad sense. In general, we can even consider attacks coming from different replay devices as unseen attacks. Typically, one-class classifiers are well suited for such outlier detection tasks. However, in practice, the performance of one-class classifiers is inferior compared to binary classifiers for known attacks, since they do not leverage useful information from the known attacks. Ideally, the PAD system should perform well in both known and unseen attack scenarios.

Clearly, there is a necessity of a method which can learn a compact one-class representation while utilizing the discriminative information from known attacks. While the collection of attacks could be difficult and costly, collecting *bonafide* samples are rather easy. A new classification strategy is required to handle the realistic scenario where a limited variety of attack classes are available.

Though one-class classifiers (*OCC*) offer a way to model the *bonafide* class, the efficient use of *OCC* requires the feature representation to be compact while containing discriminative information for PAD task. In the proposed framework, we use a CNN-based approach to learn the feature representation. A novel loss function is proposed to learn a representation of *bonafide* samples leveraging the known attack classes.

Consider a typical CNN architecture for PAD, where the output layer contains one node and the loss function used is Binary Cross Entropy (*BCE*), which is defined as follows:

$$\mathcal{L}_{BCE} = -(y \log(p) + (1 - y) \log(1 - p)) \tag{13.1}$$

where y is the ground truth, ($y = 0$ for attack and $y = 1$ for *bonafide*) and p is the probability.

When trained only with *BCE* loss, the network learns a decision boundary based on the *bonafide* and attacks present in the training set. However, it may not generalize when encountered with an unseen attack in the test time as it could be over-fitted to attacks which are "known" from the training set.

To overcome this issue, we propose the "One-Class Contrastive Loss" (*OCCL*) function which operates on the embedding layer. Proposed One-Class Contrastive Loss (*OCCL*) function is used as an auxiliary loss function in conjunction with binary cross-entropy loss. The feature map obtained from the penultimate layer of the CNN is used as the embedding. The loss function is inspired from center loss [60] and contrastive loss [24], which are usually used in the face recognition applications.

In face recognition applications, center loss is used as an additional auxiliary loss function, the task of the center loss is to minimize the distance of the embeddings from their corresponding class centers. The center loss is defined as follows:

$$\mathcal{L}_{center} = \frac{1}{2} \sum_{i=1}^{m} \|x_i - c_{y_i}\|_2^2 \tag{13.2}$$

where L_{center} denotes the center loss, m the number of training samples in a mini-batch, $x_i \in R_d$ denotes the ith training sample, y_i denotes the label, and c_{y_i} denotes the y_i^{th} class center in the embedding space.

The main issue with center loss in the PAD application is that the loss function penalizes for large intra-class distances and does not care about the inter-class distances. Contrastive center loss [52] tries to solve this issue by adding the distance between classes (inter-class) in the formulation. However, for the PAD problem, modeling the attack class as a cluster and finding a center for the attack class is not trivial. The attacks could be of different categories: 2D, 3D, and partial attacks, and it is not ideal forcing them to cluster together in the embedding space. It is only necessary to have the embeddings of attacks far from *bonafide* cluster in the embedding space. Hence, we put the compactness constraint only on the *bonafide* class, while forcing the embeddings of PAs to be far from that of *bonafide*.

To formulate the loss function, we start with the equation for contrastive loss function proposed by Lecun et al. [24].

$$
\begin{aligned}
\mathcal{L}_{\text{Contrastive}}(W, Y, X^1, X^2) =& (1 - Y)\frac{1}{2}D_W^2 \\
&+ Y\frac{1}{2}\max(0, \text{m} - D_W)^2
\end{aligned}
\tag{13.3}
$$

where W is the network weights; X^1, X^2 are the pair; and Y the label of the pair, i.e., whether they belong to the same class or not. m is the margin, and D_W is the distance function between two samples. The data is provided as pairs (X^1, X^2) and the distance function D_W can be computed as the Euclidean distance.

$$
D_W = \sqrt{\|X^1 - X^2\|_2^2}
\tag{13.4}
$$

Now, in our loss formulation, the critical difference is how we define D_W. In the original contrastive loss, D_W is the distance between samples. In our case, we need the representation of *bonafide* samples to be compact in an embedding space. At the same time, we want to maximize the distance between *bonafide* cluster and attack samples in the embedding space. This can be achieved by defining DC_W to be the distance from the center of *bonafide* class as follows:

$$
DC_W = \sqrt{\|X^i - c_{BF}\|_2^2}
\tag{13.5}
$$

where X^i is the embedding for i^{th} sample, and c_{BF} is the center of *bonafide* class in the embedding space.

The center of the *bonafide* class is updated in every mini-batch during training as follows:

$$
c_{BF} = \hat{c}_{BF}(1 - \alpha) + \alpha\frac{1}{N}\sum_{i=1}^{N}e_i
\tag{13.6}
$$

where c_{BF} and $\hat{c}_{BF}$ denote the new and old *bonafide* centers. α is a scalar which prevents sudden changes in the class centers in mini-batch. e_i denotes the difference between embeddings for the *bonafide* samples in the current mini-batch compared to the previous center, and N denotes the number of *bonafide* samples in the mini-batch.

Combining the equations, our auxiliary loss function becomes

$$
\begin{aligned}
\mathcal{L}_{OCCL}(W, Y, X) =& Y\frac{1}{2}DC_W^2 \\
&+ (1 - Y)\frac{1}{2}\max(0, \text{m} - DC_W)^2
\end{aligned}
\tag{13.7}
$$

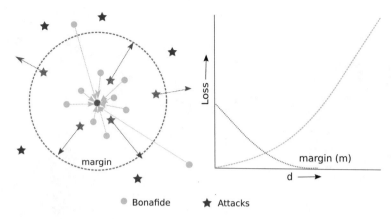

Fig. 13.4 Loss functions acting on the embedding space, left) *bonafide* representations are pulled closer to the center of *bonafide* class (green), while the attack embeddings (red) are forced to be beyond the margin. The attack samples outside the margin do not contribute to the loss, right) The loss as a function of distance from the *bonafide* center

where DC_W denotes the Euclidean distance between the samples and the *bonafide* class center, Y denotes the ground truth, i.e., $Y = 0$ for attacks and $Y = 1$ for *bonafide* (note the change in labels from the standard notation due to the ground truth convention). It is to be noted that the proposed loss function ***does not*** require pairs of samples, which is a requirement in usage of contrastive loss. This makes it easier to train the model without requiring an explicit selection of pairs during training.

This auxiliary loss makes the representation of *bonafide* compact pushing it closer to the center of *bonafide* class and penalizes attack samples which are closer than the margin m. Attack samples which are farther than the margin m are not penalized. An illustration of the loss functions acting on the embeddings of *bonafide* and attack samples is shown in Fig. 13.4.

We combine the proposed loss function with standard binary cross entropy for training. The combined loss function to minimize is given as

$$\mathcal{L} = (1 - \lambda)\mathcal{L}_{BCE} + \lambda\mathcal{L}_{OCCL} \tag{13.8}$$

where $\mathcal{L}$ denotes the total loss for the CNN. $\mathcal{L}_{BCE}$ and $\mathcal{L}_{OCCL}$ denote the binary cross entropy and one-class contrastive loss, respectively. λ denotes a scalar value to set the weight for each loss functions. In our experiments, we set the value of λ as 0.5.

The combined loss function $\mathcal{L}$ tries to learn a decision boundary between the available attacks and *bonafide* while the auxiliary loss tries to make the feature representation of the *bonafide* compact in the embedding space. We expect the decision boundary learned in this fashion to be more robust in unseen attacks compared to the network learned only with *BCE*. The embedding obtained in this manner is used with a one-class classifier for the PAD task.

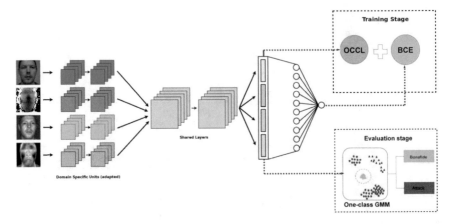

Fig. 13.5 Schematic diagram of the proposed framework. The CNN architecture is trained with two losses and then used as a fixed feature extractor with frozen weights. The one-class GMM is trained using the embeddings obtained from *bonafide* class alone

An illustration of the proposed framework is shown in Fig. 13.5. At the time of training, both losses are used, and the model corresponding to the lowest validation score is selected. It is to be noted that, at the time of CNN training, both *bonafide* and (known) attack samples are used. After the CNN training, the network weights are frozen, and the *bonafide* samples are feedforwarded to obtain the embeddings.

13.3.3.1 One-Class Gaussian Mixture Model

After the training of *MCCNN* with *BCE* and *OCCL*, the trained weights of the network are frozen, and it is used as a fixed feature extractor for the PAD task. Now that a compact representation is available, the objective is to learn a one-class classifier using the features obtained. We use one-class Gaussian mixture model for this task. The one-class GMM is a generative approach which is used for modeling the distribution of the *bonafide* class in the proposed framework.

A Gaussian mixture model is defined as the weighted sum of K multivariate Gaussian distributions as

$$p(x|\Theta) = \sum_{k=1}^{K} w_k \mathcal{N}(x; \mu_k, \Sigma_k) \tag{13.9}$$

where $\Theta = \{w_k, \mu_k, \sigma_k\}_{\{k=1,...,K\}}$ are the weights, means, and the covariance matrix of the GMM.

Expectation-Maximization (EM) [12] was used to compute the parameters of the GMM. A full covariance matrix is computed for each component, and the number of components to use was empirically selected as five ($K = 5$).

During the training phase, embeddings obtained from *bonafide* class only are used to train the one-class GMM.

In test time, a sample is first forwarded though the network to obtain the embedding x, and then fed to the one-class GMM to obtain the log-likelihood score as follows:

$$score = \log(p(x|\Theta)) \qquad (13.10)$$

Algorithm 1: Algorithm for training the proposed framework

Data: (x_i, y_i), where x_i is multi-channel input and $y_i \in 0, 1$; 0 – for attack and 1– for
  bonafide
Result: W_C – CNN weights, Θ_{GMM} – Parameters of GMM
1 **Constants** : λ – weighting factor, μ – learning rate
2 **Initialize** : C_{BF} – center of *bonafide* class, W_C – initial weights of CNN from pre-trained
  model
3 **for** *mini-batch* $\leftarrow$ *1 to P* **do**
4   Forward x_i through the CNN
5   Compute the combined loss: $\mathcal{L} = (1 - \lambda)\mathcal{L}_{BCE} + \lambda\mathcal{L}_{OCCL}$
6   Back-propagate the loss and update the weights of DSUs and FC layers
7   Update the *bonafide* center:
8     $c_{BF} = \hat{c}_{BF}(1 - \alpha) + \alpha\frac{1}{N}\sum_{i=1}^{N} e_i$
9 **end**
10 Forward x_j (*bonafide*, where $y_j = 1$) through the CNN to obtain Embeddings E_j
11 Estimate parameters of GMM from E_j:
12 $\Theta_{GMM} = (w_k, \mu_k, \Sigma_k)$
13 **Parameters** $\leftarrow (W_C, \Theta_{GMM})$

In summary, the proposed framework can be considered as a one-class classifier-based framework for PAD. The crucial distinction is that the features used are ***learned***. The loss function proposed forces the CNN to learn a compact representation for the *bonafide* class leveraging the information from known attack classes. The algorithm for training the framework is shown in Algorithm 1.

13.3.4 Implementation Details

To increase the number of samples, data augmentation using random horizontal flips with a probability of 0.5 was used in training. Adam optimizer [31] was used to minimize the combined loss function. Learning rate of 1×10^{-4} and a weight decay parameter of 1×10^{-5} were used. The network was trained for 50 epochs on GPU grid with a batch size of 32. The model corresponding to minimum validation loss in the *dev* set is selected as the best model. For the four-channel models, the MCCNN architecture has about 13.1 M parameters and about 14.5 GFLOPS. The implementation was done using PyTorch [48] library.

13.4   Experiments

In order to evaluate the effectiveness of the proposed approach, we have performed experiments in three publicly available databases, namely, *WMCA* [23], *MLFP* [1], and *SiW-M* [37] datasets. Recently published *CASIA-SURF* [67] database also consists of multi-channel data, namely, color, depth, and infrared channels with a limited set of attack instruments. However, the raw data from the sensors were not publicly available; in the publicly available version of the database, images were masked and scaled with custom preprocessing reducing the dynamic range of depth and infrared channels severely. Moreover, there was no guaranteed alignment between the channels. Therefore, we cannot use our framework with *CASIA-SURF* database due to the mentioned limitations.

13.4.1   WMCA Dataset

We have conducted an extensive set of experiments on *Wide Multi-Channel presentation Attack* (*WMCA*)[3] database, which contains a total of *1679* video samples of *bonafide* and attack attempts from *72* identities. The database contains information from four different channels collected simultaneously, namely, color, depth, infrared, and thermal channels. The data was collected using two consumer devices, Intel® RealSense™SR300 capturing RGB-NIR-Depth streams and Seek Thermal CompactPRO for the thermal channel. The database contained around 80 different PAIs constituting seven different categories of attacks: print, replay, funny eyeglasses, fake head, rigid mask, flexible silicone mask, and paper masks. The RGB visualization of the attack categories is shown in Fig. 13.6 and the different sessions in Fig. 13.7. Detailed information about the *WMCA* database can be found in the publication [23]. The statistics of the number of samples in each category and their types are shown in Table 13.1. We have made challenging protocols in the *WMCA* dataset to perform an extensive set of evaluations emulating real-world unseen attack scenarios.

13.4.1.1   Protocols in *WMCA*

To test the performance of the algorithm in known and unseen attack scenarios, we created three protocols in the *WMCA* dataset. The protocols are described below:

- **grandtest**: This is the exact same *grandtest* protocol available with *WMCA* database, here all the attack types are present in almost equal proportions in the *train*, *development*, and *evaluation* sets. The attack types and *bonafide* samples are divided into threefolds, and the client ids are disjoint across the three sets. Each

---

[3] Database available at: https://www.idiap.ch/dataset/wmca.

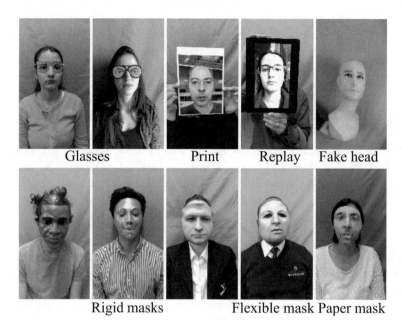

Glasses   Print   Replay   Fake head

Rigid masks   Flexible mask Paper mask

Fig. 13.6 Attack categories in *WMCA* dataset, only RGB images are shown. Print and Replay constitutes the 2D attacks and all others are 3D attacks (Image taken from [23])

Fig. 13.7 Different sessions in *WMCA* dataset, only RGB images are shown. A total of six sessions was used in the *WMCA* (Image taken from [23])

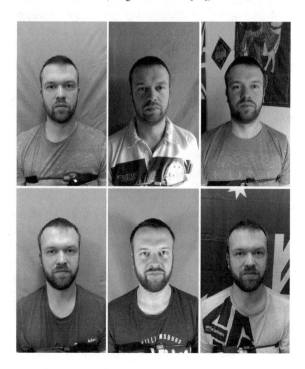

Table 13.1 Statistics of attacks in *WMCA* database

PA category	Type	#Presentations
Bonafide	–	347
Glasses	Partial	75
Print	2D	200
Replay	2D	348
Fake head	3D	122
Rigid mask	3D	137
Flexible mask	3D	379
Paper mask	3D	71
Total		1679

presentation attack instrument had a separate client id. The train, dev, eval splits were made in such a way that a specific PA instrument will appear in only onefold.

- **unseen-2D**: In this protocol, we use same splits as *grandtest* and removed all 2D attacks from *train* and *development* groups. *Evaluation* set contains only *bonafide* and 2D attacks. This emulates the performance of a system when encountered with 2D attacks which was not seen in training.

- **unseen-3D**: In this protocol, we use same splits as *grandtest* and removed all 3D attacks from *train* and *development* groups. *Evaluation* set contains only *bonafide* and 3D attacks. This emulates the performance of a system when encountered with 3D attacks which were not seen in training. This is the most challenging protocol as the model sees only the simpler 2D attacks in training and encounter challenging 3D attacks in testing.

While the *grandtest* protocol emulates the known attack scenario, other protocols emulate the unseen attack scenario. All protocols are made available publicly.

13.4.2 MLFP Dataset

MLFP dataset [1] consists of attacks captured with seven 3D latex masks and three 2D print attacks. The dataset contains videos captured from color, thermal, and infrared channels. Since channels were captured individually in different recording sessions, multi-channel approaches are not trivial. Also, the alignment of channels is not possible since they are not collected simultaneously. Hence, we only use the RGB videos from the MLFP dataset for our experiments. The database contains videos of 10 subjects wearing both print and latex masks. There are 440 videos are consisting of both attacks and *bonafide* for the RGB channel.

13.4.2.1 Protocols in *MLFP*

To emulate known and unseen attack scenarios, we created three new protocols in the *MLFP* dataset. There are two types of attacks, namely, print and mask. Only two sets, i.e., *train* and *evaluation* are created due to the small size of the dataset. We used a subset of the train set (10%) for model selection. The protocols are described below:

- **grandtest**: This protocol emulates the known attack scenario. Both the attacks are present in both *train* and *evaluation* set. However, the subjects and the PAs are disjoint across the two sets.
- **unseen-print**: In this protocol, only *bonafide* and mask attacks are present in *train* set; the *evaluation* set contains only *bonafide* and print attacks. This emulates unseen attack scenario.
- **unseen-mask**: In this protocol, only *bonafide* and print attacks are present in *train* set; the *evaluation* set contains only *bonafide* and mask attacks. This protocol also emulates unseen attack scenario.

13.4.3 SiW-M Dataset

The Spoof in the Wild database with Multiple Attack Types (*SiW-M*) [37] consists of a wide variety of attacks captured only in RGB spectra. The database consists of images from 493 subjects, and a total of 660 *bonafide* and 968 attack samples. A total of 1628 files, consisting of 13 different attack types, collected in different sessions, pose, lighting, and expression (PIE) variations. The attacks consist of various types of masks, makeups, partial attacks, and 2D attacks. The videos are available in 1080P resolution.

13.4.3.1 Protocols in *SiW-M*

To emulate unseen attack scenarios, we use the leave-one-out (LOO) testing protocols available with the *SiW-M* [37] dataset. The protocols consist of only *train* and *eval* sets. In each LOO protocol, the training set consists of 80% percentage of the live data and 12 types of spoof attacks. The evaluation set consists of 20% of *bonafide* data and the attack which was left out in the training phase. The subjects in *bonafide* sets are disjoint in *train* and *evaluation* sets. A subset of the train set (5%) was used for model selection. Additionally, we have created a *grandtest* protocol, specifically for cross-database testing which contains all the attack types in all the folds.

13.4.4 Evaluation Metrics

We report the standardized ISO/IEC 30107-3 metrics [29], Attack Presentation Classification Error Rate (APCER), Bonafide Presentation Classification Error Rate (BPCER), and Average Classification Error Rate (ACER) in the *test* set. A BPCER threshold of 1% is used for computing the threshold in *dev* set. The APCER and BPCER in both *dev* and *eval* sets are also reported. Additionally, the ROC curves for experiments are also shown in all the protocols. For the *MLFP* dataset, we report only EER in the *evaluation* set since only two sets are available. For SiW-M database, we apply a threshold selected a priori in all protocols, for computing the metrics, to be comparable with the results in [37].

13.4.5 Baselines

We have implemented three feature-based baselines and two CNN-based baselines. For a fair comparison, all the benchmarks are multi-channel methods and use the same four channels. Besides, an RGB-only CNN model is also added for comparison. A short description of the baselines along with the acronyms used is shown below:

- *MC-RDWT-Haralick-SVM*: This baseline is the multi-channel extension of the RDWT-Haralick-SVM approach proposed in [1]; the images from all channels are stacked together after preprocessing. For each channel, the image is divided into a 4×4 grid, and Haralick [25] features obtained from the RDWT decompositions are concatenated from all the grids in all channels to get the joint feature vector. The joint feature is used with a linear SVM for PAD.
- *MC-RDWT-Haralick-GMM*: Here, the feature extraction stage is same as *MC-RDWT-Haralick-SVM*; however, the classifier used is one-class GMM. Only *bonafide* samples are used in training this model. This model is added to show the performance of one-class models in unseen attack scenarios.
- *MC-LBP-SVM*: Here, again, the same preprocessing is performed on all the channels first. After this, spatially enhanced histograms of LBP representation from all the component channels are computed and concatenated to a feature vector. The features extracted are fed to an SVM for PAD task.
- *DeepPixBiS*: This is a CNN-based system [20] trained using both binary and pixelwise binary loss functions. This model only uses RGB information for PAD.
- *MC-ResNetPAD*: We reimplemented the architecture from [47] extending it to four channels, based on their open-source implementation.[4] This approach obtained the first place solution in the "CASIA-SURF" challenge. For a fair comparison, instead of using an ensemble, we used the best pre-trained model as suggested in [47].

---

[4] Available from: https://github.com/AlexanderParkin/ChaLearn_liveness_challenge.

- *MCCNN(BCE)*: This is the multi-channel CNN system described in [23], which achieved state-of-the-art performance in the *grandtest* protocol. The model is trained using Binary Cross-Entropy (*BCE*) loss only.

All the baseline methods described are reproducible, and the details about the parameters can be found in our open-source package.[5]

13.4.6   Experiments and Results in WMCA Dataset

We have tested the baselines and the proposed approach in three different protocols in *WMCA*. The proposed approach is denoted as *MCCNN(BCE+OCCL)-GMM*.

- *MCCNN(BCE+OCCL)-GMM*: Here, the *bonafide* embeddings from the *MCCNN* trained using both the losses are used to train a GMM, and in the evaluation stage, the score from the one-class GMM is used as the PAD score.

The results in each protocol are described below.

13.4.6.1   Experiments in *Grandtest* Protocol

The *grandtest* protocol emulates the known attack scenario. Table 13.2 shows the results in the *grandtest* protocol. The proposed approach outperforms the feature-based methods by a large margin as expected. The model *MC-RDWT-Haralick-GMM* trained using a one-class model achieves the worse results. It is interesting to note that the *MC-RDWT-Haralick-SVM* model, trained using the same feature as a binary classifier, performed much better. This shows one weakness of one-class classifiers in a known attack scenario, as they do not use the known attacks in training. The *MCCNN(BCE)* achieves much better performance as compared to *MC-ResNetPAD*. The *MCCNN(BCE)* trained as a binary classifier achieves the best performance in this protocol. The proposed *MCCNN(BCE+OCCL)-GMM* approach achieves comparable performance to *MCCNN(BCE)*. This indicates that the one-class GMM classifier performs on par with the binary classification, provided they are trained with compact feature representations.

13.4.6.2   Experiments in *Unseen-2D* and *Unseen-3D* Protocol

The *unseen-2D* and *unseen-3D* protocols emulate the unseen attack scenario (Table 13.3). The *unseen-3D* is the most challenging protocol since it is trained only on 2D print and replay attacks and encounters a wide variety of 3D attacks such as silicone masks, fake heads, mannequins, etc. in the *eval* set.

---

[5] Source code: https://gitlab.idiap.ch/bob/bob.paper.oneclass_mccnn_2019.

Table 13.2 Performance of the baseline systems and the proposed method in *grandtest* protocol of *WMCA* dataset. The values reported are obtained with a threshold computed for BPCER 1% in *dev* set

Method	Dev (%)		Test (%)		
	APCER	ACER	APCER	BPCER	ACER
MC-RDWT-Haralick-SVM	3.6	2.3	5.4	1.2	3.3
MC-LBP-SVM	3.6	2.3	8.5	0.6	4.6
MC-RDWT-Haralick-GMM	43.4	22.2	47.7	1.7	24.7
DeepPixBiS (RGB only) [20]	1.0	1.0	8.2	3.7	6
MC-ResNetPAD [47]	3.8	2.4	3.5	1.6	2.6
MCCNN (BCE) [23]	0.4	0.7	0.5	0	**0.2**
MCCNN (BCE+OCCL)-GMM	0.1	0.6	0.6	0.1	0.4

Table 13.3 Performance of the baseline systems and the proposed method in **unseen** protocols of *WMCA* dataset. The values reported are obtained with a threshold computed for BPCER 1% in *dev* set

Method	Unseen-2D			Unseen-3D		
	APCER	BPCER	ACER	APCER	BPCER	ACER
MC-RDWT-Haralick-SVM	0.3	0.1	0.2	66.0	0.1	33.1
MC-LBP-SVM	40.7	0.1	20.4	38.9	0.2	19.5
MC-RDWT-Haralick-GMM	0.0	0.2	**0.1**	70.8	1.9	36.4
DeepPixBiS (RGB only) [20]	77.7	0.3	39	74.7	16.3	45.5
MC-ResNetPAD [47]	4.1	0.9	2.5	92.2	6.4	49.3
MCCNN (BCE) [23]	0.0	1.0	0.5	62.0	0.0	31.0
MCCNN (BCE+OCCL)-GMM	0.3	0.6	0.5	15.4	3.9	**9.7**

Most of the approaches perform well in the *unseen-2D* protocol. This result is intuitive as these models are trained on challenging 3D attacks, detection of 2D attacks is much easier. Moreover, the 2D attacks can be easily identified in depth, thermal, and infrared channels. Even some feature-based methods perform well in this protocol, with *MC-RDWT-Haralick-GMM* method achieving the best performance. This shows the advantage of one-class model in an unseen attack scenario. The proposed approach *MCCNN(BCE+OCCL)-GMM* and *MCCNN(BCE)* baseline perform comparably in this protocol. Notably, the DeepPixBiS model achieves much worse results in this protocol. This could be because discriminating between *bonafide* and 2D attacks are harder when only RGB information is used.

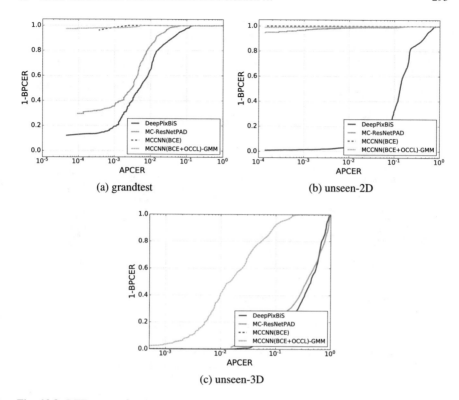

Fig. 13.8 DET curves for the *eval* sets of different protocols of *WMCA* dataset **a** *grandtest*, **b** *unseen-2D*, **c** *unseen-3D* protocol

The *unseen-3D* protocol shows important results. All the baselines show inferior performance when encountered with unseen-3D samples. This shows the failure of binary classifiers in generalizing to challenging unseen attacks. The *MCCNN(BCE)* approach, while being architecturally similar, fails to generalize when trained in the binary classification setting. With the proposed approach, performance improves to 9.7% when the one-class GMM is used on the *bonafide* representations. Since the network learns to map the *bonafide* samples to a compact cluster in the feature space, even in the presence of unseen attacks, the decision boundary learned for the *bonafide* class is robust. The unseen attacks map far from the *bonafide* cluster and hence becomes easy to discriminate from *bonafide* samples. This result is encouraging since the network has shown only 2D attacks in training, and still it manages to achieve good performance against challenging 3D attacks. The ROCs for all the protocols are shown in Fig. 13.8.

The t-SNE [38] plots of the embeddings for all protocols are shown in Fig. 13.9. Five frames from each video in the evaluation sets of the protocols are used for this visualization. While the difference between *bonafide* and attacks is clear in the *grandtest* and *unseen-2D*, difference in *unseen-3D* protocol is very evident. It can

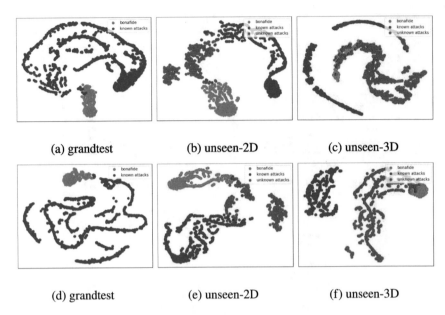

(a) grandtest   (b) unseen-2D   (c) unseen-3D

(d) grandtest   (e) unseen-2D   (f) unseen-3D

Fig. 13.9 t-SNE plots of embeddings in the protocols in *WMCA* dataset. First row (a, b, c) shows the embeddings when only *BCE* loss was used. Second row (d, e, f) shows the embeddings when both the losses are used. Embeddings of both known and unseen attacks are shown in the figures for each protocol. Grandtest protocol contains only known attacks in the test set

be clearly seen that the *bonafide* class clusters together and is far from the *bonafide* representation in the embedding space in the *unseen 3D* protocol when the proposed loss is used. Unseen attacks overlap with *bonafide* embeddings when only *BCE* is used. This clearly demonstrates the effectiveness of the proposed approach for unseen attack detection. The unseen attacks which are overlapping with the *bonafide* region are shown in Fig. 13.10. It can be seen that some video replay samples and flexible silicone 3D masks get misclassified in unseen-2D and unseen-3D protocols, respectively.

13.4.6.3 Ablation Study with Channels

To evaluate the performance of the proposed framework on different set of channels, we perform an ablation study by including a different set of channels. We used only the best performing *MCCNN(BCE+OCCL)-GMM* approach in this ablation study. In all combinations, the grayscale channel is present since it is used as a reference. This is required as the embedding from the grayscale part can be used for face recognition as well.

The acronyms for different channels are shown below:

- G: Grayscale image,

Fig. 13.10 The attack samples which are closer to *bonafide* cluster in **a** unseen-2D (Fig. 13.9e) and **b** unseen-3D (Fig. 13.9f) protocol for the proposed framework

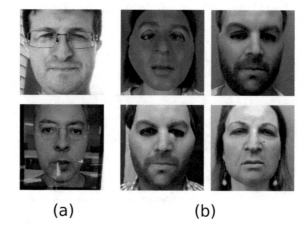

(a)   (b)

Table 13.4 Performance of the proposed framework with different combinations of channels in all protocols of *WMCA* dataset. The values reported are obtained with a threshold computed for BPCER 1% in *dev* set

Channels	Grandtest	Unseen-2D	Unseen-3D
	ACER	ACER	ACER
GDIT	0.4	0.5	9.7
GDI	1.1	11.2	23.1
GT	2.2	3.2	21.5
GD	2.3	49.4	45.4
GI	1.1	2.2	22.6

- D: Depth image,
- I: Infrared channel, and
- T: Thermal channel.

Various combinations of these channels are experimented with, and the results are tabulated in Table 13.4. It is to be noted that the channels G, D, and I come from the same device and T is coming from a different device. Usually, thermal cameras are expensive, compared to RGB-D cameras, and hence the combinations involving subsets of G, D, and I are more interesting from a deployment point of view.

From Table 13.4, it can be seen that the performance degrades as channels are removed. However, the combination GI achieves reasonable performance while considering the performance-cost ratio. The ROCs for different protocols are shown in Fig. 13.11.

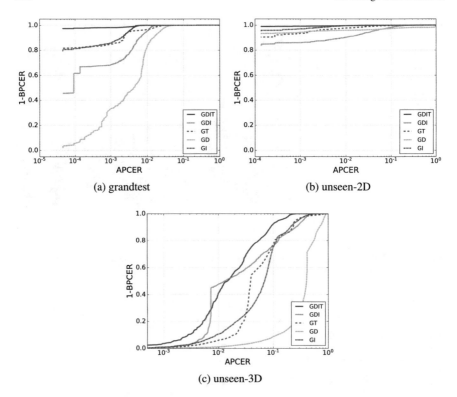

(a) grandtest   (b) unseen-2D

(c) unseen-3D

Fig. 13.11 Ablation study with different combination of channels, DET curves for the *eval* sets of different protocols of *WMCA* dataset **a** *grandtest*, **b** *unseen-2D*, **c** *unseen-3D* protocol

13.4.7 Experiments and Results in MLFP Dataset

We have used only the RGB channel for the experiments since the other channels were not captured simultaneously. For the MCCNN framework and other baselines, "R," "G," and "B" are considered as the different channels in these experiments. We have performed the experiments in the three newly created protocols and the results are tabulated in Table 13.5.

From the results in Table 13.5, it can be seen that the CNN-based approach outperforms the feature-based approaches. The MCCNN framework, with the addition of the newly proposed loss, outperforms the architecture trained with BCE only, showing the effectiveness of the proposed approach.

Even though the proposed approach performs better than the baselines, it is to be noted that the key point of the proposed approach, leveraging multi-channel information, is not utilized here. The architecture is not optimized for PAD in RGB and this experiment is performed only to show the change in performance with the new loss function. Nevertheless, the proposed approach achieves better performance as compared to the baselines in all the protocols.

Table 13.5 Performance of the proposed framework in the protocols in *MLFP* dataset. Only RGB channel was used in this experiments. The values reported are the EER in the *evaluation* set

Algorithm	Grandtest	Unseen print	Unseen mask
MC-RDWT-Haralick-SVM	9.8	12.0	32.2
MC-LBP-SVM	6.3	27.1	9.3
MC-RDWT-Haralick-GMM	27.4	40.8	21.5
DeepPixBiS (RGB only) [20]	6.3	24.8	17.5
MCCNN (BCE)	5.5	9.2	5.2
MCCNN (BCE+OCCL)-GMM	1.2	3.3	3.4

13.4.8 Experiments and Results in SiW-M Dataset

Table 13.6 shows the performance of the proposed framework, again only in the RGB scenario. CNN-based methods are much more powerful than feature-based methods in this case. It can be seen that the proposed approach achieves better performance compared to the baseline methods. The performance of the *MCCNN (BCE+OCCL)-GMM* model is better than that of the *MCCNN(BCE)* model. It can be seen that the addition of the new loss function makes the classification of unseen attacks more accurate.

13.4.9 Cross-Database Evaluations

Since we cannot carry out a cross-database evaluation between a multi-channel database and an RGB-only database, we only used the RGB channels from two datasets for the cross-database evaluation. We selected WMCA and SiW-M datasets since they are relatively large and consist of a wide variety of attacks.

Table 13.7 shows that the MCCNN model achieves comparable performance with and without the new loss. In general, performance in the cross-database setting is poor for all models. The poor performance can be due to the disparity in acquisition conditions and attack types. A larger variety of attacks makes it more difficult for the classifier to identify attacks only via RGB channels. Cross-database performance against this multitude of attacks appears to be more challenging than typical cross-database evaluations that only use 2D attacks. Using multiple channels [23] may

Table 13.6 Performance of the proposed framework in the leave-one-out protocols in *SiW-M* dataset. Only RGB channel was present in this dataset

Methods	Metrics	Replay	Print	Mask attacks					Makeup attacks			Partial attacks			Average
				Half	Silicone	Trans.	Paper	Manne.	Obfusc.	Imperson.	Cosmetic	Funny eye	Paper glasses	Partial paper	
MC-RDWT-H-SVM	ACER	17.15	16.52	22.71	22.49	24.37	8.49	10.19	42.47	8.26	25.79	33.85	24.50	9.61	20.4 ± 10.3
	EER	16.88	16.53	21.80	20.73	21.94	7.34	9.88	32.56	2.37	23.51	31.72	21.94	10.05	18.2 ± 9.0
MC-LBP-SVM	ACER	16.83	17.54	16.38	28.76	30.46	12.15	15.04	59.44	11.97	24.45	30.41	28.31	13.24	23.4 ± 12.9
	EER	15.96	16.83	16.87	28.51	29.77	10.54	12.75	52.60	1.90	24.61	28.32	26.76	11.29	21.2 ± 12.6
Auxiliary [36]	ACER	16.8	6.9	19.3	14.9	52.1	8.0	12.8	55.8	13.7	11.7	49.0	40.5	5.3	23.6 ± 18.5
	EER	14.0	4.3	11.6	12.4	24.6	7.8	10.0	72.3	10.1	9.4	21.4	18.6	4.0	17.0 ± 17.7
DTN [37]	ACER	9.8	6.0	15.0	18.7	36.0	4.5	7.7	48.1	11.4	14.2	19.3	19.8	8.5	16.8 ± 11.1
	EER	10.0	2.1	14.4	18.6	26.5	5.7	9.6	50.2	10.1	13.2	19.8	20.5	8.8	16.1 ± 12.2
DeepPixBiS [20]	ACER	13.94	8.30	6.38	16.53	35.47	4.43	4.81	55.27	5.80	18.95	37.48	43.87	3.69	19.6± 17.4
	EER	11.68	7.94	7.22	15.04	21.30	3.78	4.52	26.49	1.23	14.89	23.28	18.90	4.82	12.3± 8.2
MCCNN (BCE)	ACER	23.01	18.52	7.66	15.02	22.56	4.29	6.02	40.31	5.86	20.19	29.72	32.52	16.54	18.6 ± 11.1
	EER	17.08	11.83	7.56	12.82	16.09	0.71	6.85	25.94	2.29	16.30	18.90	22.82	13.13	13.2 ± 7.4
MCCNN(BCE+OCCL)-GMM	ACER	12.61	12.84	9.69	11.97	25.16	6.87	5.89	29.90	6.34	16.01	16.83	26.97	13.66	14.9 ± 7.8
	EER	12.82	12.94	11.33	13.70	13.47	0.56	5.60	22.17	0.59	15.14	14.40	23.93	9.82	**12.0 ± 6.9**

Table 13.7   The results from the cross-database testing between WMCA and SiW-M datasets using the grandtest protocol, only RGB channels were used in this experiment

Method	Trained on WMCA		Trained on SiW-M	
	Tested on WMCA	Tested on SiW-M	Tested on SiW-M	Tested on WMCA
MC-RDWT-Haralick-SVM	14.6	29.6	15.1	45.3
MC-LBP-SVM	26.6	45.5	19.6	38.6
MC-RDWT-Haralick-GMM	27.9	34.0	25.5	43.6
DeepPixBiS	7.5	49.1	14.7	44.4
MCCNN (BCE)	12.1	34.0	9.9	42.3
MCCNN (BCE+OCCL)-GMM	12.3	31.9	9.5	41.8

alleviate these issues. This also indicates the limitation of RGB-only methods while dealing with a wide variety of attacks.

13.5   Discussions

The experiments in the *WMCA* database clearly show that the CNN-based methods outperform the feature-based methods by a large margin. When comparing the method *MCCNN (BCE)* with the proposed method, the performance in the known attack scenario is comparable. This indicates that the proposed one-class GMM-based approach performs par with the binary classification, thanks to the embedding learned with the proposed loss function. Most approaches work well in the *unseen-2D* protocol, as it can be clearly distinguished in many channels. Furthermore, it shows that simpler attacks are easy to spot if the network is trained in challenging attacks. While the performance in the *grandtest* and *unseen-2D* protocols is comparable, the proposed method achieves a great increase in performance in the most challenging *unseen-3D* protocol. The proposed loss function forces the network to learn a compact representation for *bonafide* examples in the feature space. Both known and unknown attacks are mapped far away from the *bonafide* cluster in the feature space. The decision boundary learned from the one-class model seems robust in identifying both seen and unseen attacks in such a scenario. This finding is significant for several reasons. It is to be noted that in the *unseen-3D* protocol, the network is trained with only 2D attacks, i.e., prints and replays. The proposed method achieves excellent performance in a test set consisting of challenging 3D attacks such as custom silicone masks, paper masks, mannequins, etc. The real-world implications of this approach are very promising. The proposed method can be used to develop robust PAD systems without the requirement of having to manufacture costly presentation attacks. Depending on availability, the PAD models can be trained using easily available attacks. The proposed framework utilizes the available (known) attack categories to learn a robust representation to facilitate known and unseen attack detection. It is to be noted that the compact representation is made possible by the joint multi-channel representation used.

In practical deployment scenarios, computational or cost constraints can prevent the use of all four channels. In such a situation, models trained on available channels can be selected based on the cost-performance ratio by sub-selecting the channels. The results of the ablation study in Table 13.4 can be used to determine which channels should be used in such cases.

In a similar way, the experiments in *MLFP* and *SiW-M* databases also show that CNN-based methods outperform feature-based baselines. Although we did not use multichannel information in these experiments, the experimental results show the performance improvement with the new loss function. Using the proposed framework together with network backbones designed specifically for RGB PAD might improve the results.

The cross-database performance shows the limitations of the RGB channel when tested with a wide variety of attacks. The performance of the baselines, as well as the proposed approach, is poor when only using RGB data. This shows the challenging nature of RGB-only PAD while considering a multitude of attacks. Using multiple channels as done with the *WMCA* dataset might improve the performance.

13.6 Conclusions

Detecting face presentation attacks is often considered as a binary classification task which results in over-fitting to known attacks and results in poor generalization against unseen attacks. In this chapter, we address this problem with a new multi-channel framework that uses a one-class classifier. A novel loss function is formulated which forces the network to learn a compact yet discriminative representation for the face images. Thanks to the proposed loss function, the *bonafide* samples form a compact cluster in the feature space. A decision boundary around the representation of *bonafide* class can be obtained using the one-class model. Both known and unknown attacks are mapped far away from the *bonafide* cluster in the feature space, which can be classified by the one-class model. The proposed framework offers a new way to learn a robust PAD system from *bonafide* and available (known) attack samples. The proposed system was evaluated in the challenging datasets such as *WMCA*, *MLFP*, and *SiW-M* and was observed to surpass the baselines in both known and unseen attack scenarios. The drastic improvement in the performance in the *unseen-3D* protocol in *WMCA* shows the robustness of the proposed approach against unseen attacks thanks to the multi-channel information. The proposed method also shows an improvement even when used together with RGB channels alone. The source code and protocols to reproduce the results are made available publicly to enable further extensions of the proposed framework.

Acknowledgements Part of this research is based on work supported by the Office of the Director of National Intelligence (ODNI), Intelligence Advanced Research Projects Activity (IARPA), via IARPA R&D Contract No. 2017-17020200005. The views and conclusions contained herein are those of the authors and should not be interpreted as necessarily representing the official policies or endorsements, either expressed or implied, of the ODNI, IARPA, or the U.S. Government. The U.S. Government is authorized to reproduce and distribute reprints for Governmental purposes notwithstanding any copyright annotation thereon.

References

1. Agarwal A, Yadav D, Kohli N, Singh R, Vatsa M, Noore A (2017) Face presentation attack with latex masks in multispectral videos. In: 2017 IEEE conference on computer vision and pattern recognition workshops (CVPRW), pp 275–283. https://doi.org/10.1109/CVPRW.2017.40
2. Anjos A, Marcel S (2011) Counter-measures to photo attacks in face recognition: a public database and a baseline. In: 2011 international joint conference on biometrics (IJCB), pp 1–7. IEEE
3. Arashloo SR, Kittler J, Christmas W (2017) An anomaly detection approach to face spoofing detection: a new formulation and evaluation protocol. IEEE Access 5:13868–13882
4. Atoum Y, Liu Y, Jourabloo A, Liu X (2017) Face anti-spoofing using patch and depth-based CNNS. In: 2017 IEEE international joint conference on biometrics (IJCB), pp 319–328. IEEE
5. Bhattacharjee S, Marcel S (2017) What you can't see can help you–extended-range imaging for 3D-mask presentation attack detection. In: Proceedings of the 16th international conference on biometrics special interest group., EPFL-CONF-231840. Gesellschaft fuer Informatik eV (GI)
6. Bhattacharjee S, Mohammadi A, Marcel S (2018) Spoofing deep face recognition with custom silicone masks. In: 2018 IEEE 9th international conference on biometrics theory, applications and systems (BTAS)
7. Boulkenafet Z, Komulainen J, Akhtar Z, Benlamoudi A, Samai D, Bekhouche SE, Ouafi A, Dornaika F, Taleb-Ahmed A, Qin L et al (2017) A competition on generalized software-based face presentation attack detection in mobile scenarios. IJCB 7
8. Boulkenafet Z, Komulainen J, Hadid A (2015) Face anti-spoofing based on color texture analysis. In: 2015 IEEE international conference on image processing (ICIP), pp 2636–2640. IEEE
9. Chingovska I, Anjos A, Marcel S (2012) On the effectiveness of local binary patterns in face anti-spoofing. In: Proceedings of the 11th international conference of the biometrics special interest group, EPFL-CONF-192369
10. Chingovska I, Dos Anjos AR (2015) On the use of client identity information for face anti-spoofing. IEEE Trans Inf Forensics Secur 10(4):787–796
11. Costa-Pazo A, Bhattacharjee S, Vazquez-Fernandez E, Marcel S (2016) The replay-mobile face presentation-attack database. In: 2016 international conference of the biometrics special interest group (BIOSIG). IEEE, pp 1–7
12. Dempster AP, Laird NM, Rubin DB (1977) Maximum likelihood from incomplete data via the EM algorithm. J R Stat Soc Series B 39(1):1–38
13. Dhamecha TI, Nigam A, Singh R, Vatsa M (2013) Disguise detection and face recognition in visible and thermal spectrums. In: 2013 international conference on biometrics (ICB), pp 1–8. IEEE
14. Engelsma JJ, Jain AK (2019) Generalizing fingerprint spoof detector: learning a one-class classifier. arXiv preprint arXiv:1901.03918

15. Erdogmus N, Marcel S (2014) Spoofing face recognition with 3D masks. IEEE Trans Inf Forensics Secur 9(7):1084–1097
16. Fatemifar S, Awais M, Arashloo SR, Kittler J (2019) Combining multiple one-class classifiers for anomaly based face spoofing attack detection. In: International conference on biometrics (ICB)
17. de Freitas Pereira T, Anjos A, Marcel S (2018) Heterogeneous face recognition using domain specific units. IEEE Trans Inf Forensics Secur
18. Galbally J, Marcel S, Fierrez J (2014) Image quality assessment for fake biometric detection: application to iris, fingerprint, and face recognition. IEEE Trans Image Process 23(2):710–724
19. Gan J, Li S, Zhai Y, Liu C (2017) 3D convolutional neural network based on face anti-spoofing. In: 2017 2nd international conference on multimedia and image processing (ICMIP), pp 1–5. IEEE
20. George A, Marcel S (2019) Deep pixel-wise binary supervision for face presentation attack detection. In: International conference on biometrics
21. George A, Marcel S (2020) Can your face detector do anti-spoofing? face presentation attack detection with a multi-channel face detector. Idiap Research Report, Idiap-RR-12-2020
22. George A, Marcel S (2020) Learning one class representations for face presentation attack detection using multi-channel convolutional neural networks. IEEE Trans Inf Forensics Secur 1 (2020)
23. George A, Mostaani Z, Geissenbuhler D, Nikisins O, Anjos A, Marcel S (2019) Biometric face presentation attack detection with multi-channel convolutional neural network. IEEE Trans Inf Forensics Secur 1. DoI:https://doi.org/10.1109/TIFS.2019.2916652
24. Hadsell R, Chopra S, LeCun Y (2006) Dimensionality reduction by learning an invariant mapping. In: 2006 ieee computer society conference on computer vision and pattern recognition (CVPR'06), vol 2. IEEE, pp 1735–1742
25. Haralick RM (1979) Statistical and structural approaches to texture. Proc IEEE 67(5):786–804
26. Heusch G, George A, Geissbühler D, Mostaani Z, Marcel S (2020) Deep models and shortwave infrared information to detect face presentation attacks. IEEE Trans Biom Behav Identity Sci (T-BIOM)
27. Heusch G, Marcel S (2018) Pulse-based features for face presentation attack detection. In: 2018 IEEE 9th international conference on, special session On biometrics theory, applications and systems (BTAS), image and video forensics in biometrics (IVFIB)
28. Heusch G, Marcel S (2019) Remote blood pulse analysis for face presentation attack detection. In: Handbook of biometric anti-spoofing, pp 267–289. Springer
29. ISO/IEC JTC 1/SC 37 (2016) Biometrics: information technology—international organization for standardization. Iso standard, International Organization for Standardization
30. Jaiswal A, Xia S, Masi I, AbdAlmageed W (2019) Ropad: robust presentation attack detection through unsupervised adversarial invariance. arXiv preprint arXiv:1903.03691
31. Kingma DP, Ba J (2014) Adam: a method for stochastic optimization. arXiv preprint arXiv:1412.6980
32. Learned-Miller E, Huang GB, RoyChowdhury A, Li H, Hua G (2016) Labeled faces in the wild: a survey. In: Advances in face detection and facial image analysis. Springer, pp 189–248
33. Leys C, Ley C, Klein O, Bernard P, Licata L (2013) Detecting outliers: do not use standard deviation around the mean, use absolute deviation around the median. J Exp Soc Psychol 49(4):764–766
34. Li H, He P, Wang S, Rocha A, Jiang X, Kot AC (2018) Learning generalized deep feature representation for face anti-spoofing. IEEE Trans Inf Forensics Secur 13(10):2639–2652
35. Li H, Lu H, Lin Z, Shen X, Price B (2015) LCNN: low-level feature embedded CNN for salient object detection. arXiv preprint arXiv:1508.03928
36. Liu, Y., Jourabloo, A., Liu, X.: Learning deep models for face anti-spoofing: Binary or auxiliary supervision. In: Proceedings of the IEEE conference on computer vision and pattern recognition, pp 389–398
37. Liu, Y., Stehouwer, J., Jourabloo, A., Liu, X.: Deep tree learning for zero-shot face anti-spoofing. In: The IEEE Conference on Computer Vision and Pattern Recognition (CVPR) (2019)

38. Maaten LVd, Hinton G (2008) Visualizing data using t-SNE. J Mach Learn Res 9:2579–2605
39. Määttä J, Hadid A, Pietikäinen M (2011) Face spoofing detection from single images using micro-texture analysis. In: 2011 international joint conference on biometrics (IJCB). IEEE, pp 1–7
40. Mallat S (2016) Understanding deep convolutional networks. Philo Trans R Soc A 374(2065):20150203
41. Marcel S, Nixon M, Li S (2014) Handbook of biometric anti-spoofing-trusted biometrics under spoofing attacks. Adv Comput Vis Pattern Recognit. Springer
42. Marcel S, Nixon MS, Fierrez J, Evans N (2018) Handbook of biometric anti-spoofing: presentation attack detection. In: Marcel S, Nixon MS, Fierrez J, Evans N (eds). Springer International Publishing, 2nd edn. ISBN: 978-3319926261. https://doi.org/10.1007/978-3-319-92627-8.http://www.eurecom.fr/publication/5667
43. Mehta S, Uberoi A, Agarwal A, Vatsa M, Singh R, Crafting a panoptic face presentation attack detector
44. Mostaani Z, George A, Heusch G, Geissenbuhler D, Marcel S (2020) The high-quality wide multi-channel attack (hq-wmca) database. Idiap-RR Idiap-RR-22-2020, Idiap
45. Nikisins O, George A, Marcel S (2019) Domain adaptation in multi-channel autoencoder based features for robust face anti-spoofing. In: 2019 international conference on biometrics (ICB). IEEE, pp 1–8
46. Nikisins O, Mohammadi A, Anjos A, Marcel S (2018) On effectiveness of anomaly detection approaches against unseen presentation attacks in face anti-spoofing. In: The 11th IAPR international conference on biometrics (ICB 2018), EPFL-CONF-233583
47. Parkin A, Grinchuk O (2019) Recognizing multi-modal face spoofing with face recognition networks. In: Proceedings of the IEEE conference on computer vision and pattern recognition workshops, pp 0–0
48. Paszke A, Gross S, Chintala S, Chanan G, Yang E, DeVito Z, Lin Z, Desmaison A, Antiga L, Lerer A (2017) Automatic differentiation in pytorch. In: NIPS-W
49. Patel K, Han H, Jain AK, Ott G (2015) Live face video vs. spoof face video: Use of moiré patterns to detect replay video attacks. In: 2015 international conference on biometrics (ICB). IEEE, pp 98–105
50. Perera P, Patel VM (2019) Learning deep features for one-class classification. IEEE Trans Image Process 28(11):5450–5463
51. Pérez-Cabo D, Jiménez-Cabello D, Costa-Pazo A, López-Sastre RJ (2019) Deep anomaly detection for generalized face anti-spoofing. In: Proceedings of the IEEE conference on computer vision and pattern recognition workshops, pp 0–0
52. Qi C, Su F (2017) Contrastive-center loss for deep neural networks. In: 2017 IEEE international conference on image processing (ICIP). IEEE, pp 2851–2855
53. Radford A, Metz L, Chintala S (2015) Unsupervised representation learning with deep convolutional generative adversarial networks. arXiv preprint arXiv:1511.06434
54. Raghavendra R, Raja KB, Venkatesh S, Busch C (2017) Extended multispectral face presentation attack detection: an approach based on fusing information from individual spectral bands. In: 2017 20th international conference on information fusion (Fusion). IEEE, pp 1–6
55. Ramachandra R, Busch C (2017) Presentation attack detection methods for face recognition systems: a comprehensive survey. ACM Comput Surv (CSUR) 50(1):8
56. Shao R, Lan X, Yuen PC (2017) Deep convolutional dynamic texture learning with adaptive channel-discriminability for 3D mask face anti-spoofing. In: 2017 IEEE international joint conference on biometrics (IJCB). IEEE, pp 748–755
57. Steiner H, Kolb A, Jung N (2016) Reliable face anti-spoofing using multispectral SWIR imaging. In: 2016 international conference on biometrics (ICB). IEEE, pp 1–8
58. Wang G, Lan C, Han H, Shan S, Chen X (2019) Multi-modal face presentation attack detection via spatial and channel attentions. In: Proceedings of the IEEE Conference on computer vision and pattern recognition workshops, pp 0–0
59. Wen D, Han H, Jain AK (2015) Face spoof detection with image distortion analysis. IEEE Trans Inf Forensics Secur 10(4):746–761

60. Wen Y, Zhang K, Li Z, Qiao Y (2016) A discriminative feature learning approach for deep face recognition. In: European conference on computer vision. Springer, pp 499–515

61. Wu X, He R, Sun Z, Tan T (2018) A light CNN for deep face representation with noisy labels. IEEE Trans Inf Forensics Secur 13(11):2884–2896

62. Xiong F, AbdAlmageed W (2018) Unknown presentation attack detection with face RGB images. In: 2018 IEEE 9th international conference on biometrics theory, applications and systems (BTAS). IEEE, pp 1–9

63. Xiong X, De la Torre F (2013) Supervised descent method and its applications to face alignment. In: Proceedings of the IEEE conference on computer vision and pattern recognition, pp 532–539

64. Yang J, Lei Z, Li SZ (2014) Learn convolutional neural network for face anti-spoofing. arXiv preprint arXiv:1408.5601

65. Yosinski J, Clune J, Bengio Y, Lipson H (2014) How transferable are features in deep neural networks? In: Advances in neural information processing systems, pp 3320–3328

66. Zhang K, Zhang Z, Li Z, Qiao Y (2016) Joint face detection and alignment using multitask cascaded convolutional networks. IEEE Signal Process Lett 23(10):1499–1503

67. Zhang S, Wang X, Liu A, Zhao C, Wan J, Escalera S, Shi H, Wang Z, Li SZ (2018) Casia-surf: a dataset and benchmark for large-scale multi-modal face anti-spoofing. arXiv preprint arXiv:1812.00408

Chapter 14
Scalable Person Re-identification: Beyond Supervised Approaches

Rameswar Panda and Amit Roy-Chowdhury

Abstract Person re-identification across cameras is an important problem as it enables associating targets over a wide area, which is likely to be viewed by multiple cameras. It is an extremely active area of research today. Most of the approaches are extensively supervised, in the sense that they require significant labeling effort to train re-identification models, usually based on deep networks. However, as in other problems in computer vision, it raises the question of scalability of the approaches as the number of people to be associated grows or the size of the network grows. In this chapter, we focus on two problems that hold the potential for developing highly scalable person re-identification approaches. In the first, we focus on the problem of how to limit the labeling effort even as the number of targets in the network grows. On the challenging Market-1501 dataset, we demonstrate that with only 8% labeling, we can achieve performance very close to that with full-set labeling. In the second problem, we focus on the size of the camera network and consider how to onboard new cameras into an existing network with little to no additional supervision. We leverage upon transfer learning approaches for this purpose and demonstrate the results on a benchmark dataset. Overall, the chapter provides some research directions and initial results in pushing person re-identification beyond fully supervised approaches and lays the groundwork for future research in this area.

14.1 Introduction

Person re-identification (re-id) across cameras is an extremely important problem in computer vision as it forms the foundation for scene understanding across a wide area, which is likely to be covered by multiple cameras. Similar to most other recognition

R. Panda · A. Roy-Chowdhury (✉)
University of California, Riverside, Riverside, USA
e-mail: amitrc@ece.ucr.edu

problems, the most successful approaches have been based on supervised training phases. Labeled data across pairs of cameras are used to learn models that define the transformation between the views in two cameras, and these learned models are used to associate between images during the testing phase. However, this level of supervision hampers scalability of the problem because of the need to label quantities of data, which grows with the size of the camera network and the variety of conditions that may be encountered.

In this chapter, we discuss the possibility of significantly reducing the level of supervision in person re-identification problems without any sacrifice in performance. This will ensure that it is possible to scale re-identification problems to larger and larger networks of cameras without compromising the accuracy of the association task. We specifically consider the following questions in person re-identification in camera networks, building upon recent work in our group.

Optimal subset selection for labeling. Given unlabeled training data across a network of cameras and a similarity measure, can we select a minimal subset of images that should be labeled and from which the person re-identification models can be learned? The intuition here is that if we choose this minimal subset judiciously, the labels can be propagated using the similarity measure to the rest of the dataset. Thus, most of the labels would be obtained automatically with only a small subset of images being labeled.

On-boarding new cameras through transfer learning. Is it possible to leverage upon learned re-identification models to onboard new cameras to an existing network, again with limited to no additional supervision? We will show that, building upon transfer learning concepts, we can augment an existing network with additional cameras. Compared to a fully supervised approach to learn models, in this augmented network, it is possible to achieve similar performance with little additional supervision by building upon the transfer learning process.

The first problem mentioned above relates to scalability of person re-id as the number of people grows. The second relates to scalability as the size of the camera network grows. We start with a survey of relevant literature in person re-identification, followed by an exposition of two approaches where we demonstrate how person re-id approaches with limited supervision can be developed—what the challenges are, what possible approaches exist, and some sample results.

14.2 Related Work

A thorough recent survey on person re-identification can be found at [84]. In this chapter, we focus on a few relevant papers that are critical to our problem of person re-identification under limited supervision.

Supervised Person Re-identification. Most existing person re-id techniques are based on supervised learning. These methods either seek the best feature representation [5, 40, 49, 75] or learn discriminant metrics [23, 26, 36–38, 58, 66, 82, 86] that yield an optimal matching score between two cameras or between a gallery and a probe image. Recently, deep learning methods have shown significant performance improvement on image classification and has been applied to person re-id [11, 16, 42, 74, 76, 78, 80]. Combining feature representation and metric learning with an end-to-end deep neural network is also a recent trend in re-identification [1, 35, 77]. Considering that a modest-sized camera network can easily have dozens of cameras, these supervised re-id models will require a huge amount of labeled data which are difficult to collect in real-world settings.

Unsupervised Person Re-identification. Unsupervised learning models have received little attention in re-identification because of their weak performance on benchmarking datasets compared to supervised methods. Representative methods along this direction use either hand-crafted appearance features [12, 41, 45, 46] or saliency statistics [81] for matching persons without requiring a huge amount of labeled data. Recently, dictionary learning-based methods have also been utilized in an unsupervised setting [2, 28, 29, 43].

Open-World Person Re-Identification. Open-world recognition has been introduced in [6] as an attempt to move beyond the dominant static setting where the number of classes is not fixed in recognition. Inspired by such approaches, there have been few works in re-identification [8, 87] which try to address the open-world scenario by assuming that gallery and probe sets contain different identities of persons.

Incremental and Active Learning. In an effort to bypass tedious labeling of training data, there has been recent interest in "active learning" [64] to intelligently select unlabeled examples for the experts to label in an interactive manner. This can be achieved by choosing one sample at a time by maximizing the value of information [25], reducing the expected error [3], or minimizing the resultant entropy of the system [7]. Recently, works selecting batches of unlabeled data by exploiting classifier feedback to maximize informativeness and sample diversity [10, 17] were proposed. Specific application areas in computer vision include, but are not limited to, tracking [70], scene classification [25, 68], semantic segmentation [67], video annotation [27], and activity recognition [22]. Authors in [4] propose a scalable re-id framework using manifold smoothing. Active learning is introduced for incremental updates in [71]. Another scalable re-id framework with a human in the loop is proposed in [72]. In [14], an entropy-based selection approach is proposed for reducing manual annotation. In [50],the authors uses a dominant clustering-based approach for probe relevant set selection and utilizes it for pair selection in a dynamic setting. Transitivity is utilized in [13] for increasing performance by re-organizing the predicted assignment matrix. The method proposed in [39] also uses similar ideas in a deep learning-based framework.

Domain Adaptation. Domain adaptation, which aims to adapt a source domain to a target domain, has been successfully used in many areas of computer vision and machine learning, e.g., object classification and action recognition and speech processing. Despite its applicability in classical vision tasks, domain adaptation for re-identification still remains as a challenging and under-addressed problem. Only very recently, domain adaptation for re-id has begun to be considered [33, 34, 44, 73, 85]. However, these studies consider only improving the re-id performance in a static camera network with a fixed number of cameras. Furthermore, most of these approaches learn supervised models using labeled data from the target domain.

Budget-Constrained Learning. The problem of video analysis under budget constraints has been studied by few researchers [54, 55, 69] in the recent past; however, none look into the problem of re-identification under budget constraints. Activity detection under a computational. budget is considered in [65].

14.3   Optimal Subset Selection for Labeling

14.3.1   Problem Statement

Supervised distance metric learning-based methods for person re-id are specifically popular because of their robustness toward large color variations and fast training speed. However, like other supervised methods, metric learning algorithms have their own burden of human labeling effort especially for large camera networks [60]. The total number of training pairs assumed to be available by these algorithms increases tremendously with network size and the number of persons in each camera. Manual labeling of such a huge number of pairs is a tedious and expensive process. So naturally, a question arises: given a camera network, *can we come up with a strategy of choosing a minimal subset of image pairs for labeling without compromising on recognition performance?* This is a problem of considerable significance in the context of person re-id in multi-camera networks, especially in larger ones. However, the problem has received little attention in the literature thus far. Transitive relations among person identities across multiple cameras and their logical consequences are strongly informative properties. These properties have been explored previously for globally consistent person re-id in several existing works [9, 13, 39]. Though it may not be apparent at first, we can also exploit these transitive relations to reduce manual pairwise annotation effort. To illustrate the idea, let us consider few plausible scenarios as shown in Fig. 14.1a.

- In camera pair 1–2 and 1–3, if we know from human labeling person that pairs $P_1^1 - P_2^1$ and $P_1^1 - P_3^2$ are positive matches, then from transitivity, we can directly infer that P_2^1 and P_3^2 also have same identity. Similarly, given labels of $P_1^1 - P_2^1$ (+ve) and $P_1^1 - P_3^1$ (-ve), we can infer that $P_2^1 - P_3^1$ is negative.
- However, given that we already know labels of $P_1^1 - P_2^2$ (-ve) and $P_1^1 - P_3^1$ (-ve), we still cannot conclude anything about pair $P_2^2 - P_3^1$.

So, from the examples above, we can make a simple observation, i.e., if we do not ask an expert for the label of the third pair/s in the first two cases described above, required labeling effort will be considerably reduced. However, this seemingly simple

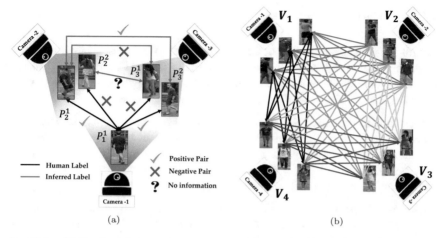

(a)   (b)

Fig. 14.1 a Motivation of our approach. Here, we have a camera network with three cameras. P_k^i represent the "i"-th person in the "k"-th camera. Now suppose, we ask the human to label the pairs $P_1^1 - P_2^1$ and $P_1^1 - P_3^2$ by asking a yes/no question. As both of them are positive matches, after we know the labels of these two pairs using transitivity property, we can correctly infer the label of $P_2^1 - P_3^2$. Similarly, if we know labels of $P_1^1 - P_2^1$ and $P_1^1 - P_3^1$, we can precisely infer that $P_2^1 - P_3^1$ is a negative match. However, knowing the labels of pairs $P_1^1 - P_2^2$ and $P_1^1 - P_3^2$ does not give us any information about the pair $P_2^2 - P_3^1$. **b Network Representation**. This figure demonstrates the representation of a camera network with four cameras as a k-partite graph with k=4. Best viewed in color

strategy implicitly makes an invalid assumption that we already have access to the pair-labels from human. Also, note that, if we arbitrarily choose subsets of pairs for labeling, there is no guarantee that we will be able to take advantage of pairwise-relations as we will end up frequently in situations like the third scenario (occurrence probability of this scenario is significantly higher than the other two). So, in order to actually reduce annotation effort using this transitivity-based approach, we have to choose image pairs in a judicious manner. Toward this objective, in this work, we first formulate this pair subset selection as a combinatorial optimization problem on edge-weighted k-partite graph. This combinatorial optimization can be represented as a binary integer program which we can solve exactly for smaller datasets using standard techniques such as branch and cut [52] and cutting plane algorithms [48]. However, as it is an NP-hard optimization problem, solving it with exact algorithms takes exponential order time and for larger datasets, it becomes intractable. So, in order to scale up the proposed methodology for large camera networks, we propose two polynomial time sub-optimal algorithms for our optimization problem. The first proposed algorithm is a pure greedy algorithm and second one is a 1/2-approximation algorithm [61].

14.3.2 Solution Overview

Graph-Based Representation of Camera Network. We represent any camera net-work as a edge weighted complete k-partite graph $G_k = (V, E)$ (see Fig. 14.1b). Below we describe how this partite graph is constructed from a camera network consisting of k cameras and total n persons across all cameras.

Vertex and Edges: Each vertex in G_k denotes a person in the camera network. To be precise, vertex $v_{k'}^i$ represents the i-th person from k'-th camera. From now on, throughout rest of the work, we will use the terms "person" and "vertex" inter-changeably. An edge $E_{k_1,k_2}^{i,j} = (v_{k_1}^i, v_{k_2}^j)$ denotes probable correspondence between i-th person in camera k_1 and j-th person in camera k_2.

Vertex Set Partitions: As per our definition, the set of all the persons in a camera network forms the vertex set V of G_k. Now in our framework, we assume the intra-camera vertices are not connected to each other, i.e., they form an independent vertex set. So, k sets of vertices from each different camera form k different partitions. More formally, $V = (V_1, V_2, ..., V_k)$ where $V_{k'} = \{v_{k'}^1, v_{k'}^2, \ldots, v_{k'}^{n_{k'}}\}$ is the set of $n_{k'}$ persons in k'-th camera. So, if we have $n_1, n_2, \ldots, n_k$ persons in camera 1, camera 2, ..., camera k, respectively, the cardinality of the set V is

$$|V| = \sum_{i=1}^{k} |V_i| = n_1 + n_2 + \ldots + n_k = n \qquad (14.1)$$

Now, G_k is a complete multipartite graph as we have probable correspondences (i.e., an weighted edge) between every pair of vertices from different partitions. So the total number of edges in the graph can be computed as follows:

$$|E| = \sum_{\substack{\forall k_1 \in \{1,2,...,k\} \\ \forall k_2 \in \{1,2,...,k\} \\ s.t.\ k_1 < k_2}} n_{k_1} n_{k_2} \qquad (14.2)$$

Edge weight: We define our edge weight function $\mathcal{F}_w : E \to \mathbb{R}$ as follows:

$$\mathcal{F}_w(E_{k_1,k_2}^{i,j}) = \mathcal{S}(v_{k_1}^i, v_{k_2}^j) \qquad (14.3)$$

where $\mathcal{S}$ is a function which computes similarity or association score between two persons $v_{k_1}^i$ and $v_{k_2}^j$. Note that our framework can be used with any kind of similarity measure. As we define our objective function later over non-negative edge weights, the proposed scheme will scale any negative-valued similarity score into a non-negative value using the sigmoid function. In this work, we compute similarity scores between a pair of shots of two persons as follows:

$$\mathcal{S}(v_{k_1}^i, v_{k_2}^j) = \frac{1}{1 + \exp(\mathcal{D}(f_{k_1}^i, f_{k_2}^j) - \mu)} \qquad (14.4)$$

where $f^i_{k_1}$, $f^j_{k_2}$ are the feature vectors of the corresponding persons $v^{k_1}_i$, $v^{k_2}_j$, respectively, $\mathcal{D}$ is a distance function giving distance between two feature vectors, and μ is a threshold.

Triangle: Complete subgraphs (or clique) of size 3 are termed as triangle in any graph. Naturally, whenever we have three persons(vertices), $v^i_{k_1}$, $v^j_{k_2}$, $v^l_{k_3}$ from three different cameras (camera k_1, camera k_2, and camera k_3), they form a triangle, $T^{i,j,l}_{k_1,k_2,k_3} = \left\{ v^i_{k_1}, v^j_{k_2}, v^l_{k_3} \right\}$. As we progress, we will see that triangles are the central objects around which our whole framework evolves.

With the initial setup in place, we can now formulate the image pair selection task as an optimization problem on our graph G_k. Let us consider first revisiting the problem statement of the budget-constrained pair selection task.

14.3.2.1 Optimization Problem

Given a labeling budget, B, and a set of training image pairs from a camera network, we have to select an optimal subset of size at most B for human annotation. The notion of "optimal subset" is *incomplete*. As seen from Fig. 14.1a, the transitive relations defined over associations between different persons (vertices) can be utilized for labeling effort reduction. Now we give that idea a concrete shape by making some specific observations in the context of our graph G_k.

- For any triangle in our graph, we have a total of three edges from which we can select for manual labeling.
- We may always want to select positive edges as they will contribute more toward reducing manual labeling effort because transitive inference in our graph always requires at least one positive edge.
- As shown in Fig. 14.1, if we have precise information about two edges in a triangle of our graph and one of them is a positive edge, then we can deterministically infer the label of the third edge. For this reason, we must always want to constrain the number of edges chosen for manual labeling in a triangle to be at most two in order to respect the budget.
- As we cannot foresee the actual labels, we have to choose that pair of edges from any triangle which will maximize the probability of getting at least one positive match.
- Also, note that any edge is a part of multiple triangles in our graph, so inference propagation can occur from different directions.

With these observations in mind, our optimization problem can be stated as follows:

- *Given a complete k-partite graph $G_k = (V, E)$ with non-negative edge weights and an integer B, choose a maximum-weight set S of edges from E such that $G'=(V,S)$ is triangle-free and $|S| \leq B$.*

14.3.2.2   An Equivalent Binary Integer Program

Our combinatorial optimization can be formulated as a binary integer programing problem as follows:

$$
\underset{\substack{x_{k_1,k_2}^{i,j} \\ \forall (i,j) \in \delta(k_1,k_2) \\ \forall k_1,k_2 \in \{1,\ldots,k\} \; s.t \; k_1 < k_2}}{\operatorname{argmax}} \left(\sum_{\substack{k_1,k_2=1 \\ k_1<k_2}}^{k} \sum_{i,j=1}^{n_{k_1},n_{k_2}} w_{k_1,k_2}^{i,j} x_{k_1,k_2}^{i,j} \right) \tag{14.5}
$$

$$
\text{subject to :} \sum_{\substack{k_1,k_2=1 \\ k_1<k_2}}^{k} \sum_{i,j=1}^{n_{k_1},n_{k_2}} x_{k_1,k_2}^{i,j} \leq B,
$$

$$
\forall (i,j) \in \delta(k_1,k_2) \; \forall k_1, k_2 \in \{1,2,\ldots,k\} \; s.t \; k_1 < k_2 \tag{14.6}
$$

$$
x_{k_1,k_2}^{i,j} + x_{k_1,k_3}^{i,l} + x_{k_2,k_3}^{j,l} \leq 2, \forall (i,j) \in \delta(k_1,k_2)
$$

$$
k_1, k_2, k_3 \in \{1,2,\ldots,k\} s.t. \; k_1 < k_2 < k_3 \tag{14.7}
$$

$$
x_{k_1,k_2}^{i,j} \in \{0,1\}, \forall (i,j) \in \delta(k_1,k_2),
$$

$$
\forall k_1, k_2 \in \{1,2,\ldots,k\} \; s.t. \; k_1 < k_2 \tag{14.8}
$$

where Eq. (14.5) represents the linear objective function, which aims to maximize the total weight of the chosen subgraph. $\delta(k_1,k_2)$ denotes the edge-set between cameras k_1 and k_2. Equations (14.6)–(14.8) are the constraints we have to satisfy. $x_{k_1,k_2}^{i,j}$ denotes the edge between i-th person in camera k_1 and j-th person in camera k_2. $x_{k_1,k_2}^{i,j}$'s are defined over all possible values of i, j, k_1, and k_2 as described above and together all possible $x_{k_1,k_2}^{i,j}$'s form the decision variable set. $w_{k_1,k_2}^{i,j}$'s are the weights of the corresponding edges and B is our labeling budget. The first constraint (14.6) dictates that we can select at most B number of edges. Equation (14.7) constrains that the subgraph formed by the selected edges be triangle-free. Equation (14.8) denotes that optimization variable be binary, where a 1 would indicate that an edge is chosen for manual labeling and 0 otherwise.

14.3.2.3   Polynomial Time Approximation Optimal Algorithms

On smaller datasets, we can easily solve our optimization problem using traditional integer programming algorithms, such as cutting plane methods [48] and Branch and Cut [52]. These methods always provide globally optimal solutions. However, as they are exponential time algorithms, we cannot employ them for larger datasets. In order to tackle this challenge, we propose two polynomial time algorithms which drastically improve scalability.

Algorithm 1. This algorithm is motivated by the observation that if we make any cut on the vertex set of a graph, the set of cut crossing edges induces a triangle-free subgraph. So if we can make a cut that maximizes the total weight of edges crossing the cut, then we may construct a approximately optimal solution using those edges. In graph theory, the max-cut problem is well studied where the objective is to find such a max-weight cut. As max-cut is also an NP-hard [51] problem, there is no known efficient algorithm for it. However, there exists a deterministic $1/2$-approximation algorithm for max-cut [20, 53]. Our first algorithm uses this $1/2$-max-cut to achieve $1/2$ approximation for our problem. After initialization steps, the Max-Cut Select algorithm constructs the subgraph G' using the top B heaviest edges in E. Then it employs the deterministic $1/2$-max-cut algorithm on G' to generate a cut $(S, V \backslash S)$. Finally, the algorithm selects the set T of edges which crosses the cut $(S, V \backslash S)$ and returns it.

Algorithm 2. Often in practice, simple greedy heuristics give better performance as compared to other theoretically superior algorithms. This perspective has motivated us to explore greedy strategies for our problem resulting in the "Greedy-Select" algorithm. Greedy-Select begins with an empty set T and iterates over the edges in decreasing weight order. In each iteration, the algorithm adds the current edge to the set T if the current edge does not form any triangle with the existing edges in T. The algorithm terminates either when we have collected B number of edges in set T or we have iterated over all the edges in the graph.

14.3.3 Sample Experimental Results

Dataset. We experiment on Market-1501 dataset [83], which is one of the largest person re-identification datasets available today. It has 32,668 images of 1501 persons taken from six cameras. We use the train-test split given in the dataset. Apart from large variations in pose and illuminations, the size of the dataset itself introduces a new level of computational challenge. For Market-1501 dataset, the optimization problem we consider has more than 4.3 millions of variables.

Settings. We use KISS metric learning method [30] for our experiments. To represent each person node in the graph, we use 29600-dimensional LOMO features [40]. For metric learning, we project the features into 100-dimensional space using PCA. We use the Euclidean metric as our distance function. In any online setting, similarity scores at any time instant can be computed using the learned metric from the previous instance. Given a budget of B, we use a portion of the budget $pB(0 < p < 1)$ [we used $p = 0.7$ for experiments] to select triangle-free edges using our optimization problem. However, in cases where the selected edges in a triangle are both negative matches, we cannot infer about the label of the third edge and we may want to gather information about it. For this reason, after the first stage of triangle-free selection, we employ a greedy top selection mechanism to exhaust the rest of the budget. We use Cumulative Matching Curves (CMC) to demonstrate recognition performance

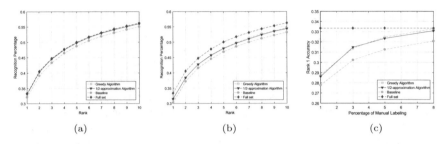

Fig. 14.2 This figure presents the comparisons of the proposed approach with baselines on the Market-1501 dataset using Configuration 2. (**a**) and (**b**) are CMC curves with 8% and 3% manual labeling, respectively. **c** Presents the plot for manual labeling effort vs. Rank-1 accuracy. Best viewed in color

at a given budget. We also provide labeling effort vs. recognition performance plots trade-off between the two.

Baseline. In this work, we use top-B edge selection as the baseline strategy. For all our experiments, we compare our method against this baseline.

Results. Figure 14.2a demonstrates re-id performance with 8% labels. From Fig. 14.2a, we observe that both of our approaches achieve full-set accuracy with this amount of labeling. While with 3% labels, the performance of the proposed approaches slightly degrades (see Fig. 14.2b). In Fig. 14.2c, we provide the manual labeling percentage vs rank-1 accuracy graph. From all these three graphs, it can be easily observed that our approach performs better than the baseline across all the conducted experiments on Market dataset [61].

14.4 On-Boarding New Cameras through Transfer Learning

14.4.1 Problem Statement

In this section, we address a very practical problem in camera networks, but which has received little attention in the person re-identification literature. *Given a camera network where the inter-camera transformations/distance metrics have been learned in an intensive training phase, how can we onboard new cameras into the installed system with minimal additional effort?* This is an important problem to address in realistic open-world re-identification scenarios, where a new camera may be temporarily inserted into an existing system to get additional information. To illustrate such a problem, let us consider a scenario with $\mathcal{N}$ cameras for which we have learned the optimal pairwise distance metrics, so providing high re-id accuracy for all camera pairs. However, during a particular event, a new camera may be temporarily

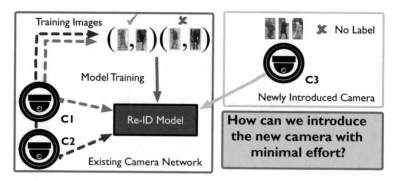

Fig. 14.3 Consider an existing network with two cameras c_1 and c_2 where we have learned a re-id model using pairwise training data from both of the cameras. During the operational phase, a new camera c_3 is introduced to cover a certain area that is not well covered by the existing two cameras. Most of the existing methods do not consider such dynamic nature of a re-id model. In contrast, we propose to adapt the existing re-id model in an unsupervised way by exploring: *what is the best source camera to pair with the new camera and how can we exploit the best source camera to improve the matching accuracy across the other camera*

onboarded to cover a certain related area that is not well covered by the existing network of $\mathcal{N}$ cameras (see Fig. 14.3 for an illustrative example). Despite the dynamic and open nature of the world, almost all work in re-identification assume a *static* and *closed* world model of the re-id problem where the number of cameras is fixed in a network. Given newly introduced camera(s), traditional re-id methods will try to relearn the inter-camera transformations/distance metrics using a costly training phase. This is impractical since labeling data in the new camera and then learning transformations with the others are time-consuming, and it defeats the entire purpose of temporarily introducing the additional camera. Thus, there is a pressing need to develop *unsupervised* learning models for re-identification that can work in such dynamic camera networks.

Transfer learning/domain adaptation [15, 32] has been successful in many classical vision problems such as object recognition [21, 24, 62] and activity classification [47, 79] with multiple classes or domains. The main objective is to scale learned systems from a source domain to a target domain without requiring a prohibitive amount of training data in the target domain. Considering a newly introduced camera as the target domain, we pose an important question in this paper: *Can the unsupervised domain adaptation be leveraged upon for re-identification in a dynamic camera network?*

Unlike classical vision problems, e.g., object recognition [62], domain adaptation for re-id has additional challenges. A central issue in domain adaptation is that from *which source to transfer*. When there is only one source of information available which is highly relevant to the task of interest, then domain adaptation is much simpler than in the more general and realistic case where there are multiple sources of information of greatly varying relevance. Re-identification in a dynamic network falls into the latter, more difficult, case. Specifically, given multiple source cameras (already installed) and a target camera (newly introduced), *how can we select the best source camera to pair with the target camera?* The problem can be easily extended to multiple additional cameras being introduced. Moreover, once the best source camera is identified, *how can we exploit this information to improve the re-identification accuracy of other camera pairs?* For instance, let us consider $\mathbf{C}_1$ being the best source camera for the newly introduced camera $\mathbf{C}_3$ in Fig. 14.3. Once the pairwise distance metric between $\mathbf{C}_1$ and $\mathbf{C}_3$ is obtained, can we exploit this information to improve the re-id accuracy across $(\mathbf{C}_2–\mathbf{C}_3)$? This is an especially important problem because it will allow us to now match data in the newly inserted target camera $\mathbf{C}_3$ with all the previously installed cameras.

14.4.2 Solution Overview

To adapt re-id models in a dynamic camera network, we first formulate a domain adaptive re-id approach based on geodesic flow kernel which can effectively find the best source camera (out of multiple installed ones) to pair with a newly introduced target camera with minimal additional effort. Then, to exploit information from the best source camera, we propose a transitive inference algorithm that improves the matching performance across other camera pairs in a network [57].

14.4.2.1 Initial Setup

Our proposed framework starts with an installed camera network where the discriminative distance metrics between each camera pairs are learned using an off-line intensive training phase. Let there be $\mathcal{N}$ cameras in a network and the number of possible camera pairs is $\binom{\mathcal{N}}{2}$. Let $\{(\mathbf{x}_i^{\mathcal{A}}, \mathbf{x}_i^{\mathcal{B}})\}_{i=1}^m$ be a set of training samples, where $\mathbf{x}_i^{\mathcal{A}} \in \mathbb{R}^D$ represents feature representation of training a sample from camera view $\mathcal{A}$ and $\mathbf{x}_i^{\mathcal{B}} \in \mathbb{R}^D$ represents feature representation of the same person in a different camera view $\mathcal{B}$.

Given the training data, we follow KISS Metric Learning (KISSME) [30] and compute the pairwise distance matrices such that distance between images of the same individual is less than the distance between images of different individuals. The basic idea of KISSME is to learn the Mahalanobis distance by considering a log likelihood ratio test of two Gaussian distributions. The likelihood ratio test between dissimilar pairs and similar pairs can be written as

$$\mathcal{R}(\mathbf{x}_i^{\mathcal{A}}, \mathbf{x}_j^{\mathcal{B}}) = \log \frac{\frac{1}{\mathcal{C}_{\mathcal{D}}} \exp(-\frac{1}{2}\mathbf{x}_{ij}^T \Sigma_{\mathcal{D}}^{-1} \mathbf{x}_{ij})}{\frac{1}{\mathcal{C}_{\mathcal{S}}} \exp(-\frac{1}{2}\mathbf{x}_{ij}^T \Sigma_{\mathcal{S}}^{-1} \mathbf{x}_{ij})} \tag{14.9}$$

where $\mathbf{x}_{ij} = \mathbf{x}_i^{\mathcal{A}} - \mathbf{x}_j^{\mathcal{B}}, \mathcal{C}_{\mathcal{D}} = \sqrt{2\pi |\Sigma_{\mathcal{D}}|}, \mathcal{C}_{\mathcal{S}} = \sqrt{2\pi |\Sigma_{\mathcal{S}}|}, \Sigma_{\mathcal{D}}$ and $\Sigma_{\mathcal{S}}$ are covariance matrices of dissimilar and similar pairs respectively. With simple manipulations,

(14.9) can be written as

$$\mathcal{R}(\mathbf{x}_i^A, \mathbf{x}_j^B) = \mathbf{x}_{ij}^T \mathbf{M} \mathbf{x}_{ij} \tag{14.10}$$

where $\mathbf{M} = \Sigma_S^{-1} - \Sigma_D^{-1}$ is the Mahalanobis distance between covariances associated to a pair of cameras. We follow [30] and clip the spectrum by an eigen-analysis to ensure $\mathbf{M}$ is positive semi-definite. Note that our approach is agnostic to the choice of metric learning algorithm used to learn the optimal metrics across camera pairs in an already installed network. We adopt KISSME in this work since it is simple to compute and has shown to perform satisfactorily on the person re-id problem.

14.4.2.2 Discovering the Best Source Camera

Our approach for discovering the best source camera consists of the following steps: (i) compute geodesic flow kernels between the new (target) camera and other source cameras; (ii) use the kernels to determine the distance between them; and (iii) rank the source cameras based on distance with respect to the target camera and choose the one with the lowest as best source camera.

Let $\{\mathcal{X}^s\}_{s=1}^{\mathcal{N}}$ be the $\mathcal{N}$ source cameras and $\mathcal{X}^T$ be the newly introduced target camera. To compute the kernels in an unsupervised way, we extend a previous method [19] that adapts classifiers in the context of object recognition to the re-identification in a camera network. The main idea of our approach is to compute the low-dimensional subspaces representing data of two cameras (one source and one target) and then map them to two points on a Grassmannian.[1] Intuitively, if these two points are close by on the Grassmannian, then the computed kernel would provide high matching performance on the target camera. In other words, both of the cameras could be similar to each other and their features may be similarly distributed over the corresponding subspaces. For simplicity, let us assume we are interested in computing the kernel matrix $\mathbf{K}^{ST} \in \mathbb{R}^{D \times D}$ between the source camera $\mathcal{X}^S$ and a newly introduced target camera $\mathcal{X}^T$. Let $\tilde{\mathcal{X}}^S \in \mathbb{R}^{D \times d}$ and $\tilde{\mathcal{X}}^T \in \mathbb{R}^{D \times d}$ denote the d-dimensional subspaces, computed using Partial Least Squares (PLS) and Principal Component Analysis (PCA) on the source and target camera, respectively. Note that we cannot use PLS on the target camera since it is a supervised dimension reduction technique and requires label information for computing the subspaces.

Given both of the subspaces, the closed loop solution to the geodesic flow kernel between the source and target camera is defined as

$$\mathbf{x}_i^{S^T} \mathbf{K}^{ST} \mathbf{x}_j^T = \int_0^1 (\psi(\mathbf{y})^T \mathbf{x}_i^S)^T (\psi(\mathbf{y}) \mathbf{x}_j^T)\, d\mathbf{y} \tag{14.11}$$

---

[1] Let d being the dimension of the subspace, the collection of all d dimensional subspaces form the Grasssmannian.

where $\mathbf{x}_i^S$ and $\mathbf{x}_j^T$ represent feature descriptor of i-th and j-th sample in source and target camera, respectively. $\psi(\mathbf{y})$ is the geodesic flow parameterized by a continuous variable $\mathbf{y} \in [0, 1]$ and represents how to smoothly project a sample from the original D-dimensional feature space onto the corresponding low-dimensional subspace. The geodesic flow $\psi(\mathbf{y})$ over two cameras can be defined as [19],

$$\psi(\mathbf{y}) = \begin{cases} \tilde{\mathcal{X}}^S & \text{if } \mathbf{y} = 0 \\ \tilde{\mathcal{X}}^T & \text{if } \mathbf{y} = 1 \\ \tilde{\mathcal{X}}^S \mathcal{U}_1 \mathcal{V}_1(\mathbf{y}) - \tilde{\mathcal{X}}_o^S \mathcal{U}_2 \mathcal{V}_2(\mathbf{y}) & \text{otherwise} \end{cases} \qquad (14.12)$$

where $\tilde{\mathcal{X}}_o^S \in \mathbb{R}^{D \times (D-d)}$ is the orthogonal matrix to $\tilde{\mathcal{X}}^S$ and $\mathcal{U}_1, \mathcal{V}_1, \mathcal{U}_2, \mathcal{V}_2$ are given by the following pairs of SVDs:

$$\mathcal{X}^{S^T} \mathcal{X}^T = \mathcal{U}_1 \mathcal{V}_1 \mathcal{P}^T, \quad \mathcal{X}_o^{S^T} \mathcal{X}^T = -\mathcal{U}_2 \mathcal{V}_2 \mathcal{P}^T \qquad (14.13)$$

With the above defined matrices, $\mathbf{K}^{ST}$ can be computed as

$$\mathbf{K}^{ST} = \begin{bmatrix} \tilde{\mathcal{X}}^S \mathcal{U}_1 & \tilde{\mathcal{X}}_o^S \mathcal{U}_2 \end{bmatrix} \mathcal{G} \begin{bmatrix} \mathcal{U}_1^T \mathcal{X}^{S^T} \\ \mathcal{U}_2^T \mathcal{X}_o^{S^T} \end{bmatrix} \qquad (14.14)$$

where $\mathcal{G} = \begin{bmatrix} \text{diag}[1 + \frac{\sin(2\theta_i)}{2\theta_i}] & \text{diag}[\frac{(\cos(2\theta_i)-1)}{2\theta_i}] \\ \text{diag}[\frac{(\cos(2\theta_i)-1)}{2\theta_i}] & \text{diag}[1 - \frac{\sin(2\theta_i)}{2\theta_i}] \end{bmatrix}$ and $[\theta_i]_{i=1}^d$ represents the principal angles between source and target camera. Once we compute all pairwise geodesic flow kernels between a target camera and source cameras using (14.14), our next objective is to find the distance across all those pairs. A source camera which is closest to the newly introduced camera is more likely to adapt better than others. We follow [59] to compute the distance between a target camera and a source camera pair. Specifically, given a kernel matrix $\mathbf{K}^{ST}$, the distance between data points of a source and target camera is defined as

$$\mathbf{D}^{ST}(\mathbf{x}_i^S, \mathbf{x}_j^T) = \mathbf{x}_i^{S^T} \mathbf{K}^{ST} \mathbf{x}_i^S + \mathbf{x}_j^{T^T} \mathbf{K}^{ST} \mathbf{x}_j^T - 2\mathbf{x}_i^{S^T} \mathbf{K}^{ST} \mathbf{x}_j^T \qquad (14.15)$$

where $\mathbf{D}^{ST}$ represents the kernel distance matrix defined over a source and target camera. We compute the average of a distance matrix $\mathbf{D}^{ST}$ and consider it as the distance between two camera pairs. Finally, we chose the one that has the lowest distance as the best source camera to pair with the target camera.

14.4.2.3 Transitive Inference for Re-identification

Once the best source camera is identified, another question that remains in adapting models is: *can we exploit the best source camera information to improve the re-*

identification accuracy of other camera pairs? Let $\{\mathbf{M}^{ij}\}_{i,j=1,i<j}^{\mathcal{N}}$ be the optimal pairwise metrics learned in a network of $\mathcal{N}$ cameras and $\mathcal{S}^\star$ be the best source camera for a newly introduced target camera $\mathcal{T}$. Motivated by the effectiveness of the Schur product in operations research [31], we develop a simple yet effective transitive algorithm for exploiting information from the best source camera. The Schur product (a.k.a. Hadamard product) has been an important tool for improving the matrix consistency and reliability in multi-criteria decision making. Our problem naturally fits to such decision making systems since our goal is to establish a path between two cameras via the best source camera. Given the best source camera $\mathcal{S}^\star$, we compute the kernel matrix between remaining source cameras and the target camera as follows:

$$\tilde{\mathbf{K}}^{\mathcal{ST}} = \mathbf{M}^{\mathcal{SS}^\star} \odot \mathbf{K}^{\mathcal{S}^\star\mathcal{T}}, \ \forall[\mathcal{S}]_{i=1}^{\mathcal{N}}, \ \mathcal{S} \neq \mathcal{S}^\star \tag{14.16}$$

where $\tilde{\mathbf{K}}^{\mathcal{ST}}$ represents the updated kernel matrix between source camera $\mathcal{S}$ and target camera $\mathcal{T}$ by exploiting information from best source camera $\mathcal{S}^\star$. The operator $\odot$ denotes the Schur product of two matrices. Equation 14.16 establishes an indirect path between camera pair $(\mathcal{S},\mathcal{T})$ by marginalization over the domain of possible appearances in best source camera $\mathcal{S}^\star$. In other words, camera $\mathcal{S}^\star$ plays a role of connector between the target camera $\mathcal{T}$ and all other source cameras.

To summarizing, to adapt re-id models in a dynamic network, we use the kernel matrix $\mathbf{K}^{\mathcal{S}^\star\mathcal{T}}$ computed using (14.14) to obtain the re-id accuracy across the newly inserted target camera and best source camera, whereas we use the updated kernel matrices, computed using (14.16), to find the matching accuracy across the target camera and remaining source cameras.

Note that although our framework is designed for unsupervised adaptation of re-id models, it can be easily extended if labeled data from the newly introduced camera become available. Specifically, the label information from the target camera can be encoded while computing subspaces. That is, instead of using PCA for estimating the subspaces, we can use Partial Least Squares (PLS) to compute the discriminative subspaces on the target data by exploiting the labeled information. PLS has shown to be effective in finding discriminative subspaces by projecting data with labeled information to a common subspace [18, 63]. This essentially leads to semi-supervised adaptation of re-id models in a dynamic camera network.

14.4.3 Sample Experimental Results

Datasets. We conduct experiments on RAiD dataset [13] to verify the effectiveness of our framework. It was collected with a view to have a large illumination variation that is not present in most of the publicly available benchmark datasets. In the original dataset, 43 subjects were asked to walk through four cameras of which two are outdoor and two are indoor to make sure there is enough variation of appearance between cameras.

Settings. The feature extraction stage consists of extracting Local Maximal Occurrence (LOMO) feature proposed in [40] for person representation. The descriptor has 26,960 dimensions. We follow [30, 56] and apply principle component analysis to reduce the dimensionality to 100 in all our experiments. To compute the distance between cameras, as well as, re-id matching score, we use kernel distance [59] (Eq. 14.15) for a given projection metric. We show results using Cumulative Matching Characteristic (CMC) curves and normalized Area Under Curve (nAUC) values, as it is common practice in re-id literature [13, 26, 28, 50, 81].

All the images for each dataset are normalized to 128×64 for being consistent with the evaluations carried out by state-of-the-art methods [5, 12, 13]. Following the literature [13, 30, 40, 57], the train and test set are kept disjoint by picking half of the available data for the training set and the rest of the half for testing. We repeated each task 10 times by randomly picking five images from each identity both for train and test time. The subspace dimension for all the possible combinations is kept 50.

Compared Methods. We compare our approach with several unsupervised alternatives which fall into two categories: (i) hand-crafted feature-based methods including CPS [12] and SDALF [5], (ii) two domain adaptation-based methods (Best-GFK and Direct-GFK) based on geodesic flow kernel [19]. For Best-GFK baseline, we compute the re-id performance of a camera pair by applying the kernel matrix, $\mathbf{K}^{S^*T}$ computed between best source and target camera [19], whereas in Direct-GFK baseline, we use the kernel matrix computed directly across the source and target camera using (14.14). The purpose of comparing with Best-GFK is to show that the kernel matrix computed across the best source and target camera does not produce optimal re-id performance in computing matching performance across other source cameras and the target camera. On the other hand, the purpose of comparing with Direct-GFK baseline is to explicitly show the effectiveness of our transitive algorithm in improving re-id performance in a dynamic network

Results. Figure 14.4 shows the results for all possible combinations on the four-camera RAiD dataset. The following observations can be made from Fig. 14.4: (i) the proposed framework for re-identification consistently outperforms all compared unsupervised methods on all three datasets by a significant margin. (ii) among the alternatives, CPS baseline is the most competitive. However, the gap is still significant due to the two introduced components working in concert: discovering the best source camera and exploiting its information for re-identification. The rank-1 performance improvements over CPS is 24.50% on RAiD dataset. (iii) Best-GFK works better than Direct-GFK in most cases, which suggests that kernel computed across the best source camera and target camera can be applied to find the matching accuracy across other camera pairs in re-identification. (iv) Finally, the performance gap between our method and Best-GFK (maximum improvement of 17% in nAUC on RAiD) shows that the proposed transitive algorithm is effective in exploiting information from the best source camera while computing re-id accuracy across camera pairs.

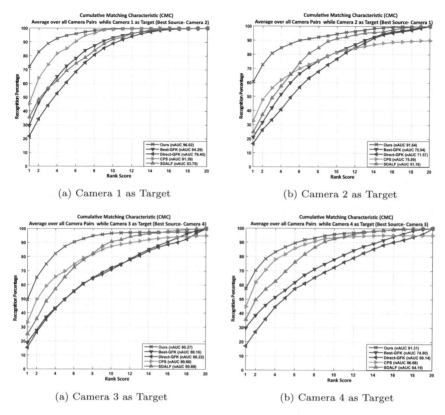

(a) Camera 1 as Target

(b) Camera 2 as Target

(a) Camera 3 as Target

(b) Camera 4 as Target

Fig. 14.4 CMC curves for RAiD dataset with four cameras. Plots (a, b, c, d) show the performance of different methods while introducing cameras 1, 2, 3, and 4, respectively, to a dynamic network. Our method significantly outperforms all the compared baselines for each case of the dynamic network. Best viewed in color

14.5 Conclusions

In this chapter, we have provided some initial research directions for ensuring scalability in person re-identification problems. We considered two aspects of scalability. First, we considered how to limit the labeling effort without any reduction in performance when compared to full-set labeling. Second, we considered how to onboard additional cameras to an existing network, with little to no additional supervision. The first is achieved by considering the similarity structure in the data through a graph-based representation and exploiting transitivity relationships in the graph. The second is achieved by building upon ideas of domain adaptation. Together, they demonstrate the ability to scale person re-identification approaches beyond fully supervised ones, building upon recent successes in these areas. We hope that this will spur future research into person re-identification under limited supervision scenarios and in situations where the network is dynamically evolving.

References

1. Ahmed E, Jones M, Marks TK (2015) An improved deep learning architecture for person re-identification. In: CVPR
2. An L, Chen X, Yang S, Bhanu B (2016) Sparse representation matching for person re-identification. Inf Sci
3. Aodha OM, Campbell NDF, Kautz J, Brostow GJ (2014) Hierarchical subquery evaluation for active learning on a graph. In: IEEE conference on computer vision and pattern recognition
4. Bai S, Bai X, Tian Q (2017) Scalable person re-identification on supervised smoothed manifold. arXiv:1703.08359
5. Bazzani L, Cristani M, Murino V (2013) Symmetry-driven accumulation of local features for human characterization and re-identification. CVIU
6. Bendale A, Boult T (2015) Towards open world recognition. In: CVPR
7. Biswas A, Parikh D (2013) Simultaneous active learning of classifiers & attributes via relative feedback. In: IEEE conference on computer vision and pattern recognition
8. Camps O, Gou M, Hebble T, Karanam S, Lehmann O, Li Y, Radke R, Wu Z, Xiong F (2016) From the lab to the real world: re-identification in an airport camera network. TCSVT
9. Chakraborty A, Das A, Roy-Chowdhury AK (2016) Network consistent data association. TPAMI
10. Chakraborty S, Balasubramanian VN, Panchanathan S (2011) Optimal batch selection for active learning in multi-label classification. In: ACM international conference on multimedia, pp 1413–1416
11. Cheng D, Gong Y, Zhou S, Wang J, Zheng N (2016) Person re-identification by multi-channel parts-based CNN with improved triplet loss function. In: CVPR
12. Cheng DS, Cristani M, Stoppa M, Bazzani L, Murino V (2011) Custom pictorial structures for re-identification. In: BMVC
13. Das A, Chakraborty A, Roy-Chowdhury AK (2014) Consistent re-identification in a camera network. In: ECCV
14. Das A, Panda R, Roy-Chowdhury A (2015) Active image pair selection for continuous person re-identification. In: ICIP
15. Daumé III H (2009) Frustratingly easy domain adaptation. arXiv:0907.1815
16. Ding S, Lin L, Wang G, Chao H (2015) Deep feature learning with relative distance comparison for person re-identification. Pattern Recognit
17. Elhamifar E, Sapiro G, Yang A, Sasrty SS (2013) A convex optimization framework for active learning. In: IEEE international conference on computer vision, pp 209–216
18. Geladi P, Kowalski BR (1986) Partial least-squares regression: a tutorial. Anal Chim Acta
19. Gong B, Shi Y, Sha F, Grauman K (2012) Geodesic flow kernel for unsupervised domain adaptation. In: CVPR
20. Gonzalez TF (2007) Handbook of approximation algorithms and metaheuristics. CRC Press
21. Gopalan R, Li R, Chellappa R (2011) Domain adaptation for object recognition: an unsupervised approach. In: ICCV
22. Hasan M, Roy-Chowdhury AK (2015) Context aware active learning of activity recognition models. In: IEEE international conference on computer vision
23. Hirzer M, Roth PM, Köstinger M, Bischof H (2012) Relaxed pairwise learned metric for person re-identification. In: ECCV
24. Jie L, Tommasi T, Caputo B (2011) Multiclass transfer learning from unconstrained priors. In: ICCV

25. Joshi AJ, Porikli F, Papanikolopoulos NP (2012) Scalable active learning for multiclass image classification. IEEE Trans Pattern Anal Mach Intell 34(11):2259–2273
26. Karanam S, Li Y, Radke RJ (2015) Person re-identification with discriminatively trained viewpoint invariant dictionaries. In: ICCV
27. Karasev V, Ravichandran A, Soatto S (2014) Active frame , location , and detector selection for automated and manual video annotation. In: IEEE conference on computer vision and pattern recognition
28. Kodirov E, Xiang T, Fu Z, Gong S (2016) Person re-identification by unsupervised\ell _1 graph learning. In: ECCV
29. Kodirov E, Xiang T, Gong S (2015) Dictionary learning with iterative laplacian regularisation for unsupervised person re-identification. In: BMVC
30. Köstinger M, Hirzer M, Wohlhart P, Roth PM, Bischof H (2012) Large scale metric learning from equivalence constraints. In: CVPR
31. Kou G, Ergu D, Shang J (2014) Enhancing data consistency in decision matrix: adapting hadamard model to mitigate judgment contradiction. Eur J Oper Res
32. Kulis B, Saenko K, Darrell T (2011) What you saw is not what you get: domain adaptation using asymmetric kernel transforms. In: CVPR
33. Layne R, Hospedales TM, Gong S (2013) Domain transfer for person re-identification. In: Proceedings of the 4th ACM/IEEE international workshop on Analysis and retrieval of tracked events and motion in imagery stream
34. Li W, Zhao R, Wang X (2012) Human reidentification with transferred metric learning. In: ACCV
35. Li W, Zhao R, Xiao T, Wang X (2014) Deepreid: deep filter pairing neural network for person re-identification. In: CVPR
36. Li Z, Chang S, Liang F, Huang TS, Cao L, Smith JR (2013) Learning locally-adaptive decision functions for person verification. In: CVPR
37. Liao S, Hu Y, Zhu X, Li SZ (2015) Person re-identification by local maximal occurrence representation and metric learning. In: CVPR
38. Liao S, Li SZ (2015) Efficient PSD constrained asymmetric metric learning for person re-identification. In: ICCV
39. Lin J, Ren L, Lu J, Feng J, Zhou J (2017) Consistent-aware deep learning for person re-identification in a camera network. In: CVPR
40. Lisanti G, Masi I, Bagdanov AD, Del Bimbo A (2015) Person re-identification by iterative re-weighted sparse ranking. TPAMI
41. Liu C, Gong S, Loy CC (2014) On-the-fly feature importance mining for person re-identification. Pattern Recognit
42. Liu J, Zha Z-J, Tian Q, Liu D, Yao T, Ling Q, Mei T (2016) Multi-scale triplet CNN for person re-identification. In: Proceedings of the 2016 ACM on multimedia conference
43. Liu X, Song M, Tao D, Zhou X, Chen C, Bu J (2014) Semi-supervised coupled dictionary learning for person re-identification. In: CVPR
44. Ma AJ, Li J, Yuen PC, Li P (2015) Cross-domain person reidentification using domain adaptation ranking SVMs. TIP
45. Ma B, Su Y, Jurie F (2012a) Bicov: a novel image representation for person re-identification and face verification. In: BMVC
46. Ma B, Su Y, Jurie F (2012b) Local descriptors encoded by fisher vectors for person re-identification. In: ECCV
47. Ma Z, Yang Y, Nie F, Sebe N, Yan S, Hauptmann AG (2014) Harnessing lab knowledge for real-world action recognition. IJCV
48. Marchand H, Martin A, Weismantel R, Wolsey L (2002) Cutting planes in integer and mixed integer programming. Discret Appl Math
49. Martinel N, Das A, Micheloni C, Roy-Chowdhury AK (2015) Re-identification in the function space of feature warps. TPAMI
50. Martinel N, Das A, Micheloni C, Roy-Chowdhury AK (2016) Temporal model adaptation for person re-identification. In: ECCV

51. Michael RG, David SJ (1979) Computers and intractability: a guide to the theory of np-completeness. WH Free. Co., San Fr
52. Mitchell JE (2002) Branch-and-cut algorithms for combinatorial optimization problems. In: Handbook of applied optimization
53. Mitzenmacher M, Upfal E (2017) probability and computing: randomization and probabilistic techniques in algorithms and data analysis. Cambridge University Press
54. Nan F, Saligrama V (2017) Adaptive classification for prediction under a budget. arXiv:1705.10194
55. Nan F, Wang J, Saligrama V (2016) Pruning random forests for prediction on a budget. In: NIPS
56. Paisitkriangkrai S, Shen C, van den Hengel A (2015) Learning to rank in person re-identification with metric ensembles. In: CVPR
57. Panda R, Bhuiyan A, Murino V, Roy-Chowdhury AK (2017) Unsupervised adaptive re-identification in open world dynamic camera networks. In: CVPR
58. Pedagadi S, Orwell J, Velastin S, Boghossian B (2013) Local fisher discriminant analysis for pedestrian re-identification. In: CVPR
59. Phillips JM, Venkatasubramanian S (2011) A gentle introduction to the kernel distance. arXiv:1103.1625
60. Roy-Chowdhury AK, Song B (2012) Camera networks: the acquisition and analysis of videos over wide areas. Synth Lect Comput Vis
61. Roy S, Paul S, Young NE, Roy-Chowdhury AK (2018) Exploiting transitivity for learning person re-identification models on a budget. In: Proceedings of the IEEE conference on computer vision and pattern recognition, pp 7064–7072
62. Saenko K, Kulis B, Fritz M, Darrell T (2010) Adapting visual category models to new domains. In: ECCV
63. Schwartz WR, Kembhavi A, Harwood D, Davis LS (2009) Human detection using partial least squares analysis. In: ICCV
64. Settles B (2012) Active learning. Synth Lect Artif Intell Mach Learn 6(1):1–114
65. Su Y-C, Grauman K (2016) Leaving some stones unturned: dynamic feature prioritization for activity detection in streaming video. In: ECCV
66. Tao D, Jin L, Wang Y, Yuan Y, Li X (2013) Person re-identification by regularized smoothing kiss metric learning. TCSVT
67. Vezhnevets A, Buhmann JM, Ferrari V (2012) Active learning for semantic segmentation with expected change. In: IEEE conference on computer vision and pattern recognition, pp 3162–3169
68. Vijayanarasimhan S, Grauman K (2011) Large-scale live active learning: training object detectors with crawled data and crowds. In: IEEE conference on computer vision and pattern recognition
69. Vijayanarasimhan S, Jain P, Grauman K (2010) Far-sighted active learning on a budget for image and video recognition. In: CVPR
70. Vondrick C, Ramanan D (2011) Video annotation and tracking with active learning. In: Advances in neural information processing systems
71. Wang H, Gong S, Xiang T (2016) Highly efficient regression for scalable person re-identification. arXiv:1612.01341
72. Wang H, Gong S, Zhu X, Xiang T (2016) Human-in-the-loop person re-identification. In: ECCV
73. Wang X, Zheng W-S, Li X, Zhang J (2015) Cross-scenario transfer person re-identification. TCSVT
74. Wu L, Shen C, Hengel AVD (2016) Personnet: person re-identification with deep convolutional neural networks. arXiv:1601.07255
75. Wu Z, Li Y, Radke RJ (2016) Viewpoint invariant human re-identification in camera networks using pose priors and subject-discriminative features. TPAMI
76. Xiao T, Li H, Ouyang W, Wang X (2016) Learning deep feature representations with domain guided dropout for person re-identification. arXiv:1604.07528

77. Xiao T, Li S, Wang B, Lin L, Wang X (2016) End-to-end deep learning for person search. arXiv:1604.01850
78. Yan Y, Ni B, Song Z, Ma C, Yan Y, Yang X (2016) Person re-identification via recurrent feature aggregation. In: ECCV
79. Yang Y, Ma Z, Xu Z, Yan S, Hauptmann AG (2013) How related exemplars help complex event detection in web videos?. In: ICCV
80. Yi D, Lei Z, Liao S, Li SZ, et al (2014) Deep metric learning for person re-identification. In: ICPR
81. Zhao R, Ouyang W, Wang X (2013) Unsupervised salience learning for person re-identification. In: CVPR
82. Zhao R, Ouyang W, Wang X (2014) Learning mid-level filters for person re-identification. In: CVPR
83. Zheng L, Shen L, Tian L, Wang S, Wang J, Tian Q (2015) Scalable person re-identification: a benchmark. In: ICCV
84. Zheng L, Yang Y, Hauptmann AG (2016) Person re-identification: past, present and future. arXiv:1610.02984
85. Zheng W-S, Gong S, Xiang T (2012) Transfer re-identification: from person to set-based verification. In: CVPR
86. Zheng W-S, Gong S, Xiang T (2013) Reidentification by relative distance comparison. TPAMI
87. Zheng W-S, Gong S, Xiang T (2016) Towards open-world person re-identification by one-shot group-based verification. TPAMI

Chapter 15
Towards Causal Benchmarking of Bias in Face Analysis Algorithms

Guha Balakrishnan, Yuanjun Xiong, Wei Xia, and Pietro Perona

Abstract Measuring algorithmic bias is crucial both to assess algorithmic fairness and to guide the improvement of algorithms. Current bias measurement methods in computer vision are based on *observational* datasets and so conflate algorithmic bias with dataset bias. To address this problem, we develop an *experimental* method for measuring algorithmic bias of face analysis algorithms, which directly manipulates the attributes of interest, e.g., gender and skin tone, in order to reveal causal links between attribute variation and performance change. Our method is based on generating synthetic image grids that differ along specific attributes while leaving other attributes constant. Crucially, we rely on the perception of human observers to control for synthesis inaccuracies when measuring algorithmic bias. We validate our method by comparing it to a traditional observational bias analysis study in gender classification algorithms. The two methods reach different conclusions. While the observational method reports gender and skin color biases, the experimental method reveals biases due to gender, hair length, age, and facial hair. We also show that our synthetic transects allow for a more straightforward bias analysis on minority and intersectional groups.

G. Balakrishnan (✉)
Massachusetts Institute of Technology, Amazon Web Services, Jamshedpur, India
e-mail: balakg@mit.edu; guha89@gmail.com

Y. Xiong · W. Xia
Amazon Web Services, Jamshedpur, India
e-mail: yuanjx@amazon.com

W. Xia
e-mail: wxia@amazon.com

P. Perona
California Institute of Technology, Amazon Web Services, Jamshedpur, India
e-mail: perona@caltech.edu

15.1 Introduction

Automated systems trained using machine learning methods are increasingly used to support decisions in industry, medicine, and government. While the performance of such systems is often excellent, accuracy is not guaranteed and needs to be assessed through careful measurements. Measuring *biases*, i.e., performance differences, across protected attributes such as age, sex, gender, and ethnicity, is particularly important for decisions that may affect peoples' lives. Unlike systems based on human judgment, where measuring and correcting biases is notoriously difficult, measuring and mitigating algorithmic bias are feasible and may become a powerful agent of progress towards more fair, accountable, and transparent institutions [1, 2].

The prevailing technique for measuring the performance of algorithms is to measure statistics like error frequencies on a test set that is sampled *in the wild*, hopefully mirroring some of the data statistics that will be encountered in the field. Studies of algorithmic bias in computer vision [3–6] have adopted this approach by adding one additional step: each image of the test set is annotated for attributes of interest (e.g., ethnicity, gender, and age), and the test set is then split into groups that have homogeneous attribute values. Comparing error rates across such groups yields predictions of bias. As an example, Fig. 15.1-top shows the results of a recent study of algorithmic bias in gender classification of face images. This type of study is called *observational*, because the independent variables (e.g., skin color and gender) are sampled from the environment, rather than controlled by the investigator.

Algorithmic bias is measured for two reasons. First, fairness: would changing a protected attribute, all else being equal, cause a systematic change in the output of the algorithm? For example, would two job applicants, that differed only by their gender or ethnicity, face predictably different outcomes [7]? The second reason for measuring bias is getting rid of it: which actions should one take to best improve the system's performance? For example, should the engineers who are in charge of developing systems A, B, and C (Fig. 15.1, top) infer that the best strategy is to add more examples of dark-skinned women to their training set? Thus, measuring algorithmic bias ultimately has one goal: revealing causal connections between attributes of interest and algorithmic performance.

Unfortunately, observational studies are ill-suited for drawing such conclusions. When one samples data in the wild, other variables may correlate with the variable of interest, and any one of the correlated variables may have an influence on the performance of the algorithm. Thus, it is difficult to impute the cause of performance differences to variations in the variable of interest—as the old saying goes: *"correlation does not imply causation."*

One simple instance of this problem is sample bias: samples in the wild may fail to represent specific combinations of variables of interest [8–10]. For example, the appearance of the parliamentarians in the PPB dataset [3] tends to be gender-stereotypical, e.g., very few males have long hair and almost no light-skinned females have short hair (Fig. 15.12, and [11]). The fact that hair length (a variable that may affect gender classification accuracy) is correlated in PPB with skin color (a variable

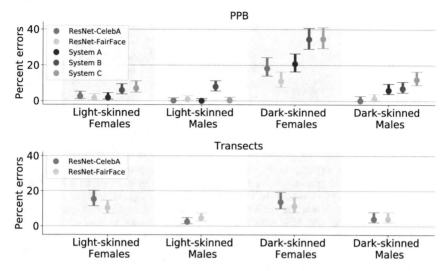

Fig. 15.1 Algorithmic bias measurements are test set dependent. (Top) Gender classification error rates of three commercial face analysis systems (System A–C) were measured in 2017 on the Pilot Parliaments Benchmark (PPB) [3], an observational dataset of portrait pictures downloaded from the websites of six national parliaments in Scandinavia and Africa. Error rates for dark-skinned females were found to be significantly higher than for other groups. We observed the same qualitative behavior when we replicated the study by training a standard classifier (ResNet-50) on two publicly available face datasets (CelebA; FairFace) and testing the two models thus obtained on a replica of the PPB dataset. (Bottom) Our experimental investigation using the Transects dataset, where the sample faces are matched across attributes, reveals a different picture of algorithmic bias (see Fig. 15.13, Sects. 15.5, and 15.6 for a more complete analysis).

of interest) complicates the analysis. In addition, the sample dataset that is used to measure bias is often not representative of the population of interest. For example, the middle-aged Scandinavians and Africans of PPB are not representatives of, say, the broad U.S. Caucasian and African-American population [12]. While observational methods do yield useful information on disparate impact within a given test set population, generalizing observational performance predictions to different target populations is hit-or-miss [13] and can negatively impact underrepresented or minority populations [14, 15]. In a nutshell, one would want a method that systematically identifies algorithmic bias while transcending the peculiarities of specific test sets.

Scientists in biology, medicine, and the social sciences are well aware of this problem and have developed practices to discover, and to control for, confounding variables. A powerful approach to discovering cause–effect relationships is the *experimental method* which involves artificially manipulating the variable of interest, while fixing all the other inputs [7, 16]. This is not easy in the case of image data, leading us to ask the question: *Can one systematically measure bias in computer vision algorithms using the experimental method?* While this is not immediately intuitive [11], we find that the answer is yes and offer a practical way forward.

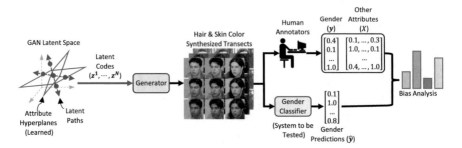

Fig. 15.2 Synopsis of our approach. A generative adversarial network (Generator) is used to synthesize "transects," or grids of images, modifying selected attributes on synthetic faces (in this example: hair length and skin tone). This is accomplished by traversing the generator's latent space in attribute-specific directions. These directions are learned using randomly sampled faces and human annotators (not shown). Human annotations on the transects provide generator-independent ground truth to be compared with algorithm output to measure algorithm errors. Attribute-specific bias measurements are obtained by comparing the algorithm's predictions with human annotations as the attributes are varied. The depicted example may study the question: *Does hair length, skin tone, or any combination of the two have a causal effect on classifier errors?* Transects exploring other attributes are shown in Figs. 15.3, 15.4, and 15.9a. The GUIs for human image annotation are shown in Fig. 15.6. Samples of image annotations are shown in Fig. 15.7

Our approach (Fig. 15.2) generates the test images synthetically, rather than sampling them from the wild, so that they are varied selectively along attributes of interest. This is enabled by recent progress in controlled and realistic image synthesis [17, 18], along with methods for collecting large amounts of accurate human annotations [19] to quantify the perceptual effect of image manipulations.

Our synthesis approach can alter multiple attributes at a time to produce grid-like matched samples of images we call *transects*. We quantify the image manipulations with detailed human annotations which we then compare with algorithm output to estimate algorithmic bias.

We evaluate our methodology with experiments on two gender classification algorithms. We first find that our transect generation strategy creates significantly more balanced data across key attributes compared to "in the wild" face datasets. Next, inspired by [3], we use this synthetic data to explore the effects of various attributes like skin color, hair length, age, and perceived gender on gender classifier errors. Our findings reveal that using an experimental method can change the picture of algorithmic bias (Fig. 15.13), which will affect the strategy of algorithm improvement, particularly concerning groups that are often underrepresented in training and test sets.

We view our work as a first step in developing experimental methods for algorithmic bias testing in computer vision which, we argue, are necessary to achieve trustworthy and actionable measurements. Much remains to be done, both in design and experimentation to achieve broadly applicable and reliable techniques. In Sect. 15.6, we discuss limitations of the current method and the next steps in this research area.

15.2   Related Work

Benchmarking in computer vision has a long history [20–22] including face recognition [6, 23–27] and face analysis [3]. Some of these studies examine biases in performance, i.e., error rates across variation of important parameters (e.g., racial background in faces). Since these studies are purely observational, they raise the question of whether the biases they measure depend on algorithmic bias or on correlations in the test data. Our work addresses this question.

A dataset is said to be biased when combinations of features of interest are disproportionately represented or, equivalently, when such features are correlated. Computer vision datasets are often found to be biased [13, 28]. Human face datasets are particularly scrutinized [5, 29–33] because methods and models trained on these data can end up being biased along attributes that are protected by the law [1]. Approaches to mitigating dataset bias include collecting more thorough examples [33], using image synthesis to compensate for the distribution gaps [32], and example resampling [34].

The machine learning community is active in analyzing biases of learning models and how one may train models where bias is mitigated [32, 35–42], usually by ensuring that performance is equal across certain subgroups of a dataset. Here, we ask a complementary question: we assume that the system to be benchmarked is *pretrained* and fixed, and we ask how to reliably measure algorithmic bias in pre-trained black-box algorithms.

Studies of face analysis systems [3, 5, 41] and face recognition systems [6, 43] attempt to measure bias across gender and skin color (or ethnicity). However, the evaluations are based on observational rather than interventional techniques—and therefore, any conclusions from these studies should be treated with caution. A notable exception is a recent study [11] using the experimental method to investigate the effect of skin color in gender classification. In that study, skin color is modified artificially in photographs of real faces to measure the effects of differences in skin color, all else being equal. However, the authors observe that generalizing the experimental method to other attributes, such as hair length, is too onerous if one is to modify existing photographs. Our goal is to develop a generally applicable and practical experimental method, where *any* attribute may be studied independently.

Recent work uses generative models to explore face classification system biases. One study explores how variations in pose and lighting affect classifier performance [29, 31, 32]. A second study uses a generative model to synthesize faces along particular attribute directions [44]. These studies rely on the strong assumption that their generative models can modify one attribute at a time. However, this assumption relies on having unbiased training data, which is almost always not practical. In contrast, our framework uses human annotations to account for residual correlations produced by our generative model.

Finally, there is research into interpreting neural networks. One strategy is to determine regions of the input that are salient, either through analysis of gradients or perturbations of the input image [45–51]. Network dissection approaches explore how particular neurons within a network affect the output, particularly in a semantic way [52, 53]. Testing with Concept Activation Vectors (TCAV) [54] provides explanations at a high level using directional derivatives to reveal the "conceptual sensitivity" of a model's prediction of a class (e.g., Smiling) to a concept. In contrast, our approach uses a synthesis model to create carefully modified input images, and human annotations to precisely quantify them.

15.3   Face Attribute Annotation in Synthetic Images

The face images used in our experiments are synthetic, and therefore, there is no real person behind each image. Thus, there is no intrinsic ground truth for face attributes such as gender, hair length, and skin tone. Such attributes are instead established by human annotators. We clarify here what we mean when we talk about face attributes in the absence of a physical ground truth.

Many attributes have both intrinsic and extrinsic manifestations. For example, "emotion" may be studied at three levels [55]: an unconscious physiological state, conscious self-perception (feelings), and emotional display (e.g., facial expression) [56]. These quantities are *intrinsic* to a person's or an animal's body and are not directly accessible to a machine. By contrast, an *extrinsic* description, i.e., the report by an onlooker of his/her perception, is more easily accessible, and this is what the machine is trained to predict.

Since we are using synthetic images, it should be clear that we are not attempting to access the intrinsic state of a person: there is no person, and there is no intrinsic gender, ethnicity, age, or emotion. However, perception of such attributes is possible. This is the same way that onlookers instinctively classify the *Venus of Milo* as "female" and Michelangelo's *David* as "male," despite the fact that they are idealized marble representations, rather than real people.

Thus, when we refer to the "age" or "gender" or any other attribute that is computed by a face analysis system from a picture, what we mean is *the algorithm's prediction of a casual observer's report of their perception of the outwards display of that attribute*. This is a bit of a mouthful, and that's why we use the abbreviated expression of "attribute," "age", or "gender." The attributes we measure from human observers are reports of subjective perceptions. However, as we find in Sect. 15.4.3, these measurements are consistent and reproducible across different observers, and so we consider statistics of such reports as objective quantities.

In our study, we discretize continuous face attributes. We have used six classes of age and skin tone, five of hair length, facial expression, and gender, etc. (see Figs. 15.6 and 15.8). This choice was made to conform with the literature, e.g., the Fitzpatrick scale of skin tone [57], and to accommodate the abilities of non-expert casual observers, the "common person," whose perception we rely on in our experiments. We make no claim to have the perfect discretization scheme; other discretization choices may be better suited in different contexts.

Gender deserves a special mention: *gender identity* is often modeled as multi-dimensional [58]. However, here we are measuring *reports of gender perception* (an extrinsic variable), rather than gender identity (the intrinsic variable), and our subjects could not reliably report beyond the traditional one-dimensional M/F dimension. Therefore, following [3] we settled for one dimension, which we discretized into five steps to accommodate different levels of confidence and ambiguity.

15.4 Method

Our framework consists of two components: a technique to synthesize sets of images with control over semantic attributes, and a procedure using these synthesized images, along with human annotators, to perform analysis of a recognition system.

In Sect. 15.4.1, we present our technique for attribute-controlled image synthesis. We introduce the concept of *transect*, a grid-like construct of synthesized images with a different attribute manipulated along each axis. A transect gives control over the joint distribution of synthesized attributes allowing us to generate *matched samples* across multiple attributes, unlike related methods that operate on only one or two attributes at a time [44, 59–61]. We then collect human annotations for each transect image to precisely quantify our modifications.

In Sect. 15.4.2, we present analyses we can perform using the annotated transects. We report a classifier's error rate, stratified along subgroups of a sensitive attribute. We also return a covariate-adjusted estimate of the *causal effect* of a binary attribute on the classifier's performance.

15.4.1 Transects: A Walk in Face Space

We assume a black-box face generator G that can transform a latent vector $\mathbf{z} \in \mathcal{R}^D$ into an image $I = G(\mathbf{z})$, where $p(\mathbf{z})$ is a distribution we can sample from. In our study, G is the generator of a pre-trained, publicly available state-of-the-art GAN ("StyleGAN2") [17, 18]. GAN latent spaces typically exhibit good disentanglement of semantic image attributes. In particular, empirical studies show that each image attribute often has a direction $\mathbf{v} \in \mathcal{R}^D$ that predominantly captures its variability [17, 53]. We base our approach on a recent study [53] for single-attribute traversals in GAN latent spaces. That method trains a linear model to predict a particular image attribute from $\mathbf{z}$ and uses the model to traverse the $\mathbf{z}$-space in a discriminative direction. Our method generalizes this idea to synthesize image grids, i.e., *transects*, spanning arbitrarily many attributes.

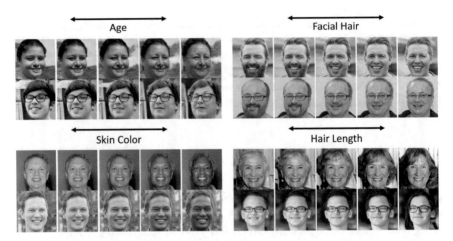

Fig. 15.3 1D transects. 1×5 sample transects synthesized by our method for various attributes. Orthogonalization was used (see Fig. 15.5)

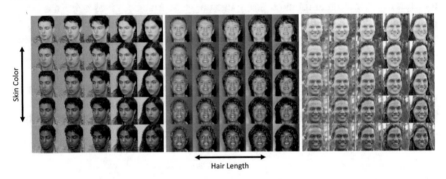

Fig. 15.4 2D transects. 5×5 transects for varying, simultaneously, hair lengths and skin tones. Multidimensional transects allow for intersectional analysis, i.e., analysis across the joint distribution of multiple attributes. Orthogonalization was used (see Fig. 15.5)

15.4.1.1 Estimating Latents-to-Attributes Linear Models

We first sample the latent space, measure the attributes at each location through human observers, and use these measurements to calculate principal axes of variation for attributes. More formally, let there be a list of N_a image attributes of interest (age, gender, skin color, etc.). As explained below, we generate an annotated training dataset $\mathcal{D}_{\mathbf{z}} = \{\mathbf{z}^i, \mathbf{a}^i\}_{i=1}^{N_z}$, where $\mathbf{a}^i$ is a vector of scores, one for each attribute, for generated image $G(\mathbf{z}^i)$. The score for attribute j, $\mathbf{a}_j^i$, may be continuous in $[0, 1]$ or binary in $\{0, 1\}$.

We produce $\mathcal{D}_{\mathbf{z}}$ as follows. First, we sample a generous number of values of $\mathbf{z}^i$ from $p(\mathbf{z})$. Second, we obtain labels $\mathbf{a}^i$ from human annotators. A related study obtains labels by only processing the generated images through a trained classifier [53]. We generally avoid this approach because any biases of the classifier due to attribute correlations—precisely the phenomena we are trying to avoid—will leak into our method.

For each attribute j, we use $\mathcal{D}_z$ to compute a $(D-1)$-dimensional linear hyperplane $h_j = (\mathbf{n}_j, b_j)$, where $\mathbf{n}_j$ is the normal vector and b_j is the offset. For continuous attributes like age or skin color, we train a ridge regression model [62]. For binary attributes, we train a Support Vector Machine (SVM) classifier [63].

When sampling from StyleGAN2 using the native latent Gaussian distribution, we noticed a bias towards generating Caucasian-looking faces which is not surprising given the fact that it was trained on Flickr-Faces-HQ (FFHQ)—a public dataset that is skewed towards that demographic (see Fig. 15.11). However, using human annotations, our method is able to partially mitigate this bias by directing sampling towards the relevant portions of the latent space (see following sections), so that it could generate a diversity of attributes. Nevertheless, training face synthesis GANs with a more diverse set of faces will be an important step in making our method more easily applicable.

15.4.1.2 Multi-attribute Transect Generation

The attribute hyperplanes may now be used to sample faces that vary along specific attributes. More formally: the hyperplane h_j specifies the subspace of $\mathcal{R}^D$ with boundary or neutral values of attribute j, and the normal vector $\mathbf{n}_j$ specifies a direction along which that attribute primarily varies. To construct a one-dimensional, length-L transect for attribute j, we first start with a random point $\mathbf{z}^i$ and project it onto h_j. We then query $L-1$ evenly spaced points along $\mathbf{n}_j$, within fixed distance limits on both sides of the h_j. Figure 15.3 presents some single transect examples (with orthogonalization, a concept introduced in the next section). We give further details on querying points in Sect. 15.4.1.4.

The 1D transect does not allow us to explore the joint space of several attributes or to fix other attributes in precise ways when varying one attribute. We generalize to K-dimensional transects in Algorithm 1 to address this. The main extensions are: (1) we project $\mathbf{z}^i$ onto the intersection of K attribute hyperplanes, and (2) we move in a K-dimensional grid in $\mathbf{z}$-space (see Fig. 15.4). Input $\mathbf{c}_k$ is a vector of decision values with respect to the hyperplane h_k, and $\mathbf{v}_k$ is a direction vector (equivalent to $\mathbf{n}_k$ here, until orthogonalization is introduced in the next section).

We are unaware of a simple closed-form solution to project $\mathbf{z}^i$ onto the intersection of arbitrarily many hyperplanes. We instead take an iterative approach: we sequentially project the point onto each hyperplane and repeat this process for some number of iterations. Repeated projections onto convex sets, the hyperplanes in our case, are guaranteed to converge to a location on the intersection of the sets [64] which, in our case, is a single point. If the hyperplanes are perfectly orthogonal, this process converges in exactly one iteration; we empirically found convergence in fewer than 50 iterations.

15.4.1.3 Orthogonalization of Traversal Directions

The hyperplane normals $\{\mathbf{n}_j\}_{j=1}^{N_a}$ are not orthogonal to one another. If we set the direction vectors equal to these normal vectors in Algorithm 1, i.e., $\mathbf{v}_j = \mathbf{n}_j$, we will likely observe unwanted correlations between attributes. We reduce this effect by producing a set of modified direction vectors such that $\mathbf{v}_j \perp \mathbf{n}_k, \forall k \neq j$ (see Algorithm 2).

Figure 15.5 illustrates the effects of orthogonalization for hair length and skin color. Without orthogonalization, the hair length transects exhibit unwanted changes in gender, with shorter hair also causing faces to appear more masculine. With orthogonalization, these unwanted changes are removed. In contrast, we see no clear difference in skin color transects with and without orthogonalization, indicating that the skin color hyperplane was already near-orthogonal to the other attribute hyperplanes.

Algorithm 1: K-attribute transect generation

Input: Generator G, tuples $\{(L_k, \mathbf{n}_k, b_k, \mathbf{v}_k, \mathbf{c}_k)\}_{k=1}^{K}$, where L_k is a transect dimension, $(\mathbf{n}_k, b_k)$ is a hyperplane, $\mathbf{v}_k$ is a direction vector, and $\mathbf{c}_k$ are signed decision values.
Output: A $L_1 \times \cdots \times L_K$ transect T^i.

$\mathbf{z}^i \sim p(\mathbf{z})$
$\mathbf{z}^{i,0} = $ projection of $\mathbf{z}^i$ onto intersection of $\{(\mathbf{n}_k, b_k)\}_{k=1}^{K}$
for $l_1 = 1 \cdots L_1$ **do**

$\quad \vdots$

$\quad$ **for** $l_K = 1 \cdots L_K$ **do**

$\quad\quad T^i(l_1, \cdots, l_K) = G(\mathbf{z}^{i,0} + \sum_{k=1}^{K} \frac{\mathbf{c}_k[l_k]}{\langle \mathbf{v}_k, \mathbf{n}_k \rangle} \frac{\mathbf{v}_k}{\|\mathbf{v}_k\|})$

Algorithm 2: Orthogonalization

Input: Vectors $\{\mathbf{n}_j\}_{j=1}^{N_a}$.
Output: Vectors $\{\tilde{\mathbf{n}}_j\}_{j=1}^{N_a}$, where $\tilde{\mathbf{n}}_j \perp \mathbf{n}_k, \forall k \neq j$

$Q, R \leftarrow$ QR-factorization of matrix $[\mathbf{n}_1, \mathbf{n}_2, \cdots, \mathbf{n}_{N_a}]$
for $i = 1 \cdots N_a$ **do**
$\quad \tilde{\mathbf{n}}_i = \mathbf{n}_i$
$\quad$ **for** $j = 1 \cdots N_a$ **do**
$\quad\quad$ **if** $i \neq j$ **then**
$\quad\quad\quad \tilde{\mathbf{n}}_i = \tilde{\mathbf{n}}_i - \frac{Q_j \cdot \langle Q_j, \tilde{\mathbf{n}}_i \rangle}{\langle Q_j, Q_j \rangle}$

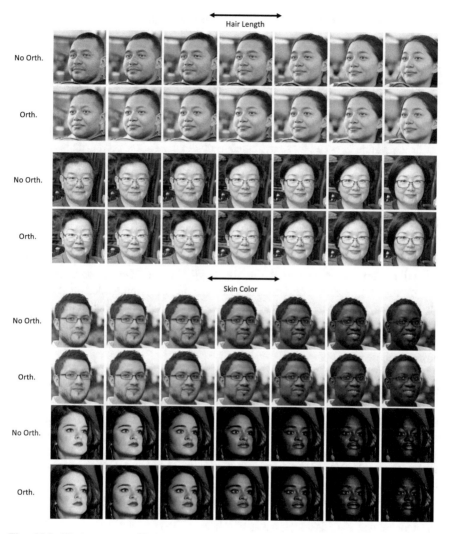

Fig. 15.5 1D transects with and without orthogonalization. Without orthogonalization (Sect. 15.4.1.2), decreasing hair length results in more masculine-looking faces. This phenomenon is not as apparent after orthogonalization (Sect. 15.4.1.3). We see only slight orthogonalization differences in the skin color transects

15.4.1.4 Setting Step Sizes and Transect Dimensions

If human annotation costs were negligible, we could simply query many grid locations with large transect dimensions L to capture subtle appearance changes over the dynamic ranges of the attributes. But given constrained resources, we set L to small values. For example, $L = 5$ for the 1D transects in Fig. 15.3 and 2D transects in Fig. 15.4, and $L = 2$ for the 3D transects in Fig. 15.9. For each attribute j, we manually set min/max signed decision values with respect to h_j and linearly interpolate L_j points between these extremes to obtain $\mathbf{c}_j$. We set per-attribute min/max values so that transects depict a full dynamic range for most random samples.

15.4.2 Analyses Using Transects

We assume a target attribute of interest, e.g., gender, and a target attribute classifier C. We will use transect images to perform bias analysis on C. Though an ideal transect will modify only selected attributes at a time, in practice, unintended attributes may also be accidentally modified. In addition, the degree to which an attribute is altered varies across transects. To measure and control these factors, we annotate each image of each transect, resulting in a second dataset $\mathcal{D}_{transect} = \{I^i, \mathbf{a}^i\}_{i=1}^{N_{images}}$ of images and human annotations.

We denote the ground truth gender of image I^i (as reported by humans) by y^i and C's prediction by $\hat{y}^i$. For ease of analysis, we discretize the remaining attributes into bins and assign an independent binary variable to each bin [65]. For instance, we may represent the "skin color" attribute with six binary variables, corresponding to the six levels shown in Fig. 15.6 (top right). For a given image, only one of these six variables would be set to 1—often called a "one-hot encoding." We denote the vector of concatenated binary covariates for image i by $\mathbf{x}^i$ and the classification error by $e^i = \ell(\hat{y}^i, y^i)$, where $\ell(\cdot, \cdot)$ is an error function.

Our first analysis strategy is to simply compare C's error rate across different subgroups in the population. Let E_j^s denote the average error of C over test samples for which covariate j is equal to $s \in \{0, 1\}$:

$$E_j^s = \frac{\sum_i e^i \mathbb{1}(\mathbf{x}_j^i = s)}{\sum_i \mathbb{1}(\mathbf{x}_j^i = s)}. \tag{15.1}$$

If the data is generated from a perfectly randomized or controlled study, the quantity $E_j^1 - E_j^0$ is a good estimate of the "Average Treatment Effect" (ATE) [66–69] of covariate j on e or the average change in e over all examples when covariate j is flipped from 0 to 1, with other covariates fixed. For example, the ATE of the "dark skin" covariate captures the average change in C's error when each person's skin tone is changed from non-dark to dark. Exactly computing the ATE from an observational dataset is not possible, because we do not observe the counterfactual case(s) for each data point, e.g., the same person with both light and dark skin tones.

Though our transects come closer to achieving an ideal controlled study than do observational datasets "from the wild" (see Sect. 15.5.3 for empirical validation), there may still be some confounding between covariates in practice (see Fig. 15.18 for an example). Since any observable confounder may be annotated in $\mathcal{D}_{transect}$, we can employ covariate-adjusted ATE estimators [70–72]. One simple covariate adjustment approach is to train a linear regression model predicting e^i from $\mathbf{x}^i$:

$$e^i = \epsilon^i + \beta_0 + \sum_j \beta_j \mathbf{x}_j^i, \tag{15.2}$$

where β's are parameters, and ϵ^i is a per-example noise term. β_j captures the ATE, the average change in e given one unit change in $\mathbf{x}_j$ holding all other variables constant, provided: (1) a linear model is a reasonable fit for the relationship between the dependent and independent variables, (2) all relevant attributes are included in the model (i.e., no hidden confounders), and (3) no attributes that are influenced by $\mathbf{x}_j$ are included in the model, otherwise these other factors can "explain away" the impact of $\mathbf{x}_j$.

An experimenter can never be completely sure that (s)he has satisfied these conditions but (s)he can strive to do so through careful consideration. Discretizing and binarizing attributes help with (1), though we still found some nonlinear influences between covariates in our experiments (see Sect. 15.5.5.1). As an example of (2), we found that earrings may be an important attribute that we did not account for in our analysis (see Fig. 15.19).

Finally, when the outcome lies in a fixed range, as is the case in our experiments with $e^i \in [0, 1]$, we use logistic instead of linear regression. β_j then represents the expected change in the *log odds* of e for a unit change in $\mathbf{x}_j$. We use such a logistic regression analysis in our experiments (see Sect. 15.5.5).

15.4.3 Human Annotation

We collect human annotations on the synthetic faces to construct $\mathcal{D}_z$ and $\mathcal{D}_{transect}$. The annotators were recruited on Amazon Mechanical Turk [19] through the AWS SageMaker Ground Truth service [73]. Annotators evaluated each image for seven attributes: gender, facial hair, skin color, age, makeup, smiling, hair length, and image fakeness. Each attribute was evaluated on a discrete scale. Each annotator evaluated each image for one attribute at a time. For each image, we collected five annotations per attribute for a total of 40 annotations per image.

We discretized each attribute using three to six levels. For example, we use the Fitzpatrick six-point scale for skin color [57] and split age into six groups ranging from children to senior citizens. For complete details about subgroups for each attribute, along with samples of our Mechanical Turk survey layouts, please see Fig. 15.6.

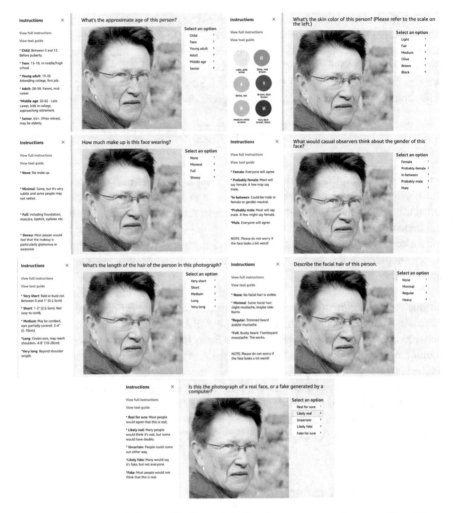

Fig. 15.6 Screenshots of the graphical user interface for seven annotations we collected from Amazon Mechanical Turk annotators using the SageMaker Ground Truth service [73]

The number of annotations that are needed by our method is rather formidable. However, we found that this is not an obstacle in practice. In our experiments, $\mathcal{D}_z$ consists of 5,000 images, and $\mathcal{D}_{transect}$ consists of 1,000 8-image transects (see examples in Fig. 15.9). The total number of annotations was thus 13,000 (images) $\times$ 8 (attributes) $\times$ 5 (annotations per image and per attribute) = 0.52M annotations. Amazon Mechanical Turk delivered on average 10–20 annotations per second, thus annotations took about 10 hours to complete over two separate sessions. Annotators were paid 1.2c per annotation, earning 10-1-5 US$ per hour.

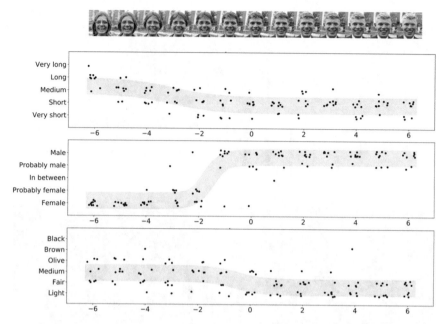

Fig. 15.7 Annotation consistency. Hair length (top), gender (middle), and skin tone (bottom) annotations on a 13-image 1D transect. This transect was annotated in a pilot experiment to fine tune our GUIs and to evaluate the consistency of the annotators and not used in our main experiment. Here nine annotations were obtained for each attribute and for each image. The annotations are shown as dots below each image. The x-axis increments one unit from one image to the next. A small amount of noise was added in x and y in order to visualize the individual annotations. The thick gray curves show the fit of a logistic function to the data. Annotations typically fall within one or two neighboring attribute levels. There are very few outliers. For a quantitative overall analysis, see Fig. 15.8

One may be concerned that annotators may not be able to give meaningful attribute annotations on synthetic images. Therefore, we explored the level of agreement of annotator responses, both in the number of pilot experiments and in the annotations we collected for the main experiment. Figure 15.7 shows the raw annotations for one 1D transect and three attributes. One may see that there are very few outlier annotations and that in most cases, annotations fall in one or two neighboring attribute levels. Figure 15.8 (top left) shows a distribution of per-image annotation standard deviations, split by attribute. One unit corresponds to the dynamic range of each attribute. For most attributes, the median annotator standard deviation is near 0.1, i.e., less than the separation between attribute levels. These observations indicate good agreement between annotators and suggest that annotations are meaningful and reproducible.

Figure 15.8 (top right) presents the distribution of mean annotator fakeness scores for the synthesized images. Only a small portion of images are deemed "Likely fake" or "Fake for sure." The realism of images is particularly important in our analysis,

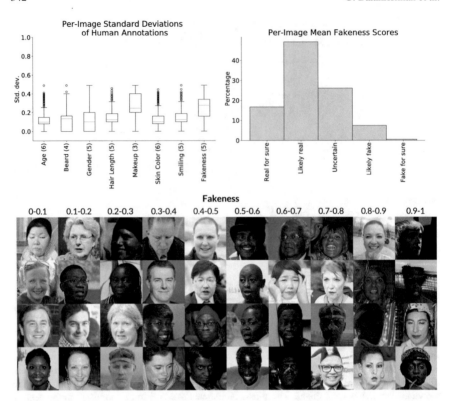

Fig. 15.8 Annotation quality and image realism. (Top left) Distributions of per-image standard deviations of human annotations for each of the attributes we considered (one unit = dynamic range of the attribute). Five annotators were asked to provide a rating for each attribute of each image. The number of rating options per attribute is indicated in brackets next to the attribute's name. The median standard deviations (red lines) are comparable to the quantization step, indicating good annotator agreement. (Top right) We asked our annotators to rate the realism of the images. The distribution of such scores is shown. Fewer than 10% of the ratings indicated fake or likely fake, suggesting that the synthetic images we randomly sampled are fairly realistic. (Bottom) We show examples of synthesized faces organized by mean human fakeness scores. Images with high fakeness scores were removed from the experiments (see Sect. 15.5.2.1)

since image artifacts can unknowingly affect the decisions of gender classifiers. In our experiments, we remove images with a fakeness score above a certain threshold (see Sect. 15.5.2.1). Figure 15.8 (bottom) shows example synthesized images organized by mean human fakeness score.

(a) **Examples of transects used in our experiments.**

| Humans: | Male | Male | Likely Female | Male | Likely Female | Male | Female | Likely Male |
| Intended: | Female | Male | Female | Male | Female | Male | Female | Male |

(b) **Human perception of the generators' manipulations.**

Fig. 15.9 Sample of three-attribute transects used in our experiments. We created 1,000 2 × 2 × 2 transects spanning skin color, hair length, and gender—four examples are shown in (**a**). We set step sizes in such a way that we obtained pale-to-dark skin tones, short-to-medium hair lengths, and M/F gender (see Sect. 15.4.1.4). Besides the intentionally modified attributes, other face attributes are held constant. For each image in each transect, we collected human annotations to measure the perceived attributes. In (**b**), we show human-annotated gender values of the bottom-left transect in (**a**) side-by-side with the generator's intended values. Humans label the first face as a male, though the generator intended to produce a female. In all our experiments, we used human perception, rather than intended generator attributes, as the ground truth

15.5 Experiments

In order to test our method on a practical application, we experiment with benchmarking bias of gender classifiers. The Pilot Parliaments Benchmark (PPB) [3], a dataset of faces of parliament members of various nations, was the first wild-collected test dataset to balance gender and skin color with the goal of fostering the study of gender classification bias across both attributes. The authors of that study found a much larger error rate on dark-skinned females as compared to other groups and conjectured that this is due to bias in the algorithms, i.e., that the performance of the algorithm changes when gender and skin color are changed, all else being equal. Our method allows us to test this hypothesis.

15.5.1 Gender Classifiers

We trained two research-grade gender classifier models, each using the ResNet-50 architecture [74]. The first was trained on the CelebA dataset [75], and the second on the FairFace dataset [76]. CelebA is the most popular public face dataset due to its size and rich attribute annotations, but is known to have severe imbalances [44]. The FairFace dataset was introduced to mitigate some of these biases.

We trained our classifiers for 20 epochs with the binary cross-entropy loss. We set the learning rate at $1e^{-4}$ for the first 10 epochs, and $1e^{-5}$ for the final 10 epochs. To avoid a baseline bias of predicting one gender over another, we enforced the likelihood of sampling male and female faces during training to be equal.

We decided not to test the commercial system for two reasons. First, reproducibility—the models we test may be re-implemented and re-trained by other researchers at any time, while commercial systems are black boxes which may change unpredictably over time. Second, our ResNet-50 models show biases comparable to those observed in the original study by [3] (see Fig. 15.2).

15.5.2 Transect Data

To produce the synthetic images for our transects, we used the generator from the StyleGAN2 architecture trained on Flickr-Faces-HQ (FFHQ) [17, 18]. This generator has both a multivariate Normal input noise space, $\mathcal{N}(\mathbf{0}, \mathbf{I})$, as well as an intermediate "style space." To train the latent space linear models (see Sect. 15.4.1.1), we sampled 5000 vectors from the noise distribution and labeled the generated images with human annotators (see Sect. 15.4.3). However, we use the *style space* as the latent space in our method, because we found it better suited for disentangling semantic attributes. We trained linear regression models to predict age, gender, skin color, and hair length attributes from style vectors. For the remaining attributes—facial hair, makeup, and smiling—we found that binarizing the ranges and training a linear SVM classifier works best.

We generated 3D transects across subgroups of skin color, hair length, and gender following the procedure described in Sect. 15.4.1.2. We use a transect size of $2 \times 2 \times 2$, with grid decision values (specified by input vector $\mathbf{c}$ in Algorithm 1) spaced to generate a pale-to-dark transition along the skin color axis, short-to-medium length along the hair length axis, and male-to-female along the gender axis. We set the decision values by trial-and-error and made them equal for all transects: $(-1.5, 1.7)$ for skin color, $(-0.5, 0)$ for hair length, and $(-1.75, 1.75)$ for gender. We generated 1000 such transects, resulting in 8000 total images. Figure 15.9 presents four example

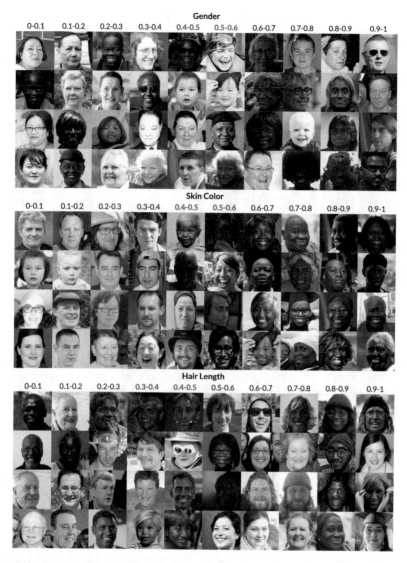

Fig. 15.10 Samples of synthesized faces, organized by mean human annotation scores. In our analysis, we omitted faces from ranges indicated in red to focus on clearly perceived females/males, light/dark skin tones, and short/long hair lengths

transects. The general characteristics of the faces—besides the intentionally modified attributes—are held constant.

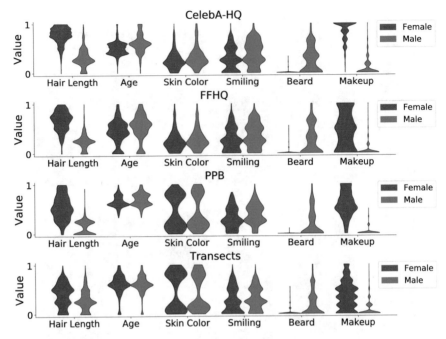

Fig. 15.11 Attribute distributions by dataset and gender groups. "Violin" plot widths are proportional to frequency counts, and each violin is scaled so that its maximum count spans the full width. Wild-collected datasets have greater attribute imbalances across gender than synthetic transects, e.g., longer hair and younger ages for women. We designed our transects to mirror PPB skin color distribution and age distributions, while mitigating hair length imbalance. Hair length versus skin color distributions are further explored in Fig. 15.12

15.5.2.1 Dataset Pruning

Not all synthesized images are ideal for our analysis. Some elicit ambiguous human responses (Fig. 15.8 top left) or are unrealistic (Fig. 15.8 top right). Furthermore, others may not belong clearly to one of our two intended categories for gender, hair length, and skin color attributes. We addressed these points by first removing any image with a mean fakeness score greater than or equal to "Likely fake" (0.75 in the normalized range of [0, 1]). We also removed faces with attribute values in the normalized subranges of [0.4, 0.6] for skin color and gender, and [0.3, 0.5] for hair length (see Fig. 15.10 for examples). After these pruning steps, we were left with 5713 images.

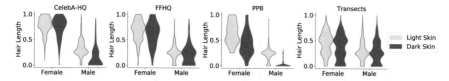

Fig. 15.12 Hair length distributions by gender and skin color groups. In the wild-collected datasets, hair length is correlated with skin color when gender is held constant. Synthetic transects may be designed to minimize this correlation

15.5.3 Comparison of Transects to Real Face Datasets

Figure 15.11 analyzes attribute distributions for the CelebA-HQ, FFHQ, and PPB datasets, along with our transects, stratified by gender. The wild-collected datasets contain significant imbalances across gender, particularly with hair length. They also have biases in age, with a larger percentage of males being older than females. An interesting correlation is that males are also more likely to smile in these data. In contrast, our transects exhibit more balance across gender. They depict more males with medium-to-long hair and fewer females with very long hair. Our transects also have a bimodal skin color distribution and an older population by design, since we are interested in mimicking those population characteristics of PPB. All datasets are imbalanced along the "Beard" and "Makeup" attributes—this is reasonable since we expect these to have strong correlations with gender.

In an ideal matched study, sets of images stratified by a sensitive attribute will exhibit the same distribution over the remaining attributes. Put simply, no other attribute should be strongly correlated with the attribute being manipulated. Figure 15.12 stratifies by skin color. We see correlations of hair length distributions and skin colors in all the wild-collected data, while the synthetic transects exhibit much better balance.

15.5.4 Analysis of Bias

We now analyze the performance of the classifiers on PPB and our transects. We verify that the classifiers exhibit similar error patterns to the commercial classifiers already evaluated on PPB [3]. Because PPB only consists of adults, we remove children and teenagers (age < 0.4 in the normalized [0, 1] scale) from our transects to make a more direct comparison, leaving us with 5335 total images.

Figure 15.1 presents classification errors split by gender (M/F) and skin color (L/D). We replicated the reported errors of the commercial classifiers in [3] and reported the errors of our classifiers on our in-house version of PPB. All classifiers perform significantly worse on dark-skinned females. Figures 15.13 and 15.14

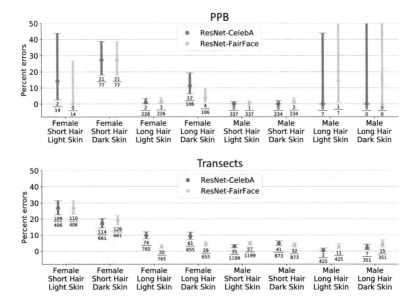

Fig. 15.13 Algorithmic errors, disaggregated by intersectional groups for wild-collected (PPB, top) and synthetic (transects, bottom). Wilson score 95% confidence intervals [77] are indicated by vertical bars, and the misclassification count and the total number of samples are written below each bar. PPB has few samples for several groups, such as short-haired, light-skinned females and long-haired males (see Fig. 15.12). Synthetic transects provide numerous test samples for all groups. The role of the different attributes in causing the errors is studied in Fig. 15.15

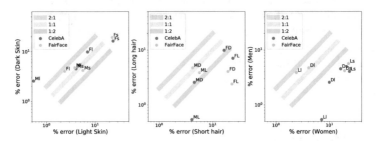

Fig. 15.14 Scatter plots of error rates using data from Fig. 15.13 (transects). Each dot compares the error rates of a pair of groups that differ by one attribute only (indicated in the label of the x and y axes). The two letters near each dot indicate the shared attributes ("M/F" indicate male and female, "D/L" indicate dark and light skin, and "s/l" indicate short and long hair). Dots falling along the equal error line indicate that skin tone has little or no effect on error. In contrast, females and persons with short hair have higher error rates

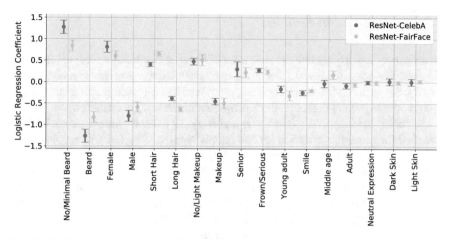

Fig. 15.15 Logistic regression coefficient values. The logistic regression model is trained to predict *absolute errors* of the gender classifiers on our transect data given attributes as input. Coefficients represent the change in *log odds* of the error for a change of 1 unit of each attribute. Larger coefficient magnitudes indicate more important variables, and positive(red)/negative(green) values correspond to variables that increase/decrease classifier error. Each attribute subgroup labeled on the *x*-axis is represented by a binary variable in the regression model, and we order attributes in this plot from large-to-small coefficient magnitudes. Error bars report standard deviations that were computed via bootstrapping 1000 times

present classification errors, stratified by gender/hair length/skin color combinations. We can make a number of broad-stroke, qualitative observations:

- The broad pattern of errors is similar across PPB and transects, with more errors on the left (females) than on the right (males).
- Transect errors are either comparable or higher than in PPB, indicating that synthetic faces can be at least as challenging as real faces. Most significantly, errors are nonzero on males, which allows the study of relative difficulties when attributes are varied.
- In PPB, there are few males with long hair and few females with short hair and light skin, making measurements unreliable for these categories. This is not a problem with transects, where faces are matched by attributes.
- Transect errors are higher when hair is shorter for women. However, hair length has a negligible effect for males (see Fig. 15.18 for a possible explanation).
- There is no consistent transect error pattern in skin tone: within homogeneous groups, changing skin tone does not seem to affect the performance of either algorithm. For example, females with long hair see no significant difference in classification error between light vs. dark skin. Looking at PPB alone, we could not make this observation, since skin tone is so strongly correlated with hair length.

15.5.5 Regression Analysis

In order to obtain a quantitative assessment of the effect (or lack thereof) of attributes on classifier error, we investigated further by calculating covariate-adjusted causal effects. For each gender classifier model, we trained an $L2$-regularized logistic regression model to predict that classifier's error conditioned on all attributes.

We discretized attributes into levels and assigned a binary variable to each level. We used the same discretization for hair length (short vs. long hair), skin color (light vs. dark skin), and gender (female vs. male) used in our experiments thus far. We used two levels for beard (no/light beard vs. beard) and makeup (no/light makeup vs makeup), three for facial expression (serious/frown vs. neutral vs. smile), and the original semantic levels for age described in Fig. 15.6. In all, this resulted in 17 input variables to our logistic regression model. We used scikit-learn's Logistic Regression function [78] and set the regularization parameter to 1.

Figure 15.15 presents coefficients for both logistic regression models. Recall that each coefficient represents the change in log odds of the classifier's error for a change of 1 unit of each covariate (see Sect. 15.4.2). Error bars depict standard deviations, obtained by bootstrapping the dataset 1000 times. A person's facial hair, gender, makeup, hair length, and age all have significant effects on classification error, and skin color has a negligible effect. Our main experimental conclusion is that observational (PPB) and experimental (transects) methods are fundamentally at odds on the causes of algorithm bias in gender classification algorithms. Observational analysis on wild-collected PPB suggests that a combination of gender and skin tone are implicated, while our experimental method using synthetic transects suggests that other attributes are far more important than skin tone.

15.5.5.1 Joint Effects of Attributes on Classification Error

Our regression analysis makes a simplifying assumption that each covariate has an independent, linear effect on classification error. The independence assumption can be a poor one. For example, Fig. 15.14-right shows that error rates vary across different intersectional groups of skin color and hair length in a way that is not simply a linear combination of each attribute.

This is also the reason we removed children and teenagers from our analysis, as these individuals tend to have different appearance characteristics from adults. Figure 15.16 illustrates this by breaking down error rates by age and gender subpopulations for two classifier decision thresholds. The difference in error rates between the genders is fairly consistent for young adults to middle-aged individuals but varies for children/teenagers and seniors. This demonstrates that age and gender have joint effects on errors.

Figure 15.17 shows faces from our synthesized transects on which the ResNet models were most incorrect. For each gender misclassification direction, we show faces on which the model predictions were farthest from the average human annotator

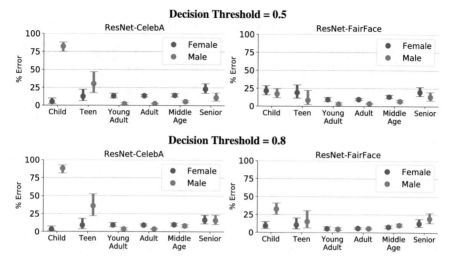

Fig. 15.16 Errors by gender and age group on our transect images. The two top plots were obtained by using a decision threshold equal to 0.5 and show a prevalence of female errors. The bottom two plots were obtained with a threshold equal to 0.8, chosen to minimize overall error. There is a non-uniform influence of age on errors. Both models tend to have lower errors for young to middle-aged adults. The differences in errors between genders are fairly consistent for adults, but differ for children, teenagers, and seniors, illustrating a combined age–gender bias in the algorithms

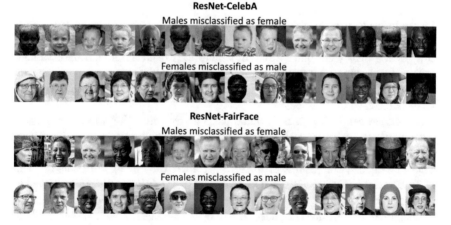

Fig. 15.17 Images with largest errors. Synthetic faces on which the classifiers most deviated from the mean human annotations

Fig. 15.18 Correlated attribute modifications. We found that our method sometimes adds a beard to a male face when attempting to only modify hair length. This is an example of an imprecise intervention which can complicate downstream bias analyses. This bias may be due to the training data itself (men with long hair tend to have facial hair) or injected by the algorithm

response. ResNet-CelebA tends to heavily misclassify young male children/babies as female, in line with the quantitative result in Fig. 15.16.

15.6 Discussion and Conclusions

15.6.1 Summary

Our study leads us to three main conclusions. First, the experimental approach to measuring algorithmic bias in computer vision is feasible. Second, the experimental approach may yield quite different conclusions from traditional observational studies. Third, when analyzing algorithmic bias, a broad spectrum of attributes and attribute combinations should be considered besides the ones of immediate interest. We examine each in detail below.

Our experimental approach is made possible by combining recent progress in image synthesis with detailed human annotations collected by crowdsourcing. Image synthesis, calibrated by human annotations, allows us to generate transects of matched samples, i.e., groups of images that differ only along attributes of interest. In contrast to the previous attribute-specific methods [11], *any* attribute may be explored, provided that it can be annotated by humans. By relying on human ground truth annotations, one does not need to rely on the synthesis method being perfect.

The experimental method and our synthesis-based experimental approach offer a number of attractive properties and advantages over traditional observational methods:

1. **Causal inferences on bias are possible**. Our method generates approximately matched samples across selected attributes, allowing for counterfactual analysis, e.g., *"Would the algorithm have made the same mistake if the same person had had a different skin color?"* Observational image data are almost never matched.

Males   Females

Fig. 15.19 Hidden confounders. There is always the possibility of a hidden confounder lurking in a dataset. As an example, we found—after already collecting annotations—that our method tends to add earrings when transitioning from dark-skinned men to dark-skinned women, a cue that a gender classifier might be used to perform disproportionately well on the latter group. Because we did not annotate this attribute, it is hidden in our analysis. Interestingly, one male in this image also has an earring; that earring becomes larger for his female counterpart

2. **Bias may be measured for underrepresented groups**. Image synthesis allows, to a great degree, uniform sampling of the space of attributes of interest—gender, skin color, and hair length in our experiments. This is very difficult to do when one relies on images that are sampled from natural distributions, which tend to be long-tailed, and therefore where some groups are underrepresented.

3. **Bias may be measured for intersectional groups**. Our method allows researchers to draw causal inferences across groups that are defined by specific attribute combinations. Single-attribute analysis may conceal biases affecting groups defined by the combination of multiple attributes [10]. Some such combinations are often vastly undersampled in natural data.

4. **Bias measurements are valid across different populations**. This is because the experimental method identifies causally linked attributes, independent of the prevalence of these attributes, i.e., the bias measurements are a property of the algorithm and not of the population on which it is used. By contrast, observational measurements do not generalize beyond the narrowly defined population where the data was collected. Furthermore, by combining appropriately the contribution of different attributes, one may predict the effects both of *disparate treatment* [79] and *disparate impact* [80] on a specific population.

5. **Accurate bias measurements may be made quickly and inexpensively**. Image synthesis is fast and inexpensive, and crowdsourced image annotation is also relatively fast and affordable. By contrast, assembling large datasets of natural images is laborious and expensive—it may take years and substantial investment, which may only be afforded by large organizations. Thus, synthetic data has the potential to democratize testing for bias.

6. **Ethical and legal concerns are greatly reduced**. Collecting face image datasets in the wild requires great care to respect the privacy and dignity of individuals, the rights of minors and other vulnerable groups, as well as copyright laws. By contrast, synthetic datasets are free from such risks because they do not depict real people.

The experimental analysis (transects) and traditional observational analysis (using PPB) diverged most significantly on the effect of skin color; the observational study was flagged as significant and the experimental method was found to be not significant in determining algorithmic bias. The experimental method reveals a number of additional sources of bias: age, hair length, and facial hair (Fig. 15.15). The two methods agree on gender. Our analysis suggests that the difference between the conclusions of the two methods is likely due to the correlation of hair length, skin color, and gender in PPB (see Figs. 15.11 and 15.12). Consequently, if one does not control for hair length, the classifiers' bias towards assigning gender on the basis of hair length is read as a bias concerning dark-skinned women. The triple correlation between hair length, gender, and skin color had been noticed in a previous study [11].

The main reason for measuring algorithmic bias is to get rid of it. Error and bias measurements guide scientists and engineers towards effective corrective measures for improving the performance of their algorithms. It is instructive to view the different predictions of the two methods through this lens. The correlational study based on PPB (Fig. 15.1) may suggest that, in order to reduce biases in our classifiers, more images of dark-skinned women should be added to their training sets. The experimental method leads engineers in a different direction. First, more training images of long-haired men and short-haired women of all races are needed. Second, correcting age bias requires more training images in the child-teen and, possibly, senior age groups.

Finally, a lesson from our study is that it is important to consider a rich number of attributes and attribute combinations, besides the one(s) of immediate interest. This is for two reasons. First, unobserved confounders can have strong effects and need to be included in the analysis. Second, the combined effect of attributes can be strongly nonlinear (see the interaction of age and gender in Fig. 15.16), and therefore an intersectional analysis [3, 8] is necessary. Selecting attributes or attribute combinations is as much of an art as a science, and therefore one has to rely on good judgment and on a healthy multidisciplinary debate to progressively reveal missing ones.

15.6.2 Limitations and Future Work

While the advantages of the experimental method are clear, our proposed method does not exempt researchers from exercising attention and good judgment. In particular, while our method greatly reduces unwanted correlations with annotated variables, it does not eliminate them completely, nor does it account for hidden confounders [81], and one will need to keep a sharp eye out for both. As an example of the first, we found that our method often adds facial hair to male faces when increasing hair length (see Fig. 15.18). This is likely a reason why our classifiers did not have higher error rates for males with longer hair (see Fig. 15.13). As an example of the second, we found that our method tends to synthesize earrings when modifying a dark-skinned face to look female (see Fig. 15.19). Depending on culture, earrings may or may not be relevant to the definition of gender. If this is an unwanted correlation, one ought to add earrings to the annotation pipeline so that it may be "orthogonalized away" by the synthesis method. Scientists building an industry-grade system for measuring face analysis bias will want to consider including a more exhaustive set of factors. A significant advantage of an approach that is based on synthetic images and human annotation is thus the following: *as soon as one residual correlation is discovered, it may be systematically annotated, compensated for in the analysis, and mitigated in the synthesis.*

A number of refinements in face synthesis will make our experimental method more practical and powerful. First, many of the faces we generated contained visible artifacts (see Fig. 15.8), which we eliminated by human annotation—even subtle artifacts can affect classifier outputs, as revealed by the literature on adversarial examples [82]. Second, we do not yet have tools to estimate the sets of physiognomies and attribute combinations that can and cannot be produced by a given generator. Current GANs are known to have difficulties in generating data outside of their training distributions. Third, we observed a bias of StyleGAN2 towards generating Caucasian faces when sampling from its latent distribution. While our method can compensate for biases through carefully oriented traversals calibrated by human annotations, it would be clearly better to start from unbiased synthesis methods. We are hopeful that these shortcomings will be incrementally resolved by a combination of training sets with increased diversity of attributes like ethnicity, gender, personal style, and age, as well as better models.

Our first-order technique for controlling synthesis can also be improved. A better understanding of the geometry of face space will hopefully yield more accurate global coordinate systems. These, in turn, will help reduce residual biases in synthetic transects, which we currently mitigate by having transects annotated by hand.

Finally, extending our method beyond gender classification to more complex tasks, such as face recognition, is not straightforward in practice and will require further study.

Acknowledgements A number of colleagues kindly read draft versions of this manuscript, providing references, insightful comments, and valuable criticisms. We are especially grateful to Frederick Eberhardt, Bill Freeman, Lei Jin, Michael Kearns, R. Manmatha, Tristan McKinney, Sendhil Mullainathan, and Chandan Singh.

References

1. Kleinberg J, Ludwig J, Mullainathany S, Sunstein CR (2019) Discrimination in the age of algorithms. Published by Oxford University Press on behalf of The John M. Olin Center for Law, Economics and Business at Harvard Law School
2. Mullainathan S (2019) Biased algorithms are easier to fix than biased people, New York Times
3. Buolamwini J, Gebru T (2018) Gender shades: intersectional accuracy disparities in commercial gender classification. In: Conference on fairness, accountability and transparency, pp 77–91
4. Brandao M (2019) Age and gender bias in pedestrian detection algorithms. arXiv:1906.10490
5. Klare BF, Burge MJ, Klontz JC, Bruegge RWV, Jain AK (2012) Face recognition performance: role of demographic information. IEEE Trans Inf Forensics Secur 7(6):1789–1801

6. Krishnapriya KS, Vangara K, King MC, Albiero V, Bowyer K (1904) Characterizing the variability in face recognition accuracy relative to race. arXiv:07325(4):2019
7. Bertrand M, Mullainathan S (2004) Are Emily and Greg more employable than Lakisha and Jamal? A field experiment on labor market discrimination. Am Econ Rev 94(4):991–1013
8. Kearns M, Neel S, Roth A, Wu ZS (2017) Preventing fairness gerrymandering: auditing and learning for subgroup fairness. arXiv:1711.05144
9. Kearns M, Neel S, Roth A, Wu ZS (2019) An empirical study of rich subgroup fairness for machine learning. In: Proceedings of the conference on fairness, accountability, and transparency, pp 100–109
10. Kearns M, Roth A (2019) The ethical algorithm: the science of socially aware algorithm design. Oxford University Press
11. Muthukumar V, Pedapati T, Ratha N, Sattigeri P, Wu C-W, Kingsbury B, Kumar A, Thomas S, Mojsilovic A, Varshney KR (2018) Understanding unequal gender classification accuracy from face images. arXiv:1812.00099
12. Lohr S (2018) Facial recognition is accurate, if you're a white guy. New York Times
13. Torralba A, Efros AA, et al (2011) Unbiased look at dataset bias. In: CVPR, vol 1, p 7
14. Merkatz RB, Temple R, Sobel S, Feiden K, Kessler DA (1993) Working group on women in clinical trials. women in clinical trials of new drugs–a change in food and drug administration policy. N Engl J Med 329(4):292–296
15. Simon V (2005) Wanted: women in clinical trials
16. Pearl J (2009) Causality. Cambridge University Press
17. Karras T, Laine S, Aila T (2019) A style-based generator architecture for generative adversarial networks. In: Proceedings of the IEEE conference on computer vision and pattern recognition, pp 4401–4410
18. Karras T, Laine S, Aittala M, Hellsten J, Lehtinen J, Aila T (2019) Analyzing and improving the image quality of stylegan. arXiv:1912.04958
19. Buhrmester M, Kwang T, Gosling SD (2016) Amazon's mechanical turk: a new source of inexpensive, yet high-quality data?
20. Barron JL, Fleet DJ, Beauchemin SS (1994) Performance of optical flow techniques. Int J Comput Vis 12(1):43–77
21. Bowyer K, Phillips PJ (1998) Empirical evaluation techniques in computer vision. IEEE Computer Society Press
22. Fei-Fei L, Fergus R, Perona P (2004) Learning generative visual models from few training examples: An incremental bayesian approach tested on 101 object categories. In: 2004 conference on computer vision and pattern recognition workshop. IEEE, pp 178–178
23. Phillips PJ, Grother P, Micheals R, Blackburn DM, Tabassi E, Bone M (2003) Face recognition vendor test 2002. In: 2003 IEEE international soi conference. Proceedings (Cat. No. 03CH37443). IEEE, p 44
24. Phillips PJ, Wechsler H, Huang J, Rauss PJ (1998) The feret database and evaluation procedure for face-recognition algorithms. Image Vis Comput 16(5):295–306
25. Phillips PJ, Yates AN, Hu Y, Hahn CA, Noyes E, Jackson K, Cavazos JG, Jeckeln G, Ranjan R, Sankaranarayanan S, et al (2018) Face recognition accuracy of forensic examiners, superrecognizers, and face recognition algorithms. Proc Natl Acad Sci 115(24):6171–6176
26. Grother P, Ngan M, Hanaoka K (2018) Ongoing face recognition vendor test (frvt) part 1: verification. Tech. Rep, National Institute of Standards and Technology
27. Grother PJ, Ngan ML, Hanaoka KK (2018) Ongoing face recognition vendor test (frvt) part 2: identification. Technical report
28. Ponce J, Berg TL, Everingham M, Forsyth DA, Hebert M, Lazebnik S, Marszalek M, Schmid C, Russell BC, Torralba A, et al (2006) Dataset issues in object recognition. In: Toward category-level object recognition. Springer, pp 29–48
29. Albiero V, Krishnapriya KS, Vangara K, Zhang K, King MC, Bowyer KW (2020) Analysis of gender inequality in face recognition accuracy. In: Proceedings of the IEEE winter conference on applications of computer vision workshops, pp 81–89

30. Drozdowski P, Rathgeb C, Dantcheva A, Damer N, Busch C (2020) Demographic bias in biometrics: a survey on an emerging challenge. IEEE Trans Technol Soc
31. Kortylewski A, Egger B, Schneider A, Gerig T, Morel-Forster A, Vetter T (2018) Empirically analyzing the effect of dataset biases on deep face recognition systems. In: Proceedings of the IEEE conference on computer vision and pattern recognition workshops, pp 2093–2102
32. Kortylewski A, Egger B, Schneider A, Gerig T, Morel-Forster A, Vetter T (2019) Analyzing and reducing the damage of dataset bias to face recognition with synthetic data. In: Proceedings of the IEEE conference on computer vision and pattern recognition workshops, pp 0–0
33. Merler M, Ratha N, Feris RS, Smith JR (2019) Diversity in faces. arXiv:1901.10436
34. Li Y, Vasconcelos N (2019) Repair: removing representation bias by dataset resampling. In: Proceedings of the IEEE conference on computer vision and pattern recognition, pp 9572–9581
35. Alvi M, Zisserman A, Nellåker C (2018) Turning a blind eye: explicit removal of biases and variation from deep neural network embeddings. In: Proceedings of the European conference on computer vision (ECCV), pp 0–0
36. Corbett-Davies S, Goel S (2018) The measure and mismeasure of fairness: a critical review of fair machine learning. arXiv:1808.00023
37. Das A, Dantcheva A, Bremond F (2018) Mitigating bias in gender, age and ethnicity classification: a multi-task convolution neural network approach. In: Proceedings of the European conference on computer vision (ECCV), pp 0–0
38. Hébert-Johnson U, Kim MP, Reingold O, Rothblum GN (2017) Calibration for the (computationally-identifiable) masses. arXiv:1711.08513
39. Hendricks LA, Burns K, Saenko K, Darrell T, Rohrbach A (2018) Women also snowboard: overcoming bias in captioning models. In: European conference on computer vision. Springer, pp 793–811
40. Khosla A, Zhou T, Malisiewicz T, Efros AA, Torralba A (2012) Undoing the damage of dataset bias. In: European conference on computer vision. Springer, pp 158–171
41. Lu B, Chen J-C, Castillo CD, Chellappa R (2019) An experimental evaluation of covariates effects on unconstrained face verification. IEEE Trans Biom Behav Identity Sci 1(1):42–55
42. Ryu HJ, Adam H, Mitchell M (2017) Inclusivefacenet: improving face attribute detection with race and gender diversity. arXiv:1712.00193
43. Patrick G, Ngan M, Hanaoka K (2019) Face recognition vendor test (frvt) part 3: demographic effects. IR 8280, NIST. https://doi.org/10.6028/NIST.IR.8280
44. Denton E, Hutchinson B, Mitchell M, Gebru T (2019) Detecting bias with generative counterfactual face attribute augmentation. arXiv:1906.06439
45. Chang C-H, Creager E, Goldenberg A, Duvenaud D (2018) Explaining image classifiers by counterfactual generation. In: ICLR
46. Dabkowski P, Gal Y (2017) Real time image saliency for black box classifiers. In: Advances in neural information processing systems, pp 6967–6976
47. Fong RC, Vedaldi A(2017) Interpretable explanations of black boxes by meaningful perturbation. In: Proceedings of the IEEE international conference on computer vision, pp 3429–3437
48. Goyal Y, Wu Z, Ernst J, Batra D, Parikh D, Lee S (2019) Counterfactual visual explanations. In Proceedings of the 36th international conference on machine learning, vol 97. PMLR, pp 2376–2384
49. Selvaraju RR, Cogswell M, Das A, Vedantam R, Parikh D, Batra D (2017) Grad-cam: visual explanations from deep networks via gradient-based localization. In: Proceedings of the IEEE international conference on computer vision, pp 618–626
50. Simonyan K, Vedaldi A, Andrew Z (2013) Visualising image classification models and saliency maps, deep inside convolutional networks
51. Sundararajan M, Taly A, Yan Q (2017) Axiomatic attribution for deep networks. In: Proceedings of the 34th international conference on machine learning, vol 70. JMLR. org, pp 3319–3328
52. Bau D, Zhu J-Y, Strobelt H, Zhou B, Tenenbaum JB, Freeman WT, Torralba A (2019) Gan dissection: visualizing and understanding generative adversarial networks
53. Zhou B, Bau D, Oliva A, Torralba A (2018) Interpreting deep visual representations via network dissection. IEEE Trans Pattern Anal Mach Intell

54. Kim B, Wattenberg M, Gilmer J, Cai C, Wexler J, Viegas F, Sayres R (2018) Interpretability beyond feature attribution: quantitative testing with concept activation vectors (TCAV). ICML
55. Anderson DJ, Adolphs R (2014) A framework for studying emotions across species. Cell 157(1):187–200
56. Darwin C, Prodger P (1998) The expression of the emotions in man and animals. Oxford University Press, USA
57. Fitzpatrick TB (1988) The validity and practicality of sun-reactive skin types i through vi. Arch Dermatol 124(6):869–871
58. Egan SK, Perry DG (2001) Gender identity: a multidimensional analysis with implications for psychosocial adjustment. Dev Psychol 37(4):451
59. Shen Y, Gu J, Tang X, Zhou B (2019) Interpreting the latent space of GANS for semantic face editing. arXiv:1907.10786
60. Singla S, Pollack B, Chen J, Batmanghelich K (2019) Explanation by progressive exaggeration. arXiv:1911.00483
61. Xiao T, Hong J, Ma J (2018) Elegant: exchanging latent encodings with GAN for transferring multiple face attributes. In: Proceedings of the European conference on computer vision (ECCV), pp 168–184
62. Hoerl AE, Kennard RW (1970) Ridge regression: biased estimation for nonorthogonal problems. Technometrics 12(1):55–67
63. Cortes C, Vapnik V (1995) Support-vector networks. Mach Learn 20(3):273–297
64. Youla DC (1987) Mathematical theory of image restoration by the method of convex projections. In: Image recovery: theory and application, pp 29–77
65. Gelman A, Hill J (2006) Data analysis using regression and multilevel/hierarchical models. Cambridge University Press
66. Angrist JD, Imbens GW (1995) Identification and estimation of local average treatment effects. Technical report, National Bureau of Economic Research
67. Heckman JJ, Vytlacil EJ (2001) Instrumental variables, selection models, and tight bounds on the average treatment effect. In: Econometric evaluation of labour market policies. Springer, pp 1–15
68. Oreopoulos P (2006) Estimating average and local average treatment effects of education when compulsory schooling laws really matter. Am Econ Rev 96(1):152–175
69. Rubin DB (2006) Matched sampling for causal effects. Cambridge University Press
70. Pocock SJ, Assmann SE, Enos LE, Kasten LE (2002) Subgroup analysis, covariate adjustment and baseline comparisons in clinical trial reporting: current practiceand problems. Stat Med 21(19):2917–2930
71. Robinson LD, Jewell NP (1991) Some surprising results about covariate adjustment in logistic regression models. In: International statistical review/Revue Internationale de Statistique, pp 227–240
72. Willan AR, Briggs AH, Hoch JS (2004) Regression methods for covariate adjustment and subgroup analysis for non-censored cost-effectiveness data. Health Econ 13(5):461–475
73. xx
74. He K, Zhang X, Ren S, Sun J (2016) Deep residual learning for image recognition. In: Proceedings of the IEEE conference on computer vision and pattern recognition, pp 770–778
75. Ziwei L, Ping L, Xiaogang W, Xiaoou T (2015) Deep learning face attributes in the wild
76. Kärkkäinen K, Joo J (2019) Fairface: face attribute dataset for balanced race, gender, and age. arXiv:1908.04913
77. Wilson EB (1927) Probable inference, the law of succession, and statistical inference. J Am Stat Assoc 22(158):209–212
78. Pedregosa F, Varoquaux G, Gramfort A, Michel V, Thirion B, Grisel O, Blondel M, Prettenhofer P, Weiss R, Dubourg V, Vanderplas J, Passos A, Cournapeau D, Brucher M, Perrot M, Duchesnay E (2011) Scikit-learn: machine learning in python. J Mach Learn Res 12:2825–2830
79. Mendez MA (1980) Presumptions of discriminatory motive in title vii disparate treatment cases. Stanford Law Rev 1129–1162

80. Rutherglen G (1987) Disparate impact under title vii: an objective theory of discrimination. Virginia Law Rev 1297–1345
81. VanderWeele TJ, Shpitser I (2013) On the definition of a confounder. Annals Stat 41(1):196
82. Szegedy C, Zaremba W, Sutskever I, Bruna J, Erhan D, Goodfellow I, Fergus R (2013) Intriguing properties of neural networks. arXiv:1312.6199

Chapter 16
Strategies of Face Recognition by Humans and Machines

Jacqueline G. Cavazos, Géraldine Jeckeln, Ying Hu, and Alice J. O'Toole

Abstract Face recognition by machines has improved markedly over the last decade. Machines now perform some face recognition tasks at the level of untrained humans and forensic face identification experts. In this chapter, first we review recent work on human and machine performance on face recognition tasks. Second, we consider the benefits of statistically fusing human and machine responses to improve performance. Third, we review strategic differences in how humans with various levels of expertise approach face identification tasks. We conclude by considering the challenging problem of human and machine performance on recognition of faces of different races. Understanding how humans and machines perform these tasks can lead to more effective and accurate face recognition in applied settings.

16.1 Introduction

Human face recognition can be impressive. With remarkable speed and accuracy, we can recognize the face of a long-missed friend across a dimly lit room. Failures of face recognition are common also. These range from the embarrassing mistakes we make every day to the highly consequential mistakes made by eyewitnesses in forensic settings. Understanding how human face recognition operates and when it is likely to succeed is especially critical in these latter cases.

In the era of machine-based face recognition, it is incumbent on researchers to better understand the circumstances in which humans versus machines will identify faces more accurately. Moreover, it is important to consider the possibility that human–machine collaborations could provide a "better" (more accurate) option in cases of consequence. Recent research indicates that strategy differences among humans [1–3], machines, and between humans and machines can be exploited to

---

J. G. Cavazos · G. Jeckeln · Y. Hu · A. J. O'Toole (✉)
The University of Texas at Dallas, Dallas, USA
e-mail: otoole@utdallas.edu

improve system performance through statistical fusion [4, 5]. The progress made on this question in recent years is the subject of this review. In particular, we will focus on the diverse range of strategies and skills found among humans and on how these diverse skills can be used intelligently in forming collaborations among humans and between humans and machines.

This chapter is organized as follows. First, we note the range of face identification skills found in humans, for both untrained individuals and professional forensic examiners. Second, we review the progress of machines on face identification from the perspective of published comparisons between human and machine accuracy over the last decade. Third, we evaluate results from human–human and human–machine fusion studies that examine the benefits of combining individual face identification judgments in efforts to improve accuracy. Fourth, we take a look at the different strategies in face identification utilized by both humans and machines. Finally, we will consider how human and machine accuracies are affected by the race/ethnicity of the face to be recognized. For humans, the race of both the perceiver and the person being perceived affects accuracy. This phenomenon is called the "other-race effect". We will see a version of this other-race effect for machines developed in different parts of the world and tested on faces of different races [6]. The other-race effect, at least for humans, may reflect a difference in recognition strategy that might benefit from careful use of fusion.

16.2   Identification Accuracy: Human Face Recognition

The study of human face recognition by psychologists has a history that goes back to the early 1980s and before. Over the last five decades, psychologists have measured human performance across a broad range of tasks. An overarching theme in behavioral research on face recognition is the wide range of performance across individuals and across groups of people. At one extreme, prosopagnosics show highly selective impairments for face recognition [7]. At the other end of the spectrum, super-recognizers show high accuracy across multiple face recognition tasks [8]. In between these extremes, there is wide person-to-person variation in face recognition skills. The behavioral literature makes it clear that "human performance" on face recognition cannot be categorized as a monolithic skill that operates equivalently across all individuals.

16.2.1   Face Identification Performance of Untrained Humans

Despite the wide person-to-person variation in face recognition abilities, untrained human subjects share some common trends in accuracy. For example, humans are experts at recognizing and identifying *familiar*, or well-known, identities. This expertise spans across photometric changes, including illumination, pose, and expression. It can also span changes in appearance, such as hair color, aging, and disguise. Human expertise for familiar faces has encouraged a false belief that humans are experts for *all* faces [9]. However, previous studies have shown that this expertise does not extend to *unfamiliar* faces (See Johnston and Edmonds [10] for a review of familiar versus unfamiliar face recognition). When asked to identify or recognize faces of unfamiliar or unknown identities, humans are susceptible to errors at an alarming rate. These accuracy differences for recognizing familiar versus unfamiliar faces raise the question of whether trained professional human observers are also prone to these susceptibilities.

16.2.2 Performance of Trained Versus Untrained Humans

Until recently, remarkably little was known about the accuracy of face identification professionals relative to untrained observers. The National Research Council report, Strengthening Forensic Science in the United States: A Path Forward [11], changed that state of affairs. The first comprehensive scientific study of the performance of professional forensic facial examiners was performed in 2015 [3]. In this study, forensic facial examiners were assessed, along with other experts in biometric systems and untrained university students. The stimuli used in this experiment were selected carefully to be highly challenging for both humans and previous-generation face recognition algorithms. This study also examined face identification performance across different stimulus exposure durations (2 s and 30 s) and stimulus orientations (upright and inverted). Forensic facial examiners performed consistently superior to motivated controls and university students. This experiment, however, imposed conditions that are not reflective of typical forensic laboratory conditions. In forensic labs, identification decisions can require days or weeks to be completed and are made with the assistance of image measurement and manipulation tools [12]. Accordingly, the results reported in [3] should be considered a lower bound estimate of the accuracy of examiners in the practice of their casework.

To address the question of performance under conditions more typical of identification decisions in a forensic laboratory, a Black Box test of forensic facial examiners was conducted in 2017 [4]. This test used the most challenging images from [3]. In the experiment, forensic facial examiners, forensic facial reviewers, professional forensic fingerprint examiners, super-recognizers, and university students were tested. Results confirmed the advantage of facial examiners, facial reviewers, and super-recognizers over fingerprint examiners and students. They also showed that the performance of individuals in all of the groups varied widely (See Fig. 16.1). The researchers proposed that the problem of individual variability in the accuracy of the high expertise group might be ameliorated by the use of face recognition software, if it can be tailored to address the weaknesses of individual examiners (see Section "Fusion: Humans and Machines").

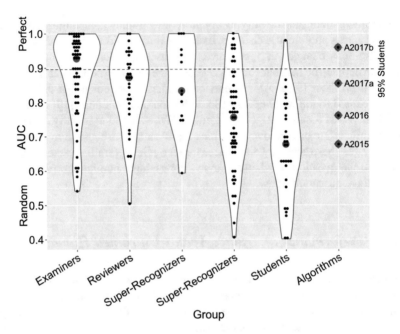

Fig. 16.1 Human and machine accuracy from [4]. Black dots indicate AUCs of individual participants; red dots are group medians. In the algorithms column, red dots indicate algorithm accuracy. Face specialists (facial examiners, facial reviewers, and super-recognizers) surpassed fingerprint examiners, who surpassed the students. The violin plot outlines are estimates of the density for the AUC distribution for the subject groups. The dashed horizontal line marks the accuracy of a 95th percentile student. For the facial examiner group, 53% were above the 95th percentile of students; for the facial reviewers, this proportion was 36%; for super-recognizers it was 46 %; and for fingerprint examiners, it was 17%. All algorithms perform in the range of human performance

16.3   Identification Accuracy: Machines Versus Humans

The study of machine-based face recognition has a long history as well that progresses from simple pattern recognition algorithms a few decades ago to algorithms that are now neurally inspired and operate with hundreds of millions of parameters. In this section, we will review the development of machines. Our focus is on measuring progression in terms of the difficulty of the stimuli an algorithm is capable of recognizing. Here we see a progression from identification of controlled images (i.e., constrained frontal images, with minimal photometric variation) to highly uncontrolled images (i.e., with wide photometric and appearance variation).

Controlled face recognition (2005–2013). In 2005, computer-based face recognition systems began to close the gap between human and machine performance. The progress of machines can be tracked by examining the difficulty of the problems on which performance was tested. Over the last two decades, large-scale evaluations of state-of-the-art face recognition algorithms, open to international competitors from academics and industry, have provided a look at the progression of task difficulty expected at any given time by a state-of-the-art face recognition algorithm [13–16]. Systematic comparisons between humans and algorithms on these tests started in 2005 (cf., for a review [17]).

In 2007, the state of the art for machines was to determine whether pairs of frontal images showed the same identity or different identities when one image was taken under controlled illumination (passport-style) and the second image was taken under uncontrolled indoor illumination. At that time, the best algorithms surpassed untrained humans on face pairs prescreened to be challenging for the machines [18]. By 2012, machine recognition had progressed to matching pairs of images with unconstrained variation in illumination (indoor and outdoor) and expression. In one test, image pairs were divided into easy, more challenging, and extremely challenging categories, based on the performance of a baseline algorithm. Human–machine comparisons at matching the identity of faces across pairs of images showed that machines were far better than humans on the easy and moderately challenging pairs [19]. On the extremely challenging pairs, machines and humans were equally matched. In no case were humans superior to the best algorithms.

To summarize, by 2012, on the problem of identifying faces from constrained frontal images, the best face recognition algorithms performed as well as or better than humans. However, it is important to consider the fact that the human perceivers tested for these comparisons were not familiar with the people in the images.

Uncontrolled face recognition (2014–present). Deep convolutional neural networks (DCNNs) trained for face recognition first appeared in 2014. These algorithms are capable of recognizing faces from uncontrolled images captured "in the wild" (e.g., [20–26]). Performance on datasets such as Labeled Faces in the Wild [27, 28], IJB-A [29], and Mega-Face [30] indicates that machines are beginning to attack the problem of generalized face recognition, which involves recognizing faces across changes in image parameters and appearance.

One factor underlying the success of DCNNs is that they are trained with millions of images of thousands of individuals captured "in the wild". Notably, the algorithm learns an identity from exposure to many images of the person. These images should span multiple pose, illumination, and expression conditions, as well as appearance variables (e.g., age, make-up, hairstyles, facial hair, glasses). Accurate labels are also critical and are provided by online crowd-sourcing or from social media applications.

How do DCNNs compare to humans? This question was addressed in [4] (see Section "Trained vs. Untrained Human Performance"). In that study, four DCNNs, developed between 2015 and 2017, took the same challenging face identification test administered to human participants. The performance of the algorithms appears along side that of the humans in Fig. 16.1. The results demonstrated that the DCNNs identified faces within the range of human accuracy. In addition, the accuracy of the algorithms increased steadily over time. Remarkably, the most recent DCNN scored slightly above the median of the forensic facial examiners.

Next, we consider the potential for fusing identification judgments to achieve more accurate performance.

16.4   Fusion

It is well known that combining multiple individual judgments can yield a response closer to the ground truth than the individual judgments themselves. This concept, known as the *wisdom-of-crowds* in psychology and *fusion* in engineering, derives its potential from the diversity of cognitive strategies employed by independent responders [31, 32]. For instance, valuable information derived from distinct face processing computations (e.g., feature analysis versus configuration analysis) can be fused in order to increase accuracy on a face identification decision. In what follows, the face identification test employed requires a decision about whether two images picture the same person or different people. We refer to this task as a *face identity matching* test. In this section, we begin with a brief overview of methods employed for fusion and then discuss the effects of fusing judgments on face identity matching tests across individual humans and between humans and machines.

16.4.1   Crowd-Sourcing Methods

Substantial improvements in face identity matching have been achieved with different crowd-sourcing methods, including the "majority vote decision rule", simple response averaging, and Partial Least Square Regression (PLS-R). In general, human crowds of varying sizes are generated by randomly sampling participants who have completed a given task individually. When the task consists of dichotomous response options, a joint decision is determined for each test item by selecting the response (e.g., "same" or "different") on which 50% or more of the individuals within the simulated crowd converge. Otherwise, individual ratings from a similarity scale (1: very dissimilar; 7: very similar) or certainty scale (1: sure different identity; 7: sure same identity) are averaged for each item independently. In addition, fusion can be performed through PLS-R, where similarity ratings from individual systems are used to predict whether two face images depict the same identity or different identities. The model first generates a weight for each of the individual systems and then combines the individual system judgments to produce an estimated similarity rating (See [5]).

16.4.2   Fusion: Human Participants

Wisdom-of-crowds effects on difficult face-matching tasks have been found across different groups of observers, including professional forensic face examiners, fingerprint examiners, super-recognizers, other forensic professionals, and untrained university students [3, 4]. Specifically, accuracy is found to increase as a function of the number of individual responders sampled. Performance typically reaches near-

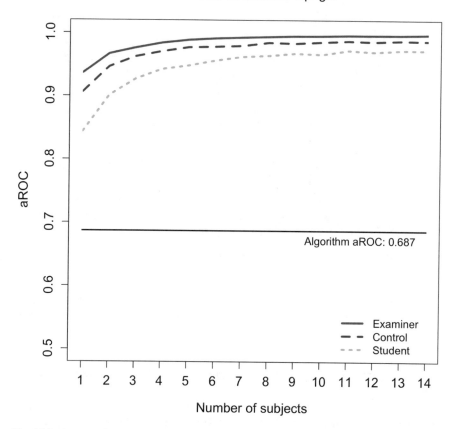

Fig. 16.2 Accuracy increases as more judgments are fused. Clear improvements for aggregating judgments are seen within all three groups of subjects. For all groups, these improvements plateau well before the maximum sample size: at asymptote, average aROCs for groups of roughly eight raters were close to perfect for all three groups (examiners = 0.997; controls = 0.987; students = 0.973) [3]

perfect accuracy at a sample size of eight individual human participants [1–3]. This is illustrated in Fig. 16.2. Although face-matching response combinations can be computed by averaging individual ratings across individual test items, similar effects have been achieved through the interactive collaboration of human participants [1, 33]. Nonetheless, it has been shown that "social" collaboration (i.e., people working together) does not increase performance beyond the benefits of simple response averaging [1].

In a recent study, researchers aggregated the judgments of multiple groups of human individuals by means of simple response averaging (See Figs. 16.3 and 16.4). Performance was substantially better for fused human judgments than for individuals working on the task alone. In addition, the findings demonstrate that fusion helped

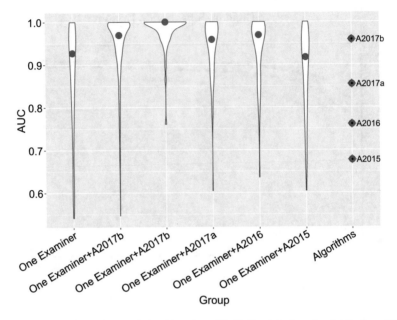

Fig. 16.3 Fusion of examiners and algorithms from [4]. Violin plots show the distribution of AUCs for each fusion test. Red dots indicate median AUCs. Fusing one examiner and A2017b is more accurate than fusing two examiners

to stabilize performance by boosting the scores of lower performing individuals and decreasing variability [4].

Although fusion boosts performance in comparison to systems operating individually, it is important to note that the best combinations of human subjects require people from high-performing groups (e.g., forensic facial examiners) [4] (See Figs. 16.1 and 16.4). For instance, perfect performance (AUC = 1) was achieved by sampling forensic facial examiners ($n = 4$) or super-recognizers ($n = 3$), but not by sampling subjects from other groups (i.e., forensic facial reviewers, fingerprint examiners, or untrained university students). Previous research has also shown that benefits from fusion for forensic facial examiners are amplified in conditions for which this group outperforms untrained groups. For example, forensic examiners benefit from fusion when the identity-diagnostic information is mostly in the face, and not when other biometric information in the body would be helpful (cf., [34, 35]).

16.4.3 Fusion: Humans and Machines

Similar wisdom-of-crowds effects have been found by fusing similarity ratings of multiple computer-based face recognition algorithms [5, 36]. The results reveal that greater boosts in performance are obtained by combining systems that differ maxi-

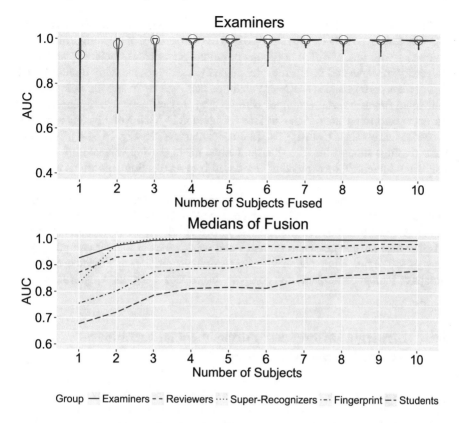

Fig. 16.4 For all groups, combining judgments by simple averaging is effective [4]. The violin plots (top panel) show the distribution of AUCs for fusing examiners. Red circles indicate median AUCs. The median AUC reaches 1.0 for fusing four examiners or fusing three super-recognizers (bottom panel). The median AUC of fusing 10 students was 0.88, substantially below the median AUC for individual examiner accuracy [4]

mally in computational strategies and that also occupy a range of individual performance levels [5]. This is sometimes preferable to combining similar, high-performing systems [5]. These findings led researchers to investigate ways to exploit the diverse strategies employed by face recognition algorithms and human participants in order to achieve optimal accuracy. Specifically, researchers have combined the similarity ratings of human subjects and computer-based face recognition algorithms on sets of difficult face-image pairs. These fusions have been accomplished through a variety of crowd-sourcing methods.

In an early study, similarity scores produced by seven face recognition algorithms and untrained human participants were used to predict the match status of difficult face-image pairs by means of weighted combinations of similarity scores derived with PLS-R. This study showed that human and machine collaborations can help achieve near-perfect face-matching performance [5].

More recently, researchers investigated the performance of state-of-the-art deep DCNNs developed between 2015 and 2017. The performance of these algorithms was compared with that of different groups of human participants, including untrained university students and forensic professionals [4]. This recent study demonstrated that perfect performance (AUC = 1) could be achieved by fusing the ratings of an individual forensic facial examiner with image-pair similarity scores generated by the best performing state-of-the-art face recognition DCNN. Notably, this accuracy exceeded that obtained by fusing two forensic examiners (See Fig. 16.3) [4]. Together, these findings support the idea that the human raters provide valuable information that is not attainable through computer-based face recognition algorithms and vice versa.

In sum, combining the judgments of independent systems (e.g., human participants and/or computer-based face recognition algorithms) yields significant boosts in the face identification accuracy. These effects result from crowd-sourcing methods that take advantage of the strengths and weaknesses of diverse computational strategies. The following section will discuss strategic differences studied in different groups of human participants and computer-based face recognition algorithms.

16.5   Strategic Differences: Forensic Facial Examiners Versus Untrained Humans

In the previous section, we presented studies demonstrating accuracy differences in face identification across various groups of human participants. These accuracy differences could be based either on skill-level differences in applying a single strategy or on the use of different face identification strategies. For the former, it is possible that all humans use the same strategy, but that some of them are simply better at applying it. For the latter, it is possible that different people approach the task in different ways. Fusion should succeed best when strategies differ. Therefore, fusion offers a useful method for probing the strategies of individuals and groups of individuals.

In addition to accuracy level differences, the face identification literature also makes it clear that perceivers use various strategies. These strategies differ in the regions of the face people focus on, the comparison methods they use, and/or in the way they report their responses. Studies examining strategic differences between professional forensic facial examiners and untrained individuals are limited. Here, we list and review four findings from two recent studies that offer evidence that strategy differs for forensic facial examiners and untrained human subjects [3, 35].

First, the most common example of strategic differences used by humans pertains to whether a person uses configural or feature-based facial analysis. Configural information in faces is based on the relational configuration of facial features (e.g., spacing between eyes) [37]. Feature-based processing focuses on isolated analyses of single features (e.g., eyes, nose). In psychology, one method for distinguishing the two strategies is to invert the images presented to a human subject. This is thought to disrupt the configural processing while allowing feature-based analysis to proceed [38]. Examiners are less impaired by inverting the faces than untrained people, indicating less reliance on configural processing [3]. This result is somewhat at odds with findings from face memory tests of untrained people, in which higher performers rely more on configural processing than feature-based processing. Here, forensic examiners show excellent results with a strategy that is non-optimal for an untrained subject.

Second, in the same set of tests [3], forensic facial examiners showed a greater accuracy advantage over untrained groups of subjects when they viewed images for 30s, but not when they viewed face images for only 2s. This indicates that examiners require more analysis time to surpass untrained groups.

The third strategic difference between professional examiners and untrained humans comes from how these groups respond to the test items to generate an identification decision. Examiners' overall accuracy advantage is disproportionately due to their ability to accurately determine when identities differ, rather than to their ability to verify that two images show the same identity. This is due in part to the cautious way examiners use the response scales, by contrast to untrained groups [35, 39].

Fourth, it is also known that forensic facial examiners use information in faces more effectively than the information from bodies [35]. Specifically, when the quality of identity information in the face versus the body was manipulated, untrained humans spontaneously used identity information from the body if the face provided only limited identity information [34]. Forensic facial examiners do not take similar advantage of body information for identification. One possible explanation for examiners "ignoring" this information may be the rigorous training they receive on facial analysis. This may enhance their ability to conduct a detailed examination of the internal facial features, but inhibit their flexibility in the use of external information when these features are of limited use [35].

In summary, various strategies are used by examiners, students, and computer algorithms. In the next section, we illustrate a special case of strategic differences that may underlie face identification performance for both humans and machines. This is the case of identification when people/machines learn from faces of a particular race and are tested on faces of a different race.

16.6   Other-Race Effects in Humans and Machines

Human face recognition accuracy is influenced by both the race of the perceiver and the race of the person to be recognized. Formally, this "other-race effect" is defined as a tendency for greater recognition accuracy for "own-race" faces compared to "other-race" faces [40, 41]. The robust effects of race on the human face recognition accuracy have been studied and found across multiple racial/ethnic and age subgroups and with different experimental designs [41].

The implications of these effects, however, extend beyond the confounds of a laboratory setting. The other-race effect has resulted in identification errors in law enforcement and in the judicial system (see [42–44] for a review of this effect in the justice system). For example, according to the Innocence Project, an organization founded to exonerate innocent convicted individuals, 41% of cases exonerated by DNA evidence involving eyewitness misidentifications involved some form of cross-race misidentification (Innocence Project 2015). This well-documented weakness in the justice system makes it a problem worth exploring. Moreover, despite the decades of research on the effect, little is known about how this phenomenon impacts face recognition algorithms. With the increased use of these algorithms for security applications, it is imperative to understand the extent to which race plays a role in algorithm accuracy. In this section, we describe evidence for the other-race effect in humans and discuss its possible causes. We also present evidence from the limited, but essential, research on how race impacts the performance of face recognition algorithms.

Beginning with humans, the other-race effect has been studied and found across multiple demographic groups. Race influences recognition accuracy in adults and has been found in children as young as five years old [45–47]. There is evidence that a perceptual expertise advantage with own- versus other-race faces emerges in infancy [48, 49]. The impact of race on identification accuracy has been reported in non-typically developing populations, including individuals with schizophrenia [50] and autism spectrum disorder [51, 52]. Notably, the other-race effect has also been found using multiple experimental paradigms, including face identity matching tasks [53], eyewitness lineups [54, 55], old/new or yes/no memory conditions [40, 56], name learning [57], and lineup constructor tasks [58]. Although stimuli and experimental parameters differ considerably across these studies, the consistent impact of race on face recognition/identification accuracy makes the other-race effect of pivotal interest to scientists.

Given the clear negative effects of the other-race effect, several studies have also examined strategies for mitigating the effect. For example, participant awareness of the other-race effect [56] and viewing caricatured images of other-race faces [59] have been shown to reduce its effects. Developmental studies show that perceptual training in infants can prevent the other-race effect from emerging [60]. The effects of perceptual and individuation training on other-race faces have been studied in children [61] and adults [62–64]. Importantly, these studies also show that the mitigating effects of these interventions are temporary.

16.6.1 Theories of the Other-Race Effect

Given its robust effects and its potential for serious consequences in applied settings, a substantial amount of research has focused on investigating the causes of the other-race effect [65–67]. From this research, two principal models have emerged as explanations for the other-race effect: the perceptual expertise model and the social-cognitive model.

The perceptual expertise model states that increased experience with own-race faces improves face recognition accuracy selectively, or at least preferentially, for own-race faces [41, 68]. The precise mechanisms influenced by expertise, however, remain open to debate (see [67]). One possibility is that expertise improves one's ability to analyze faces configurally. Configural processing is thought to improve recognition accuracy over feature-based processing [37]. Several studies show that own-race faces are processed configurally, whereas other-race faces are processed with a more feature-based strategy [67, 69–71]. Notably, the degree to which perceptual expertise with faces improves recognition accuracy is still uncertain. Although there is evidence to support the benefits of experience with faces of a certain race, the amount of this experience explains only a small portion of the variance of the other-race effect [41]. Other studies have suggested that the type, rather than amount, of experience is crucial to other-race face recognition accuracy. Thus, it seems that qualitative experience, as opposed to mere quantitative exposure, may be responsible for promoting configural holistic processing [72], and ultimately for improving accuracy.

Support for the expertise model comes also from studies that manipulate experience with other-race faces. For example, [47] found a reduced other-race effect in children and teenagers with increased other-race experience. Monoracial British–White children showed a typical other-race effect and had greater recognition accuracy for own-race (Caucasian) faces than other-race faces (Chinese, Malay, and African Black). In contrast, multiracial Chinese Malaysian children showed an other-race effect, but only for faces of the race they reported less contact with (African Black). Increased experience with other-race faces has also been shown to reverse completely the effects of race [73]. For example, one study showed that Korean adults, adopted by European parents when the children were between the ages of 3–9 years old, showed a reversed other-race effect (greater recognition accuracy for Caucasian faces than Asian faces) [73]. Interestingly, the accuracy for Korean adults on Caucasian faces was comparable to native French-born Caucasian participants. These results provide support for the influence of experience and how this other-race experience can mitigate the other-race effect.

The social-cognitive model suggests that the other-race effect is due to the human tendency to perceive individuals in terms of social groups [67]. "In-groups" consist of people we think as similar to ourselves (demographically and socially). "Out-groups" are people we think of as different from ourselves along these same dimensions. This social-cognitive model is explained by the tendency to think of people in our in-groups as individuals and people in our out-groups categorically, rather than as individuals [74, 75]. As an account of the other-race effect, the social-cognitive model applies as follows. Because we think of out-group members (e.g., other-race individuals) as categories rather than individuals, we attend only to superficial facial attributes (i.e., skin color or hair color). These features are insufficient to individuate people.

Next, we consider the other-race effect for algorithms. Although human performance might be influenced by social perception and motivation, these are clearly not situations of concern for an algorithm. For this reason, we have limited our discussion to the perceptual expertise model. For a discussion of additional other-race effect models on human face recognition accuracy, see [65–67]. As noted, the data used for training algorithms has a strong effect on algorithm performance. As we will see, level of expertise indexed by the quality of training data, including own- and other-race faces, influences algorithm performance.

16.6.2   Other-Race Effect in Machines

Given the robust effect of race on human face recognition, there is a long history of studies examining its effects on recognition algorithms and training models [6, 76, 77]. In an early study (1991), an auto-associative neural network model of face recognition was trained with a "majority" and "minority" race of faces. Majority and minority were defined simply by the imbalance of training data for two races of face [76]. Results showed that the network represented novel (untrained) faces from the majority race more accurately than novel face from the minority race [76]. Additionally, to test whether the associative network created representations of minority-race faces that were less individuated than representations of majority-race faces, the average inter-face similarity was calculated. Inter-face similarity for the minority race was greater than inter-face similarity for majority-race faces. This result is consistent with the old adage that other-race faces "all look alike".

Moving forward in time, neural network models, circa 2002, provided a look at plausible underlying computational mechanisms of a perceptual expertise theory of the other-race effect [6]. Using extant models, the researchers found an other-race effect only for a subset of face recognition algorithms. These were algorithms in which the representational system was developed through experience that warped the perceptual space in a way that was sensitive to the overall structure of the model's experience with faces of different races. These models are in some ways analogous to developmental psychology models of language that invoke a "critical period" mechanism during development. This posits that experience with different languages structures basic aspects of the perceptual space to favor the language sounds we hear most often—usually phonemes from our native language. For faces, this distinguishes the impact of experience based on when it occurs and potentially indicates that the other-race effect requires experience with other-races to be timed to be early in life to coincide with the critical period, during which visual feature sets develop.

In one early study, circa 2011, researchers examined differences in face recognition algorithms developed in different countries [77]. In that study, 13 algorithms from the Face Recognition Vendor Test 2006 (FRVT-2006) served as test algorithms. [18]. Eight of these algorithms were developed in Western countries (United States, Germany, France) and five algorithms were developed in East Asian (China, Japan, Korea). The algorithms developed in the west were fused together to create a single "Western" algorithm. Similarly, the algorithms from Eastern countries were fused to create an "East Asian" algorithm. The fused Western and East algorithms were then tested on controlled and uncontrolled Caucasian and East Asian faces. Machine accuracy mirrored the classic other-race effect found in humans. The Western fused algorithm was more accurate on Caucasian faces and the East Asian fused algorithm was more accurate on East Asian faces. This other-race effect was attributed to the likelihood that algorithms trained in East Asia would be trained mostly with East Asian faces, whereas algorithms trained in Western counties would be trained mostly with Caucasian faces. In this era, additional work looking more generally at the performance of algorithms with multiple demographic categories (race, gender, age) show analogous challenges across demographic categories [78].

To compare human and machine recognition, the researchers used a condensed version of the experiment with a smaller number of face pairs. Human participants demonstrated a classic other-race effect with a slight, but non-significant, advantage for Caucasian faces. Additionally, both algorithms were better at recognizing Caucasian face pairs. However, the Caucasian face advantage was larger for the Caucasian algorithm than the East Asian algorithm—again evidence for an other-race effect. This overall Caucasian advantage was attributed to the likelihood that the researchers who submitted algorithms to the FRVT-2006 anticipated that the majority of the faces used to test the algorithms would be Caucasian. These results demonstrate that origin of the algorithm and the training data used for the algorithm both play a crucial role in building better representations of own-vs-other-race faces. If these algorithms are to be used in diverse populations, it may be imperative that their training data reflect such diversity.

The most recent class of face recognition algorithms, based on DCNNs, became state of the art, circa 2015. These algorithms also perform differently as a function of the race of the face [79, 80] (See [80] for review on the topic up to 2019). A comprehensive and thorough test of race bias in face recognition algorithms, published in 2019 by the NIST [79], showed a variety of bias effects for different algorithms depending on performance measures (e.g., false alarm rate). Although the meta-data on the race of faces used in these tests were limited, and perhaps not directly comparable to older more controlled studies, the study nonetheless shows that the problems of race for face recognition have not been solved. A more systematic approach to measuring race bias in face recognition algorithms is needed [80]. The challenges in this endeavor include understanding data driven factors (e.g., population distributions in training and test data) and application factors (e.g., threshold setting) for these algorithms. A comprehensive analysis and discussion of these factors, and their potential impact on performance across race is available in [80].

16.7   Closing Thoughts

In summary, we have examined the diversity of ways in which humans and machines approach the problem of face identification. Understanding this diversity can be a prerequisite to developing and implementing fusion strategies for humans, machines, and between humans and machines. These can ultimately improve the accuracy of face identification in security and law enforcement applications that require the best possible accuracy.

References

1. Jeckeln G, Hahn CA, Noyes E, Cavazos JG, O'Toole AJ (2018) Wisdom of the social versus non-social crowd in face identification. B J Psychol
2. White D, Burton AM, Kemp RI, Jenkins R (2013) Crowd effects in unfamiliar face matching. Appl Cogn Psychol 27(6):769–777
3. White D, Phillips PJ, Hahn CA, Hill M, O'Toole AJ (2015) Perceptual expertise in forensic facial image comparison. Proc R Soc B 282(1814):20151292
4. Phillips PJ, Yates AN, Hu Y, Hahn CA, Noyes E, Jackson K, Cavazos JG, Jeckeln G, Ranjan R, Sankaranarayanan S et al (2018) Face recognition accuracy of forensic examiners, superrecognizers, and face recognition algorithms. In: Proceedings of the National Academy of Sciences, p 201721355
5. O'Toole AJ, Abdi H, Jiang F, Phillips PJ (2007) Fusing face-verification algorithms and humans. IEEE Trans Syst Man Cybern Part B (Cybern) 37(5):1149–1155
6. Furl N, Phillips PJ, O'Toole AJ (2002) Face recognition algorithms and the other-race effect: computational mechanisms for a developmental contact hypothesis. Cogn Sci 26(6):797–815
7. Damasio AR, Damasio H, Van Hoesen GW (1982) Prosopagnosia anatomic basis and behavioral mechanisms. Neurology 32(4):331
8. Russell R, Duchaine B, Nakayama K (2009) Super-recognizers: people with extraordinary face recognition ability. Psychon Bull Rev 16(2):252–257
9. Young AW, Burton AM (2017) Are we face experts? trends in cognitive sciences
10. Johnston RA, Edmonds AJ (2009) Familiar and unfamiliar face recognition: a review. Memory 17(5):577–596
11. Carl Metzgar CSP, A (2012) Strengthening forensic science in the us: a path forward. Prof Saf 57(1):32
12. FI, SW, Group et al (2012) Guidelines for facial comparison methods
13. Phillips PJ, Moon H, Rizvi SA, Rauss PJ (2000) The feret evaluation methodology for face-recognition algorithms. IEEE Trans Pattern Anal Mach Intell 22(10):1090–1104
14. Phillips PJ, Flynn PJ, Scruggs T, Bowyer KW, Chang J, Hoffman K, Marques J, Min J, Worek W (2005) Overview of the face recognition grand challenge. In: IEEE computer society conference on Computer vision and pattern recognition CVPR 2005, vol 1. IEEE, pp 947–954
15. Phillips PJ, Scruggs WT, O'Toole AJ, Flynn PJ, Bowyer KW, Schott CL, Sharpe M (2010) Frvt 2006 and ice 2006 large-scale experimental results. IEEE Trans Pattern Anal Mach Intell 32(5):831–846
16. Phillips PJ, Beveridge JR, Draper BA, Givens G, O'Toole AJ, Bolme D, Dunlop J, Lui YM, Sahibzada H, Weimer S (2012) The good, the bad, and the ugly face challenge problem. Image Vis Comput 30(3):177–185
17. Phillips PJ, O'Toole AJ (2014) Comparison of human and computer performance across face recognition experiments. Image Vis Comput 32(1):74–85
18. Phillips PJ, Jiang F, Ayyad J, Pénard N et al (2007) Face recognition algorithms surpass humans matching faces over changes in illumination. IEEE Trans Pattern Anal Mach Intell 9:1642–1646
19. O'Toole AJ, An X, Dunlop J, Natu V, Phillips PJ (2012) Comparing face recognition algorithms to humans on challenging tasks. ACM Trans Appl Percept (TAP) 9(4):16
20. Sun Y, Wang X, Tang X (2014) Deep learning face representation from predicting 10,000 classes. In: Proceedings of the IEEE conference on computer vision and pattern recognition, pp 1891–1898
21. Sun Y, Wang X, Tang X (2013) Hybrid deep learning for face verification. In: Proceedings of the IEEE international conference on computer vision, pp 1489–1496

22. Parkhi OM, Vedaldi A, Zisserman A et al. (2015) Deep face recognition. BMVC 1(3):6
23. Chen J-C, Ranjan R, Kumar A, Chen C-H, Patel VM, Chellappa R (2015) An end-to-end system for unconstrained face verification with deep convolutional neural networks. In: Proceedings of the IEEE international conference on computer vision workshops, pp 118–126
24. Sankaranarayanan S, Alavi A, Castillo C, Chellappa R (2016) Triplet probabilistic embedding for face verification and clustering. arXiv:1604.05417
25. Taigman Y, Yang M, Ranzato M, Wolf L (2014) Deepface: closing the gap to human-level performance in face verification. In: Proceedings of the IEEE conference on computer vision and pattern recognition, pp 1701–1708
26. Schroff F, Kalenichenko D, Philbin J (2015) Facenet: a unified embedding for face recognition and clustering. In: Proceedings of the IEEE conference on computer vision and pattern recognition, pp 815–823
27. Huang GB, Mattar M, Berg T, Learned-Miller E (2008) Labeled faces in the wild: a database forstudying face recognition in unconstrained environments. In: Workshop on faces in 'Real-Life' images: detection, alignment, and recognition
28. Kumar N, Berg AC, Belhumeur PN, Nayar SK (2009) Attribute and simile classifiers for face verification. In: 2009 IEEE 12th international conference on computer vision, pp 365–372. IEEE
29. Whitelam C, Taborsky E, Blanton A, Maze B, Adams JC, Miller T, Kalka ND, Jain AK, Duncan JA, Allen K et al (2017) Iarpa janus benchmark-b face dataset. In: CVPR workshops, pp 592–600
30. Miller D, Brossard E, Seitz S, Kemelmacher-Shlizerman I (2015) Megaface: a million faces for recognition at scale. arXiv:1505.02108
31. Kittler J (1998) Combining classifiers: a theoretical framework. Pattern Anal Appl 1(1):18–27
32. Condorcet MD (1785) Essay on the application of analysis to the probability of majority decisions. Paris: Imprimerie Royale
33. Dowsett AJ, Burton AM (2015) Unfamiliar face matching: pairs out-perform individuals and provide a route to training. B J Psychol 106(3):433–445
34. Rice A, Phillips PJ, Natu V, An X, O'Toole AJ (2013) Unaware person recognition from the body when face identification fails. Psychol Sci 24(11):2235–2243
35. Hu Y, Jackson K, Yates A, White D, Phillips PJ, O'Toole AJ (2017) Person recognition: qualitative differences in how forensic face examiners and untrained people rely on the face versus the body for identification. Vis Cognit 25(4–6):492–506
36. Czyz J, Kittler J, Vandendorpe L (2002) Combining face verification experts. In: 16th International conference on pattern recognition, 2002. Proceedings, vol 2. IEEE, pp. 28–31
37. Maurer D, Le Grand R, Mondloch CJ (2002) The many faces of configural processing. Trends Cognit Sci 6(6):255–260
38. Yin RK (1969) Looking at upside-down faces. J Exp Psycholy 81(1):141
39. Norell K, Läthén KB, Bergström P, Rice A, Natu V, O'Toole A (2015) The effect of image quality and forensic expertise in facial image comparisons. J Forensic Sci 60(2):331–340
40. Malpass RS, Kravitz J (1969) Recognition for faces of own and other race. J Pers Soc Psychol 13(4):330
41. Meissner CA, Brigham JC (2001) Thirty years of investigating the own-race bias in memory for faces: a meta-analytic review. Psychol Public Policy Law 7(1):3
42. Sporer SL (2001) The cross-race effect: beyond recognition of faces in the laboratory. Psychol Public Policy Law 7(1):170
43. Wells GL, Olson EA (2001) The other-race effect in eyewitness identification: what do we do about it? Psychol Public Policy Law 7(1):230
44. Wilson JP, Hugenberg K, Bernstein MJ (2013) The cross-race effect and eyewitness identification: how to improve recognition and reduce decision errors in eyewitness situations. Soc Issues Policy Rev 7(1):83–113
45. Pezdek K, Blandon-Gitlin I, Moore C (2003) Children's face recognition memory: more evidence for the cross-race effect. J Appl Psychol 88(4):760

46. Anzures G, Kelly DJ, Pascalis O, Quinn PC, Slater AM, De Viviés X, Lee K (2014) Own-and other-race face identity recognition in children: the effects of pose and feature composition. Dev Psychol 50(2):469
47. Tham DSY, Bremner JG, Hay D (2017) The other-race effect in children from a multiracial population: a cross-cultural comparison. J Exp Psychol 155:128–137
48. Sangrigoli S, De Schonen S (2004) Recognition of own-race and other-race faces by three-month-old infants. J Child Psychol Psychiatry 45(7):1219–1227
49. Kelly DJ, Quinn PC, Slater AM, Lee K, Ge L, Pascalis O (2007) The other-race effect develops during infancy: evidence of perceptual narrowing. Psychol Sci 18(12):1084–1089
50. Pinkham AE, Sasson NJ, Calkins ME, Richard J, Hughett P, Gur RE, Gur RC (2008) The other-race effect in face processing among African American and Caucasian individuals with schizophrenia. Am J Psychiatry 165(5):639–645
51. Wilson CE, Palermo R, Burton AM, Brock J (2011) Recognition of own-and other-race faces in autism spectrum disorders. Q J Exp Psychol 64(10):1939–1954
52. Yi L, Quinn PC, Feng C, Li J, Ding H, Lee K (2015) Do individuals with autism spectrum disorder process own-and other-race faces differently? Vis Res 107:124–132
53. Megreya AM, White D, Burton AM (2011) The other-race effect does not rely on memory: evidence from a matching task. Q J Exp Psychol 64(8):1473–1483
54. Evans JR, Marcon JL, Meissner CA (2009) Cross-racial lineup identification: assessing the potential benefits of context reinstatement. Psychol Crime Law 15(1):19–28
55. Jackiw LB, Arbuthnott KD, Pfeifer JE, Marcon JL, Meissner CA (2008) Examining the cross-race effect in lineup identification using caucasian and first nations samples. Canadian Journal of Behavioural Science/Revue canadienne des sciences du comportement 40(1):52
56. Hugenberg K, Miller J, Claypool HM (2007) Categorization and individuation in the cross-race recognition deficit: toward a solution to an insidious problem. J Exp Soc Psychol 43(2):334–340
57. Hayward WG, Favelle SK, Oxner M, Chu MH, Lam SM (2017) The other-race effect in face learning: using naturalistic images to investigate face ethnicity effects in a learning paradigm. Q J Exp Psychol 70(5):890–896
58. Brigham JC, Ready DJ (1985) Own-race bias in lineup construction. Law Hum Behav 9(4):415–424
59. Rodríguez J, Bortfeld H, Gutiérrez-Osuna R (2008) Reducing the other-race effect through caricatures. In: 8th IEEE international conference on automatic face & gesture recognition, 2008. FG'08. IEEE, pp. 1–5
60. Heron-Delaney M, Anzures G, Herbert JS, Quinn PC, Slater AM, Tanaka JW, Lee K, Pascalis O (2011) Perceptual training prevents the emergence of the other race effect during infancy. PLoS One 6(5):e19858
61. Xiao WS, Fu G, Quinn PC, Qin J, Tanaka JW, Pascalis O, Lee K (2015) Individuation training with other-race faces reduces preschoolers-implicit racial bias: A link between perceptual and social representation of faces in children. Dev Sci 18(4):655–663
62. Tanaka JW, Pierce LJ (2009) The neural plasticity of other-race face recognition. Cognit Affect Behav Neurosci 9(1):122–131
63. Matthews CM, Mondloch CJ (2018) Improving identity matching of newly encountered faces: effects of multi-image training. J Appl Res Mem Cognit 7(2):280–290
64. Cavazos JG, Noyes E, O'Toole AJ (2018) Learning context and the other-race effect: strategies for improving face recognition. Vis Res
65. Sporer SL (2001) Recognizing faces of other ethnic groups: an integration of theories. Psychol Public Policy Law 7(1):36
66. Hugenberg K, Young SG, Bernstein MJ, Sacco DF (2010) The categorization-individuation model: an integrative account of the other-race recognition deficit. Psychol Rev 117(4):1168
67. Young SG, Hugenberg K, Bernstein MJ, Sacco DF (2012) Perception and motivation in face recognition: a critical review of theories of the cross-race effect. Person Soc Psychol Rev 16(2):116–142
68. Diamond R, Carey S (1986) Why faces are and are not special: an effect of expertise. J Exp Psychol: Gen 115(2):107

69. Rhodes G, Brake S, Taylor K, Tan S (1989) Expertise and configural coding in face recognition. B J Psychol 80(3):313–331
70. Rhodes G, Hayward WG, Winkler C (2006) Expert face coding: configural and component coding of own-race and other-race faces. Psychon Bull Rev 13(3):499–505
71. Tanaka JW, Kiefer M, Bukach CM (2004) A holistic account of the own-race effect in face recognition: evidence from a cross-cultural study. Cognition 93(1):B1–B9
72. Bukach CM, Cottle J, Ubiwa J, Miller J (2012) Individuation experience predicts other-race effects in holistic processing for both caucasian and black participants. Cognition 123(2):319–324
73. Sangrigoli S, Pallier C, Argenti A-M, Ventureyra V, de Schonen S (2005) Reversibility of the other-race effect in face recognition during childhood. Psychol Sci 16(6):440–444
74. Bodenhausen GV, Macrae CN, Hugenberg K (2003) Activating and inhibiting social identities: implications for perceiving the self and others. In: Foundations of social cognition: a festschrift in honor of Robert S. Jr. Erlbaum, Wyer
75. Millon T, Lerner MJ, Weiner IB (2003) Handbook of psychology: personality and social psychology. Wiley, New Jersey
76. O'Toole A. J, Deffenbacher K, Abdi H, Bartlett JC (1991) Simulating the "other-race effect" as a problem in perceptual learning. Connect Sci 3(2):163–178
77. Phillips PJ, Jiang F, Narvekar A, Ayyad J, O'Toole AJ (2011) An other-race effect for face recognition algorithms. ACM Trans Appl Percept (TAP) 8(2):14
78. Klare BF, Burge MJ, Klontz JC, Bruegge RW, Jain AK (2012) Face recognition performance: role of demographic information. IEEE Trans Inf Forensics Sec 7(6):1789–1801
79. Grother P, Ngan M, Hanaoka K (2019) Face recognition vendor test (frvt) part 3: demographic effects. National Institute of Standards and Technology
80. Cavazos JG, Phillips PJ, Castillo CD, O'Toole AJ (2020) Accuracy comparison across face recognition algorithms: where are we on measuring race bias? IEEE transactions on biometrics, behavior, and identity science

Chapter 17
Evaluation of Face Recognition Systems

Patrick Grother and Mei Ngan

While face recognition research has been perennial and popular since its inception, there has been a marked escalation in this research in recent years due to the confluence of several factors, primarily the development of advanced machine learning algorithms, free and robust software implementations thereof, ever faster GPU processors for running them, vast web-scraped face image databases, open performance benchmarks, and a vibrant face recognition literature.

The new algorithms were largely been developed to exhibit invariance to pose, illumination, and expression variations that characterize photojournalism and social media images. The initial research [1, 2] employed large numbers of images of relatively few ($\sim 10^4$) individuals to learn invariance. Inevitably much larger populations ($\sim 10^7$) were employed for training [3, 4] but the benchmark, LFW with an EER metric [5], represents an easy task: one-to-one verification at very high false match rates. While a larger scale identification benchmark duly followed (Megaface [6]) its primary metric, rank one hit rate, contrasts with the high threshold discrimination task required in large-population governmental applications of face recognition, namely credential de-duplication, law enforcement, and intelligence searches. There identification into galleries containing from up to 10^8 individuals must be performed using very few images per individual and stringent thresholds adopted to afford very low false positive identification rates.

Technological advances have led to massive reductions in recognition errors. The largest independent published benchmark, NIST's Face Recognition Vendor Test [7], has documented that the leading commercial algorithms now produce twenty times fewer false negatives in 2018 than they did in 2013 on an identical portrait image identification task with gallery size 1.6 million. This contrasts with progress in the period 2010–2013 when error rates fell only by a factor of two on the same

P. Grother (✉) · M. Ngan
National Institute of Standards and Technology, Gaithersburg, USA
e-mail: patrick.grother@nist.gov

task. These gains are realized at both high and low false positive identification rates, showing that CNN-based algorithms can exhibit both invariance *and* discrimination. This demonstrates that the major commercial developers have integrated and, in many cases, replaced existing algorithms with those based on convolutional neural networks (CNNs).

This chapter covers topics in evaluation including some that are often ignored in research settings yet which present operational concerns. These are low false positive rates, impostor distribution stability, and measurement of computational resources.

17.1 Introduction

Progress in face recognition has always been underpinned by evaluation, i.e., quantification of accuracy to assess whether an algorithm is superior to its progenitors and to its competitors. Most evaluation is done within the laboratories of the developers and is necessarily tightly bound to research. In university settings, peer-reviewed journals invariably insist that accuracy be measured against published benchmarks. In commercial settings, however, the results rarely see the light of day. In both cases, developers have historically been frustrated by the lack of suitable image datasets. This constraint has lifted in recent years as face imagery can be taken from the internet, in many cases from identity-labeled photographs residing on photography, news, and, particularly, social media websites. These are available in large part due to the advent of the digital camera, particularly on the smartphone, and critically, the internet as a distribution mechanism.

Regardless of the quantity of available data, researchers usually utilize the data in an iterative build–test–refine process, in which error cases are isolated and algorithmically addressed. This process may expose ground truth identity errors in the dataset. It may also lead to specialization and impede generalization of the algorithm to other datasets.

17.1.1 Objectives of Evaluation

The goals behind evaluation of face recognition algorithms have remained consistent despite the rapid-changing pace of the technology. Fundamentally, independent evaluation of face recognition software on operational data allows for fair, repeatable, and relevant assessment of performance of core capability. Capability testing establishes criteria for whether the technology is viable and what's possible, understanding what the technology limits are, and assessing whether application and technical requirements can be met. Comparative testing of algorithms affords an understanding of which technologies work and likewise and which technologies don't work. This supports end-users in their procurement decisions and also motivates the advancement of new technology. Longitudinal monitoring of performance gives a quantitative measurement of performance gains and influences upgrade schedules and contract re-competes. Face recognition errors incur cost, for example, failure to enroll (FTE) imposes additional time, procedures, modalities, and processes, so measurement of such quantities supports shaping operational expectations if the software is deployed. Characterizing performance allows procedures to mitigate risk, for example, finding that FTE is very high prompts environmental redesign to regulate ambient light or other causes for failure.

17.2 Verification

This section gives metrics for verification. Identification is dealt with later. The verification task is the fundamental biometric operation—to determine whether two images are of the same face or not. Verification is now by far the most common application of biometrics, being widely deployed in applications such as access control and authentication, particularly on mobile phones. It quantifies the ability to answer the question are two samples from the same person or not.

Verification involves a claim of identity with a verification sample being compared with a particular (prior) reference sample. The comparison can be either a genuine or impostor comparisons, the former are often known as authentic or mated comparisons. It is reasonable and sufficient for much algorithm development to proceed by running verification tests—the LFW [5] task being the most famous—to quantify core algorithmic efficacy.

17.2.1 Metrics

Nomenclature: This section defines *false non-match rate* (FNMR) *false match rate* (FMR). In decision theory, these are generically the Type I and II error rates, respectively. These are the fundamental sample-comparison *matching* error rates. The academic literature, particularly, often uses the terms *false accept rate* (FAR) and *false reject rate* (FRR) but industrial practice, and formally standardized biometrics performance tests, reserve these terms to represent *transactional* error rates defined over the entire interaction of a human with a biometric system that potentially involves multiple presentations, samples, and comparisons.

Scores versus distances: For the purposes of core algorithm evaluation, however, the matching error rates are the key accuracy indicators. Algorithms are assumed to produce comparison scores, which in commercial face recognition are conventionally non-negative similarity scores. In academic settings, algorithms usually produce *distances* (with metric properties). For the discussion here, scores are assumed to be similarity scores, with higher-is-more-similar semantics. Distances may be converted to similarities via monotone transformations such as $1/(1 + d)$ or negation $U - d$, with U constant, usually large.

False non-match computation: Given N_G scores from genuine comparisons, FNMR is the proportion of genuine scores below threshold, T:

$$\text{FNMR}(T) = 1 - \frac{1}{N_G} \sum_{i=1}^{N_G} H(s_i - T) \qquad (17.1)$$

where the step function H(x) is 1 if $x \geq 0$ and 0 otherwise. The inequality supports the commercial convention of authenticating a claimant if the score equals the threshold. It is common for algorithms to fail to produce a template from some input images, particularly in non-cooperative wild images where detection may be difficult. These so-called failures to enroll outcomes must be accounted for when comparing algorithms. One means of doing this is to assign low scores, e.g., 0, to any comparison involving that template. This simulates false rejection of a user and increases FNMR. Alternatively, failure rates can be included explicitly. For example, consider N_G pairs of images each comprised of a portrait and a wild image, with template generation failure rates of F_1 and F_2, respectively. A generalized false non-match rate can be computed as

$$\text{GFNMR}(T) = F_1 + F_2 + (1 - F_1)(1 - F_2)\text{FNMR}(T) \qquad (17.2)$$

Modified formulae are necessary for human-in-the-loop tests to model, for example, a policy where a subject does not continue in a test if they cannot enroll.

False match computation: Scores from N_I impostor comparisons are used in the false match rate (FMR) computation, which states the proportion of impostor comparisons yielding scores at or above T:

$$\text{FMR}(T) = \frac{1}{N_I} \sum_{i=1}^{N_I} H(s_i - T) \qquad (17.3)$$

In cases where an algorithm fails to produce a template from an input image, a low score is again assigned as the result of any comparison involving that template. This practice actually benefits (reduces) FMR.

17.2.2 Error Tradeoff Characteristics

FNMR quantifies inconvenience of users; FMR quantifies the likelihood that an impostor can falsely match an identity. As low FMR values are needed in strong authentication processes, the most important depiction of biometric performance then is to plot FNMR(T) vs FMR(T). Such a plot is an error tradeoff characteristic, an example of which is shown in Fig. 17.1. Note that, when computed empirically, the DET points

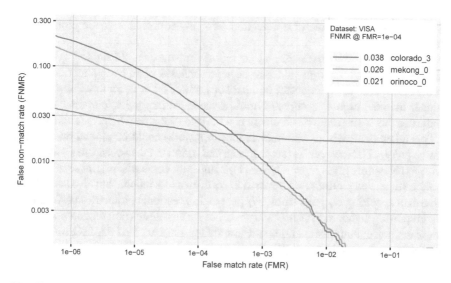

Fig. 17.1 The traces show error tradeoff characteristics for three algorithms applied to the comparison of visa photographs. The inset legend affords comparison at one particular operating point

are connected via horizontal and vertical lines, and therefore the common term DET *curve* is incorrect.

Relation of DET and ROC: The term receiver operating characteristic (ROC) is conventionally a plot of true match rate (TMR = 1 - FNMR) against FMR and as such is only trivially different to the DET. However the DET should be preferred because for highly accurate biometrics, TMR values become close to 1 and visualization of difference in that range becomes difficult. A logarithmic FNMR axis spreads plots out, readily revealing a factor of two reduction in FNMR for example.

Axis scaling: The term detection error tradeoff characteristic (DET) is used for this plot—the word "detection" hailing from target detection in radar applications—although it has been claimed by the speech recognition community for the special case where both axes are probit transformed such that Normal genuine and impostor score distributions yield straight lines on the plot. Practically researchers should use appropriate scaling—linear, probit, log-log, log-linear—to afford good visualization of the error rates. All such plots may be called DETs.

The importance of low FMR: It is unfortunately common in the literature for error tradeoff characteristics to be plotted with FMR on a linear axis. This confines the range of typical operational values, [0.0001, 0.01], to a very narrow region of the visible plot, emphasizing only irrelevant FMR values, [0.1, 1]. This issue motivated the development of the improved LFW benchmark[8] specifying much greater numbers of impostor comparisons than the 6000 mandated in the original LFW design[5].

The DET should usually therefore be reported with transformed FMR axis—often logarithmic as in Fig. 17.1—unless a linear range can be justified. FMR should span an interval from as low as the data can support statistically to at least 0.1. Figure

17.1 presents an example of a detection error tradeoff characteristic plotted on a logarithmic scale.

Uncertainty: The test itself, and the reported results, should support quantification of FMR at low enough values to be representative of some application. Given scores from a set of N_I impostor comparisons, FMR $(T_0) = 0$, for any threshold, T_0, higher than the highest observed impostor score. At that point, binomial statistics indicates that we can be 95% confident that the true FMR is below $3/N_I$. This assumes independent comparisons and implies that DET plots should be plotted only on the range $[3/N_I, 1]$. However, as it is common and useful to cross-compare all available images (N images gives $N_I = N(N-1)/2$ impostor comparisons), these will not be independent, and plotting FMR down to $3/N_I$ is then less reliable but also less expensive than acquiring N_I pairs from $2N_I$ people. International biometrics performance and reporting standard ISO/IEC 19795-1. For further discussion on components of variance, uncertainty estimation, and test sizing, see Part 1 of the ISO/IEC 19795 performance testing standard [9].

17.2.3 Population-Specific Error Rates

It is generally the case that face recognition accuracy varies with subject-specific covariates such as sex, age, and race. In those common operational situations where images with such variations are sent to an algorithm configured with a single threshold, the FNMR and FMR values will *both* vary. While it is common to report DETs for sub-populations A, B, C..., such plots omit essential information, namely whether the DETs are shifted vertically, horizontally, or in some combination. Possible remedies to this are to show genuine and impostor distributions as a function of threshold or to tabulate or show linked DET points corresponding to fixed thresholds. A variant of the latter is shown in Fig. 17.2.

While for all algorithms, the blue line lies below the red line, which can be naively interpreted as males are easier to recognize than females, examination of points of equal threshold shows that the difference is more to do with higher false match rates, extending beyond an order of magnitude in one case.

Demographically matched impostors It is conventional in tests of biometric algorithms for the experimental design to use zero-effort impostors, meaning that an impostor pair is used without regard to biographic (or biometric) information. However, because the face is a genetically linked trait, the distribution of impostor scores produced by comparing faces of subjects of the same sex and national origin is markedly higher. This is depicted in Fig. 17.3.

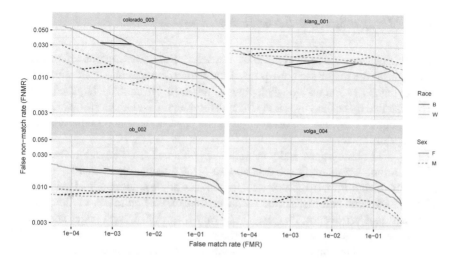

Fig. 17.2 DETs For four algorithms for four groups male and female, black and white per associated biographic metadata. The DETs are linked by gray lines corresponding to with three thresholds. These reveal the DETs are not simply displaced vertically but rather that female–female impostor comparisons almost always yield higher false match rates than male–male. Not shown are points for male–female impostors. This kind of figure can be used to show excursions in *both* FNMR and FMR across some covariate. Here the covariates, sex, and race are discrete and binary

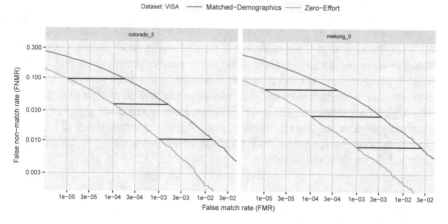

Fig. 17.3 For two algorithms, each configured with three thresholds, the black lines connect DET points corresponding to zero-effort and demographically matched impostors. The genuine comparisons are identical in both cases, so the DETs are shifted horizontally. The shifts from both algorithms are large and are of operational concern because, for example, an attacker would not steal a passport from a subject of different sex and national origin

17.2.4 Image-Specific Error Rates

Similarly, image-specific covariates such as resolution and compression affect recognition. Indeed the major factors—pose, illumination, expression—have given their names to databases and driven whole lines of research. Given that adequate lighting can be installed by design, invariably the most influential parameter on actual recognition outcomes has been the orientation of the head in one photograph relative to that in a prior image.

Using wild photographs and yaw estimates obtained from an automated pose-estimation tool, we quantify the dependence of face recognition accuracy on yaw. The ability of algorithms to compensate for viewing angle is summarized in Fig. 17.4 which shows false non-match rate as a function of yaw angle, θ, of the face in enrollment and verification images. These vary over $\pm 90°$. Each panel encodes a false non-match rate FNMR for an algorithm at a particular threshold. This is set to give a false match rate of 0.001 for images of the frontal pose, i.e., those with $|\theta| \leq 15$. The FNMR values are generally lowest for frontal pairs, then for pairs with the same yaw angle, and they increase with difference in yaw.

While figures [10] equivalent to Figure 17.4 have appeared in the literature since at least 2004, the effect of yaw on FMR has gone largely unreported. The assumption may be that faces at different poses would simply not match, but it is worth checking in, for example, a phone unlocking application that a non-frontal presentation does not present a false match hazard. This is evident in Fig. 17.5 which shows how FMR itself varies with the pair of yaw angles. This figure is relevant in applications where a global threshold is set and pose varies widely. In all panels, the center cell has FMR = 0.001, by design. The results for other yaw angles show different behaviors. First, the more accurate algorithms often have weak dependence of FMR on yaw angles (prevalence of gray). Others give consistently low FMR when angles differ (prevalence of blue) consistent with an inability to match. A final class of algorithms gives *higher* FMR when yaw angles differ (prevalence of red in the periphery). This is typically unexpected and undesirable. Such figures are not relevant if a specific pair of poses can be forced by design and capture-time checking, in which case a dedicated threshold could be set.

17.2.5 Summary Statistics

The default, and recommended, way to state verification accuracy is in terms of FNMR at threshold set to achieve a certain FMR, typically FMR = 0.001, for automated border control gates for example. This serves as a standard and simple way to compare core algorithm recognition capability. However, it is very common in the academic literature to quote summary statistics:

- **Equal Error Rate**: The EER is obtained from a DET by finding the threshold where FMR is equal to FNMR. The metric was popular even before LFW [5] adopted essen-

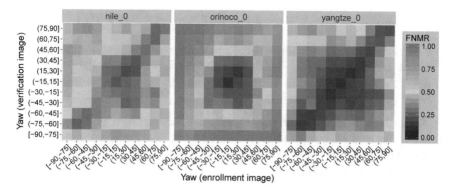

Fig. 17.4 For three commercial algorithms evaluated in 2017, the figure shows the effect of yaw between the enrollment image and the verification image on FNMR. The threshold is fixed in all cells to that value that gives FMR = 0.001 on near frontal images (the center cell). Three behaviors are evident: same-pose invariance (left), pose sensitivity (center), and substantial cross-pose invariance (right)

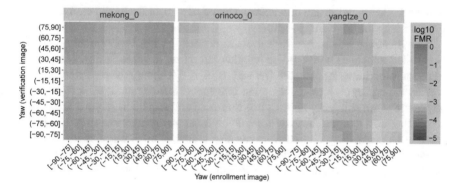

Fig. 17.5 For three commercial algorithms evaluated in 2017, the figure shows the effect of yaw between the enrollment image and the verification image on FMR. The threshold is fixed in all cells to that value that gives FMR = 0.001 on near frontal images. The elevated false positive rates when both images have large yaw angles may present a vulnerability

tially the same quantity as its benchmark metric. The EER should be deprecated because it usually corresponds to an operationally high and unrealistic FMR. Usually the operating FMR of relevance is much lower than the FNMR, for example, when FMR = 0.001 and FNMR = 0.01.

- **Area Under the Curve** The AUC is the area under the ROC, integrating FNMR over FMR on [0, 1]. This verification metric corresponds to the expected rank 1 recognition rate in identification trials where the gallery size is two and scores are computed independently. It is favored by radiologists and in other human studies where scores are reported in very discrete scales.

- **Half total error rate** The HTER is (FMR + FNMR) / 2 is one special case of decision cost function

$$\text{DCF}(T) = (1 - P)\ C_{\text{FNM}}\ \text{FNMR}(T)\ +\ P\ C_{\text{FM}}\text{FMR}(T) \qquad (17.4)$$

which expresses the (monetary) cost of errors committed by an verification system, where proportion P are impostors comparisons and C_x quantify the cost of false non-match and match errors. The problem with HTER is that the two errors almost never have equal priors and costs.

While these metrics are well defined, easy to compute, and reasonably common, they all drive research away from the low-FMR regime most useful to the marketplace uses of face recognition. They should be deprecated in favor of citing FNMR at some low FMR, e.g., FMR ≤ 0.001.

17.3   Identification

Background: The largest segment of the face recognition marketplace is identification, with identification and duplicate detection of standardized cooperative portrait images being the largest applications.

Identification, which involves no claim of identity, involves searching a set of N enrolled templates with a probe template, to return matches, if any. Many identification algorithms implement 1:N search as N 1:1 comparisons, followed by a sort operation. In operations, there are two modes of operation: First is to return any entries with similarity above some threshold and second to return a fixed number of the most similar entries, essentially a k-nearest neighbor search. Neither of these tasks necessarily require an exhaustive search of all N enrollments, and it is incumbent therefore on an evaluation to not prescribe nor constrain the underlying implementation, for example, by requiring the algorithm to return all N comparison scores.

Fundamentally, a probe is searched against an enrollment database of existing identities and a candidate list of potential matching subjects is retrieved. Applications are generally differentiated by the kinds of images being compared (e.g., driving license photos vs. surveillance photos), enrolled population sizes, and the prior probability that a search has an enrolled mate. In most applications, the core accuracy of a facial recognition algorithm is the most important performance variable.

17.3.1   Closed Versus Open-Universe Identification

In one-to-many identification, unknown imagery is searched against a gallery of N individuals previously collected, and the goal is to determine the identity of the query subject.

In a *closed-universe*, every subject queried against the gallery is known to have a mate (i.e., exists in the gallery). While this might apply to situations such as the one-to-many search on a cruise ship where everyone being searched is known to have boarded the ship, closed-universe applications occur very infrequently. Because only mated searches are conducted, closed-universe tests generally report a rank-based hit rate, ignoring scores produced by the algorithm, neglecting to look at false positives, and frequently encouraging optimization of the wrong quantity when commonly, high threshold (low false positive rate) discrimination is required in large-population government applications of face recognition.

Closed-universe experiments are often seen in academic tests [6]. The University of Washington's Megaface Challenge [6] is an open academic face recognition challenge that contains a one-to-many protocol which evaluates algorithmic capability to retrieve the correct hit against galleries with up to 1 million distractors. The protocol is closed-universe as it only conducts searches that are known to exist in the gallery. As such, the challenge reports identification hit rate as a function of distractor size and rank but does not report identification performance against false positive identification rate.

Most applications are naturally *open-universe*, where some proportion of query subjects will not have a corresponding mate entry among the previously enrolled identities. In high search volume applications where most searches do not have a mate (e.g., casino surveillance for cheats), the emphasis is not controlling false positive outcomes. In low-volume applications, for example, bank robbery investigation, false positives will be investigated manually.

From a testing perspective, open-universe is accomplished by running both mated searches and non-mated searches. The non-mated searches afford the ability to generate a false positive identification rate against a threshold. Face identification algorithms must minimize both false positive errors where an unenrolled person is mistakenly returned and false negative errors where an enrolled person is missed. This is critical whenever the proportion of non-mated searches is naturally large particularly in the canonical "watch-list" surveillance application where a large majority of individuals in the field of view are not enrolled and the system should return nothing. So a face recognition system looking for terrorism suspects in a crowded railway station must be configured to produce few false positives, no more than what can be sustainably adjudicated by trained reviewers who would determine the veracity of the candidate match and then initiate action. If, on the other hand, the proportion of unmated searches is low, as is the case, for example, with patrons entering their gymnasium, the system must be configured to tolerate a few false positives, i.e., to admit the infrequent non-customer who attempts access.

False negative identification performance: It is necessary to report accuracy in terms of *both* false negative identification rate quantifying how often enrollees are not recognized and the false positive identification rate stating how often algorithms incorrectly issue false alarms.

Outputs from the mated searches are used in the false negative identification rate (FNIR) computation. FNIR is defined as the number of mated searches which fail to produce the enrolled mate in the top R ranks with score above threshold, T. FNIR is therefore known as a miss rate. Its value will generally increase with the size of the enrolled database, N, because the recognition algorithm is tasked with assigning a low score to *all* $N - 1$ non-mated enrollments. Thus, for each of M mated searches, the algorithm returns L candidates with hypothesized identities and similarity scores. If the identity of the search face is ID_i and that of the r-th candidate is ID_r, then

$$\text{FNIR}(N, R, T) = 1 - \frac{1}{M} \sum_{i=1}^{M} \sum_{r=1}^{R} H(s_{ir} - T) \, \delta(ID_i, ID_r) \qquad (17.5)$$

where s_{ir} is the r-th highest score from the i-th search, the step function H(x) is 1 if $x \geq 0$ and 0 otherwise, and the function $\delta(x, y)$ is 1 if $x = y$ and 0 otherwise.

In cases where an algorithm fails to produce a template from a probe input image—the FNIR computation conventionally proceeds by assigning a low score, $-\infty$, and high rank, $L + 1$, simulating a miss. For non-mate searches, this treatment actually improves FPIR.

As discussed below, FNIR varies with population size N and the number of candidates examined to rank, R that is above threshold, T.

Effect of Gallery Size As the world population increases in size, face recognition databases are also getting larger. It is known that face recognition performance degrades as the population size being searched increases. With an increased population size, comes an increased chance of finding another person in the gallery who looks similar to the search probe. This has an impact on the false match rate and how often a similar-looking person is incorrectly matched. How algorithmic performance degrades with increasing population size is a primary challenge in any application where individuals are enrolled at a greater rate than they are unenrolled. Therefore, reporting accuracy as a function of the population size being searched is essential as this will inform end-users of face recognition systems, who each have their own unique operating points along the population size spectrum. A desirable property of a good face recognition algorithm is slow degradation in accuracy as the population size increases. Figure 17.6 presents algorithm accuracy plotted against enrolled population size. The key point about the graph is the lines are quite flat, and the lines drift up quite slowly on a logarithmic scale as the enrolled population size increases. This substantiates the viability for the face recognition industry, where searches are conducted with usable accuracy performance as the enrolled population goes up to state and nation sizes.

Effect of candidate list length In investigative applications, a human reviewer may be employed to look for mates that have been displaced from rank one by higher scoring non-mates. A plot of FNIR(R) is then the relevant metric stating how often mates are not contained in, say, the top R = 50 candidates. Figure 17.7. The complement of this quantity the cumulative match characteristic is perhaps the most commonly reported metric: $\text{CMC}(N, R) = 1 - \text{FNIR}(N, R)$.

False positive identification rate: Scores from the non-mated searches are used in the false positive identification rate (FPIR) computation, which states the proportion of non-mate searches yielding *any* candidates at or above a threshold T:

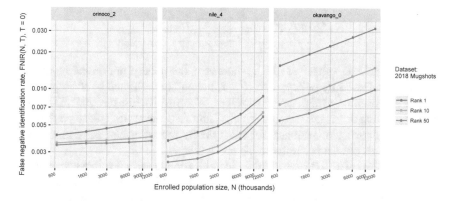

Fig. 17.6 Figure shows rank-based miss rates versus enrolled population, FNIR(N, 0), estimated over mated searches. The threshold of zero means the computation ignores scores, as a mate may be at rank 1 but have a low score

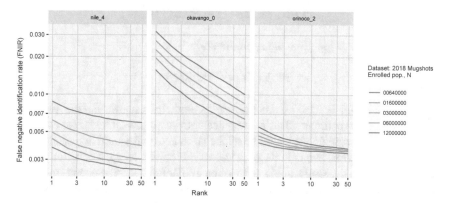

Fig. 17.7 Miss rate, FNIR(N, R, 0), showing the value to an investigator of manually reviewing candidate lists to rank, R

$$\text{FPIR}(T) = \frac{1}{N_I} \sum_{i=1}^{N} H(s_i - T) \qquad (17.6)$$

In cases where an algorithm fails to produce a template from an input image, a low score is again assigned as the result of any comparison involving that template. This practice actually benefits (reduces) FPIR.

An alternative quantity expressing false positive accuracy is "selectivity". This term hails from the fingerprint recognition literature, particularly for latent print identification. In operations, it is defined as the expected number of candidates a human reviewer would have to examine before a mate is found. The definition, in algorithm testing, is related but different: the expected number of candidates returned in a non-mated search. Recall, FPIR, is the proportion of *searches* returning *any* non-mate candidates. Selectivity expresses how many:

$$\text{SEL}(T, R) = \frac{1}{N_I} \sum_{i=1}^{N} \sum_{r=1}^{R} H(s_{ir} - T) \tag{17.7}$$

Selectivity takes on values on [0,R], where R is the upper limit on the number of candidates reviewed by policy, with $R \leq L$ the number returned by the algorithm. For many algorithms, at high threshold values, false positives are rare and $\text{SEL}(T) \rightarrow \text{FPIR}(T)$. When this is not the case, an algorithm is concentrating false positives in the results of particular searches, for example, by returning images of any person wearing glasses or beards similar to the probe image.

Relationship of FPIR and FMR In identification, a false positive occurs when at least one comparison produces a false positive. This requires correct rejection of all N non-mated enrollments and if these are independent comparisons, then a verification trial producing an estimate of false match rate FMR(T) could be used to estimate FPIR(T); thus,

$$\text{FPIR}(T) = 1 - (1 - \text{FMR}(T))^{N} \tag{17.8}$$

The binomial expansion gives the approximation

$$\text{FPIR}(T) = N\,\text{FMR}(T) \tag{17.9}$$

This formula is widely used by practitioners in biometrics supporting the use of the verification DET to estimate identification accuracy as follows. Given, for example, Fig. 17.1, the horizontal axis would be relabeled as FPIR and would be rescaled by a factor of N. However, two caveats are necessary: First, it's an approximation which overstates FPIR and closely holds only when $N\,\text{FMR}(T) \ll 1$. Second and more seriously, this model is incorrect for implementations that do not implement 1:N search as N 1:1 comparisons. This is evident in Fig. 17.8 and discussed now.

Linked DET When assessing an algorithm's performance on images from different populations or on images with different properties, it is very easy to naively look at DET and draw conclusions about error rates when lines lie above or beneath one another. Because most operational systems always operate at a fixed threshold, it is necessary to compare points of equal operating threshold between the different populations. Figure 17.8 presents DET plots for identification in mugshot images for searches into five different enrolled population sizes. The three black lines join points of equal threshold. Horizontal lines represent an increase in $\text{FPIR}(T)$ and vertical lines represent an increase in $\text{FNIR}(T)$. In the rightmost plot, $\text{FPIR}(T)$ is almost independent of N, and the binomial model (Eq. 17.8) and its linear approximation (Eq. 17.9) do not hold. This is consistent with 1:N search not being implemented as N 1:1 comparisons. The implication of this is that identification algorithms should be tested as such and not simulated using scores produced in verification recognition trials.

Aging Face recognition accuracy is undermined by aging, the progressive essentially irreversibly effects of soft and hard-tissue changes to the anatomy. This has been

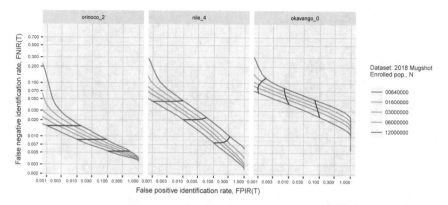

Fig. 17.8 Identification DET plots showing FNIR vs FPIR for five enrolled population sizes (colored traces). The black lines show accuracy varies at fixed thresholds: Horizontal lines correspond to linear growth in false positives with N, per binomial theory; approximately vertical lines show an algorithm attempting to produce false positives independent of N. This can be achieved by stabilizing the distribution of the highest impostor score

quantified using longitudinal analysis applied to recognition scores produced from long-run repeat images of a large set of individuals. The application of mixed-effect models [11] to aging allows for imbalanced sets of irregularly sampled data to be modeled as a shared population component (reflecting that everybody ages) and an individual-specific "random-effect" (allowing subjects to vary around that). This approach, which relies on multiple encounters of the same individual, such as those seen in frequent traveler systems, yields an aging rate, i.e., how rapidly the score is expected to degrade in average subjects. Of course, this depends on the recognition algorithm. It can handle, additionally, the effect of age and aging—where a five-year time lapse in a fifteen-year old is wholly different than in a fifty-year old. The sensitivity of algorithms to aging is evident in Fig. 17.9 which shows the decline in scores. While the figure is not a replacement for mixed-effect analysis (because imbalance can give misleading results), it does expose the better algorithms. For example, aging can be quantified in terms of how quickly scores degrade. Comparison between algorithms can be achieved by normalizing, e.g., to standard deviations per decade.

In summary, while most evaluations concentrate on quantification of error rates, changes in the whole score distribution reveals systematic effects. Mixed-effect modeling of scores and error rates [12] is the recommended approach to quantifying aging.

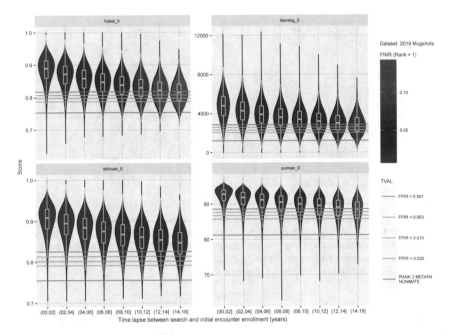

Fig. 17.9 In a fixed gallery of the N = 3 million, the violin plots show mate scores (on native ranges) returned from 10.9 million searches, broken out by time-lapse in years. As faces age, they become progressively less similar to the initial enrolled face. The fill color encodes FNIR (1) i.e., rank one miss rate; the horizontal lines show thresholds corresponding to various FPIR values and the score of rank 2 non-mates. Plots showing native scores can reveal systematic effects rather than just those causing recognition errors confined to the tails of distribution

17.3.2 Enrollment Gallery Composition

Most face identification experiments proceed by placing N images of N people in a gallery. For pure algorithm evaluation, this is the recommended approach. However, operational reality departs from this simple case because, over time, subjects are encountered multiple times, e.g., as part of a visa application process, and these images are added to the gallery.

Recognition accuracy is improved when $K > 1$ images are available. Figure 17.10 shows the effect of providing multiple images for each identity to three different face recognition algorithms and the impact it has on accuracy. In all cases, FNIR reduces with K; persons with multiple enrollments are more readily recognized. However, for two algorithms, there are an order of magnitude increases in FPIR, showing that non-mate searches match multiple-image enrolees spuriously.

Key to the mitigation of this is how multiple images are employed. Referring to Fig. 17.11, there are two approaches:

- **Event-based, unconsolidated gallery**: Here images are added to the gallery without regard for whether the person already exists or not. Under this model, there

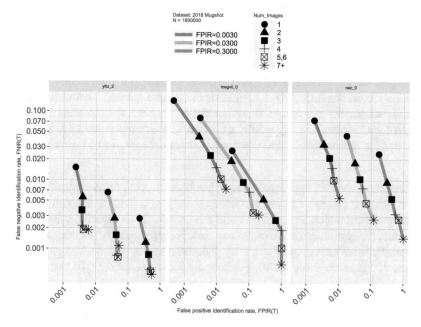

Fig. 17.10 The effect of providing multiple images for each identity to three face recognition algorithms and the impact it has on accuracy. Plotted is identification miss rates vs. false positive rates at four operating thresholds. The enrolled population size is fixed at 1.6 million

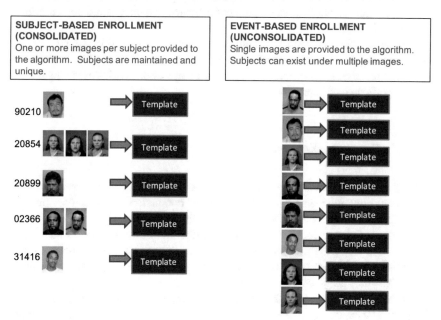

Fig. 17.11 This figure presents examples of how enrollment data is stored in a subject-based, consolidated gallery versus an event-based, unconsolidated gallery

can be multiple images of the same person in the gallery. Templates are generated from single images independently and are treated as different identities. Administratively, there might be record-keeping that associates same-person images, but the underlying face recognition algorithm is not aware of this.

- **Subject-based, consolidated gallery**: Unique identities of people are maintained, and a record contains $K \geq 1$ images of the unique subject. This person-centric model affords the face recognition algorithm an opportunity to fuse feature vectors, select the best image using quality assessment, or to simply extract features from K images independently and then arrange to effect score-level fusion during the search.

The recommended approach—which has been adopted in identification benchmarks [13]—is to construct a consolidated gallery by providing the algorithm with K images from which feature extraction can proceed in an algorithm-defined way, for example, by concatenation of feature vectors (common), by selecting the "best" images, by some metric, or by some template-level fusion scheme. The consolidated approach was used for Fig. 17.10. When an algorithm consolidates by concatenating the individual feature vectors, the resulting template will have size linear in K. In the best-image and fusion approaches, size is independent of K. Both schemes exist commercially with the most accurate algorithm in a recent benchmark [7] performing template-level fusion.

There are algorithm effects associated with the different enrollment gallery types. For example, Fig. 17.12 presents a DET comparing performance for two state-of-the-art face recognition algorithms on both consolidated and unconsolidated enrollment galleries.

How an enrollment gallery is built and maintained is an operational decision, and depending on the face recognition algorithm, there may be varying performance gains. Testing algorithms on different gallery compositions will provide insight into optimal performance scenarios.

17.3.3 Database Segmentation

Database segmentation is a phenomenon that occurs when a gallery is somehow heterogeneous containing images with mixtures of population or image-specific properties. This will often cause a face recognition algorithm to easily distinguish images with one set of characteristics from another, rather than with the actual facial identities of the subject. Database segmentation occurs naturally whenever the enrollment gallery is not homogenous, and a number of factors including subject age, race, image quality, beards, and glasses can affect this.

Figure 17.13, as excerpted from [14], shows an example of database segmentation. The particular experiment uses images from the LFW [5] and IJB-A [15] photojournalism datasets as probes and mates, enrolled with 80 million distractor images scraped from social media websites (Web-Face). The differences in properties between the

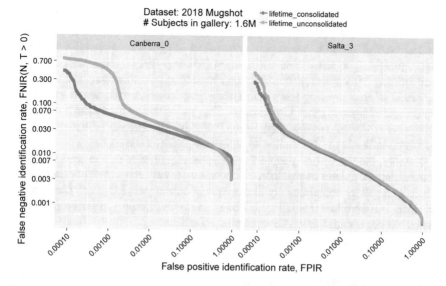

Fig. 17.12 This figure presents FNIR vs FPIR on the same set of searches made against a consolidated versus unconsolidated enrollment database

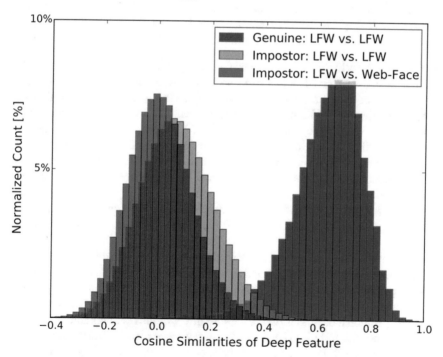

Fig. 17.13 As excerpt from [14], distributions of cosine similarities of the genuine pairs, within-dataset impostor pairs and between-dataset impostor pairs for the combinations of LFW and Web-Face

photojournalism images versus social media images lead to a database segmentation effect, which can be clearly seen by the left-shifted distribution of similarity scores for imposter distribution when photojournalism images are compared with social media images. The left-shifted distribution indicates that the algorithm can more easily distinguish imposters (due to image characteristics rather than the identities of the faces) when comparing LFW against Web-Face images, versus when LFW images are compared against other LFW images. This undesirable effect causes the actual gallery size to be an overstatement of the "effective" gallery size. As such, this can lead to undesirable overestimation of algorithm performance accuracy.

Notably, the number of people in that gallery is unknown. The test design, to make as large a gallery as possible, eschews the normal design goal of constructing a balanced gallery containing equal numbers of entries per individual. Care should be taken to ensure that the gallery is homogenous with respect to image properties to prevent or minimize database segmentation effects.

17.4 Computational Efficiency

In high applications in which the volumes of enrollments or searches are large, or in which there is a rapid response requirement, the duration of the algorithms operation becomes important. Therefore, evaluators should measure and report the following.

- **Feature extraction time**: Operational face recognition systems usually must localize a face and extract features from an image in less than one second—this is necessary for human satisfaction and operational throughput. While additional or exotic hardware may be fielded to achieve this, there will be marketplace pressures to avoid doing so. Similarly, the use of GPU processors, while ubiquitous for the training of algorithms, is (empirically) not needed for sub-second feature extraction from contemporary face recognition algorithms. In addition, many feature extraction algorithms have not or cannot be parallelized when operating on a single face image.
  Developers should measure feature extraction time. This may be achieved by wrapping the appropriate function call with high-resolution timers (such as the std::chrono facility in C++).
- **Search duration and complexity**: Recent test results show that 1:N search speeds can span up to three orders of magnitude [16]. Given the implications for hardware procurement, it becomes essential to measure speed and to only invest in slow algorithms if they offer measurable accuracy advantages. Further, given very large operational databases, the scalability of algorithms is important. It has been reported previously [16, 17] that search duration can scale sublinearly with enrolled population size N. Further, there has been considerable recent research on indexing, exact [18] and approximate nearest neighbor search[18, 19], and fast-search [20]. Figure 17.14 shows search duration as a function of database size for linear and sublinear algorithms.

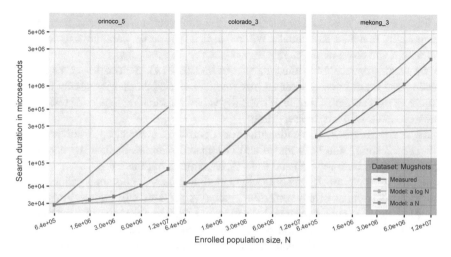

Fig. 17.14 For three algorithms, the duration of search is a function of enrolled population size. The times are measured on a single c. 2016 CPU. On each plot are the actual durations (in red and as points) and then projections based on linear (blue) and logarithmic (green) growth. Those algorithms that give sublinear growth have accuracy approaching that of linear versions from the same developer

- **Fast-search data structure construction and maintenance**: While most algorithms implement linear search with a linear data structure, those that achieve sublinear search durations generally require pre-processing and organization of the enrollment data structure prior to conducting search. This may not be an inexpensive operation. Researchers should report the duration of gallery construction operations and the expected complexity (linear, quadratic, etc).
- **Insertion and deletion** Likewise, developers should report the durations of functions for inserting and deleting gallery entries and their dependence on N.

17.5 Summary and Recommendations

We summarize with a set of recommendations:

▷ RECOMMENDATION 1: To encourage explicit consideration of the impostor distribution and low FMR, face recognition algorithm developers should report verification metrics as the primary indicators of accuracy. Researchers should not report just rank-based metrics from identification trials.

▷ RECOMMENDATION 2: When reporting verification results, prefer DETs over ROCs, i.e. plot FNMR instead of true match rate TMR.

▷ RECOMMENDATION 3: When reporting verification results, plot DETs and, absent an overriding reason, emphasize lower false match rates by using a logarithmic scale for the FMR axis showing FMR ≪ 1.

▷ RECOMMENDATION 4: Given adequate data construct impostor pairs from demographically matched individuals. When that is not possible, compute and comment on within and cross-demographic impostor score distributions.

▷ RECOMMENDATION 5: While the series of ISO/IEC 19795 biometrics performance testing and reporting standards are not free, biometric laboratories should invest in copies as they contain a wealth of carefully vetted information that guide test design, measurement, analysis and reporting.

▷ RECOMMENDATION 6: For identification algorithms, conduct open-universe identification testing. Run both mated and non-mated searches against the enrollment gallery. Report both FNIR(T) and FPIR(T), for a range of thresholds.

▷ RECOMMENDATION 7: Database segmentation causes the nominal gallery size to be an overstatement of the "effective" gallery size. This will underestimate FNIR. Unless a galleries represents a real situation, or other should not be an overt mixture, or if that's not possible, should be balanced a uniformly sampled from a represent a real-application, or other and should n Prevent database segmentation by using images with uniform properties.

▷ RECOMMENDATION 8: Populate with galleries with equal numbers of images per person. Avoid imbalanced galleries. When multiple images are available per person, use them as probe images.

▷ RECOMMENDATION 9: Measure duration of all elemental functions, including detection, feature extract, comparison, and search. For fast-search data structures report duration of construction and search and the expected complexity of those operations (e.g. $O(\sqrt{N})$.

References

1. Taigman Y, Yang M, Ranzato M, Wolf L (2014) Deepface: closing the gap to human-level performance in face verification. In: Proceedings of the 2014 IEEE conference on computer vision and pattern recognition, ser. CVPR '14. IEEE Computer Society, Washington, DC, USA, pp 1701–1708. https://doi.org/10.1109/CVPR.2014.220
2. Parkhi OM, Vedaldi A, Zisserman A (2015) Deep face recognition. In: British machine vision conference
3. He K, Zhang X, Ren S, Sun J (2016) Deep residual learning for image recognition. In: 2016 IEEE conference on computer vision and pattern recognition (CVPR), pp 770–778
4. Schroff F, Kalenichenko D, Philbin J (2015) Facenet: a unified embedding for face recognition and clustering. CoRR, arxiv:1503.03832
5. Huang GB, Ramesh M, Berg T, Learned-Miller E (2007) Labeled faces in the wild: a database for studying face recognition in unconstrained environments. University of Massachusetts, Amherst. Tech. Rep. 07–49
6. Kemelmacher-Shlizerman I, Seitz SM, Miller D, Brossard E (2015) The megaface benchmark: 1 million faces for recognition at scale. CoRR. arxiv:1512.00596
7. Grother P, Ngan M, Hanaoka K (2018) The face recognition vendor test 2018 (frvt). National Institute of Standards and Technology, Gaithersburg, Marlyand, Tech. Rep. NIST Interagency Report, 2018, to be published
8. Liao S, Lei Z, Yi D, Li SZ (2014) A benchmark study of large-scale unconstrained face recognition. In: IEEE international joint conference on biometrics, pp 1–8

9. WG, ET Mansfield ET (2015) ISO/IEC 19795-1 biometric performance testing and reporting: principles and framework, international standard ed., JTC1: SC37. http://webstore.ansi.org
10. Gross R, Matthews I, Baker S (2004) Appearance-based face recognition and light-fields. IEEE Trans Pattern Anal Mach Intell 26(4):449–465
11. Best-Rowden L, Jain AK (2018) Longitudinal study of automatic face recognition. IEEE Trans Pattern Anal Mach Intell 40(1):148–162
12. Beveridge JR, Givens GH, Phillips PJ, Draper BA (2009) Factors that influence algorithm performance in the face recognition grand challenge. Comput Vis Image Underst 113(6):750–762
13. Maze B, Adams J, Duncan JA, Kalka N, Miller T, Otto C, Jain AK, Niggel WT, Anderson J, Cheney J, Grother P (2018) Iarpa janus benchmark—c: face dataset and protocol. In: 2018 international conference on biometrics (ICB), pp 158–165
14. Wang D, Otto C, Jain AK (2017) Face search at scale. IEEE Trans Pattern Anal Mach Intell 39(6):1122–1136
15. Klare BF, Klein B, Taborsky E, Blanton A, Cheney J, Allen K, Grother P, Mah A, Burge M, Jain AK (2015) Pushing the frontiers of unconstrained face detection and recognition: Iarpa janus benchmark a. In: 2015 IEEE conference on computer vision and pattern recognition (CVPR), pp 1931–1939
16. Grother P, Ngan M, Hanaoka K, Boehnen C, Ericson L (2017) The 2017 iarpa face recognition prize challenge (frpc). National Institute of Standards and Technology, Gaithersburg, Marlyand. Tech. Rep. NIST Interagency Report 8197. https://doi.org/10.6028/NIST.IR.8197
17. Grother P, Ngan M (2014) Interagency report 8009, performance of face identification algorithms. In: Face recognition vendor test (FRVT)
18. Ishii M, Imaoka H, Sato A (2017) Fast k-nearest neighbor search for face identification using bounds of residual score. 2017 12th IEEE international conference on automatic face & gesture recognition (FG 2017). IEEE Computer Society, Los Alamitos, CA, USA, pp 194–199
19. Babenko A, Lempitsky V (2016) Efficient indexing of billion-scale datasets of deep descriptors. In: The IEEE conference on computer vision and pattern recognition (CVPR)
20. Johnson J, Douze M, Jégou H (2017) Billion-scale similarity search with gpus. CoRR. arxiv:1702.08734

Index

Printed in the United States
by Baker & Taylor Publisher Services